T0327546

HYDROSTATIC TRANSMISSIONS AND ACTUATORS

HYDROSTATIC TRANSMISSIONS AND ACTUATORS

OPERATION, MODELLING AND APPLICATIONS

Gustavo Koury Costa

Department of Mechanics – Federal Institute of Education, Science and Technology, Recife, Pernambuco, Brazil

Nariman Sepehri

Department of Mechanical Engineering – University of Manitoba, Winnipeg, Manitoba, Canada

Library of Congress Cataloging-in-Publication Data

Costa, Gustavo K.
    Hydrostatic transmissions and actuators / Gustavo K. Costa, Department of Mechanics Federal Institute of Education, Science and Technology, Recife, Pernambuco, Brazil, Nariman Sepehri, Department of Mechanical Engineering–University of Manitoba, Winnipeg, Manitoba, Canada.
        pages cm
    Includes bibliographical references and index.
    ISBN 978-1-118-81879-4 (hardback)
    1. Oil hydraulic machinery. 2. Fluid power technology. 3. Hydrostatics. 4. Power transmission. 5. Actuators.
I. Sepehri, Nariman. II. Title.
    TJ843.C67 2015
    620.1′06–dc23
                                                                                    2015022962

A catalogue record for this book is available from the British Library.

ISBN: 9781118818794

Set in 10/12pt TimesLTStd by SPi Global, Chennai, India

1   2015

To Flávia Maria, my beloved wife

Gustavo Koury Costa

To Aresh, Parisa and Anoush, the joys of my life

Nariman Sepehri

Contents

Preface

The need for transmitting mechanical power has always been present in many fields of engineering. Take, for example an automobile where power must be transferred either from the engine to the wheels or from the driver's foot to the wheels during braking. This book focuses on two specific power transfer situations: (a) transmission between two rotating shafts, and (b) transmission between a rotating shaft and a hydraulic actuator. The power-conveying medium, in both cases, is a hydraulic fluid.

For Whom This Book Has Been Written

This book has been written for undergraduate students but will also be useful to practical engineers and junior graduate students who need to have introductory knowledge on the subject of hydrostatic power transmissions and actuation. The pre-requisites for the reader are minimal; no more than a little knowledge about power hydraulics and a basic understanding of calculus and physics are necessary. The book has been constructed in such a way that students do not need to refer other sources of information to understand the text. Every effort has been made to derive most of the equations found in the text. Only a few of the many formulas found in the book do not have a formal development because of either the degree of complexity involved or their straightforward nature. To help solidifying the concepts, we have also included a list of exercises at the end of most chapters.

Book Organization

The book is organized in a way that caters to different audiences with varied backgrounds. For students who do not have a strong foundation in fluid power, this book is best read from cover to cover. On the other hand, a practical engineer who wants to learn how to calculate the efficiency of hydrostatic transmissions, can go straight to Chapters 3 and 4. However, we have made every attempt to follow a logical and progressive way of exposing the theme, having in mind the students who will read the book from the first chapter to the last chapter. In that sense, we have followed an approach whereby the reader obtains a complete overview of the subject matter in the first chapter. And then in subsequent chapters, details are provided so that when the reader arrives to the end of the book, he or she will have acquired a solid and concise knowledge of the exposed themes.

In terms of the subjects explored in each chapter, there is a clear division following the overall exposition given in Chapter 1. From Chapter 2 to Chapter 5, we focus on power transmissions between rotating shafts (hydrostatic transmissions). Chapters 6 and 7 concentrate on hydrostatic actuators. Finally, Chapter 8 focuses on conventional and new applications of both hydrostatic transmissions and actuators.

To ensure practicality, most of the examples in the book use catalogue data from manufacturers. While great care has been taken in the reproduction of illustrations and/or information taken from manufacturers, inadvertent typographical errors or omissions may have occurred. In some particular cases where catalogue data are unavailable, or when it is not strictly necessary to provide this data, we follow a purely theoretical approach in presenting the concepts. The overall idea behind using real catalogue data is to introduce the students to real-world applications as they work through the examples of the book.

In what follows, we briefly describe the subjects covered in every chapter of this book.

Chapter 1 presents an introduction to hydrostatic transmissions and actuators. Since applications now using mechanical power transmissions constitute a potential field for hydrostatic transmission usage, we review the subject of mechanical transmissions first. Hydraulic components are gradually introduced in this chapter.

Chapter 2 reviews some basic definitions and concepts about hydraulics, such as fluid compressibility and viscosity, pressure losses and internal flows in hydraulic circuits.

Chapter 3 focuses on hydrostatic pumps and motors. After exploring the fundamental aspects of pumps and motors in general, a succinct description of some representative models is given. The definition of efficiency takes up a considerable portion of the chapter, given its importance in hydrostatic transmissions. We also explore the basics of digital displacement and floating cup technologies.

Chapter 4 explores the steady-state operation of hydrostatic transmissions. After reading this chapter, the student will be able to create a basic design of a typical hydrostatic transmission.

Chapter 5 complements the steady-state analysis of hydrostatic transmissions carried out in Chapter 4 by exploring the transient regime. In this chapter, the student has the chance to study the oil compressibility effects, introduced in Chapter 2, that occur when the hydrostatic transmission is subject to dynamic loads.

Chapter 6 focuses on the theme of hydrostatic and electrohydrostatic actuators. Several circuit designs are described in detail. The chapter closes with a description of the common pressure rail technology and its relation with hydraulic transformers.

Chapter 7 introduces the dynamic analysis of hydrostatic actuators. A nonlinear analysis is carried out, and the equations describing the model are solved numerically.

Chapter 8 puts together current and potential applications of hydrostatic transmissions and actuators. Each application is described in detail, so that students can have a good knowledge of the benefits and drawbacks of the hydrostatic technology in every case.

Appendix A lists the several ISO hydraulic symbols used in the book.

Appendix B contains the necessary mathematical tools for a complete understanding of the book. Special emphasis is given to the solution of second-order linear differential equations, where the method of the Laplace transform is briefly presented.

Appendix C reviews the basics of fluid dynamics with a special emphasis on the Navier–Stokes equations, which are developed in detail. For students who are not familiar with the theme, this appendix constitutes a sufficient basis for the subjects covered in the book.

Some examples given in this book require a numerical solution. In this case, the reader can find the source code for the corresponding computer programs written in Scilab script language[1] www.wiley.com/go/costa/hydrostatic. The parts of the book for which a computer script is available have been marked with the download icon ⬇.

We have done our best to make the text as clear and rich as possible to the student, and it is our most sincere desire that this book contributes to the understanding and the development of this very important field of fluid power engineering.

---

[1] Scilab is a free programming environment available at http://www.scilab.org (April 2014).

Acknowledgements

Primarily, the authors wish to thank the reviewers for their comments during the preparation of this book as they have definitely impacted its quality. This book was conceived during the postdoctoral studies of Gustavo Koury Costa, in the Fluid Power and Telerobotics Research Laboratory at the University of Manitoba, Canada. In this aspect, he is grateful for the support of the University of Manitoba for the post-doctoral position offering and the Federal Institute of Science and Technology of Pernambuco (Brazil) which, together with the Capes Foundation (Brazil), provided the financial aid during his stay in Winnipeg. He also acknowledges the help of Maria Auxiliadora Nicolato from Capes, for her assistance during his stay in Canada; Luciana Lima Monteiro, for her friendship and for the promptness to help with the necessary paperwork in Brazil; Robert and Madeline Blanchard for their warm welcome and support; Paulo Lyra and Carlos Alberto Brayner for helping with the post-doctoral application; and Arthur Fraser, who did not hesitate to fly from Scotland to Winnipeg to visit. Last but not least, he acknowledges the support of the colleagues from the Department of Mechanics of his home institution, who agreed to take over his classes while he was away.

Nariman Sepehri is grateful to the University of Manitoba for providing the infrastructure, Natural Sciences and Engineering Research Council of Canada (NSERC) for providing continuous support of his research in fluid power systems and controls, and all his past and present graduate students from whom he always receives considerable help and education. Finally, both authors are grateful to Shari Klassen of the University of Manitoba for her editorial work and comments on the writing style of this book.

Gustavo Koury Costa
Nariman Sepehri
August 2015

About the Companion Website

This book's companion website www.wiley.com/go/costa/hydrostatic provides you with additional resources to further your understanding, including:

- A solutions manual
- Scilab scripts
- Links to useful web resources

1

Introduction to Power Transmission

The term *power transmission* refers to a collection of devices assembled to transmit power from one physical point to another. In this chapter, we describe the most common types of power transmissions and introduce the subject of hydrostatic transmissions and actuators. This chapter is divided into six parts:

- Mechanical transmissions
- Hydrodynamic transmissions
- Hydrostatic transmissions
- Hydromechanical transmissions
- Mechanical actuators
- Hydrostatic actuators

It is important to mention that there are other types of power transmissions. For example, an electricity gridline is a type of power transmission – from the generator to the final user. However, when mechanical energy is involved (kinetic and potential), the aforementioned types are the most representative.

The majority of this chapter deals with the topic 'transmissions', with a smaller portion dedicated to 'actuators', as actuators can be seen as a special type of hydrostatic transmission where the motor is replaced by a hydraulic cylinder. We start with a basic concept common to both mechanical and hydrostatic transmissions: the transmission ratio.

1.1 Transmission Ratio

1.1.1 Generalities

Figure 1.1 illustrates a typical situation where a power transmission can be applied. The input shaft is rotating with an angular speed ω_i and is connected to a prime mover (such as

Hydrostatic Transmissions and Actuators: Operation, Modelling and Applications, First Edition.
Gustavo Koury Costa and Nariman Sepehri.
© 2015 John Wiley & Sons, Ltd. Published 2015 by John Wiley & Sons, Ltd.
Companion Website: www.wiley.com/go/costa/hydrostatic

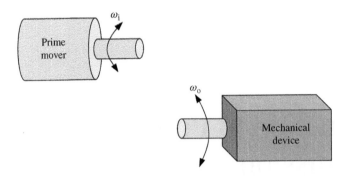

Figure 1.1   Typical situation requiring a power transmission

an electric motor or an engine) whose output power is P_i. We connect the input shaft to an output (driven) shaft that must rotate at an angular speed ω_o. The angular speed of the driven shaft may be greater or lesser than the angular speed of the input shaft or even have an opposite direction in relation to the input shaft's angular speed.

When mechanical transmissions, such as gearboxes, belts and chains, are considered, the spatial arrangement of the driving and driven shafts is of paramount importance because it dictates the technology to be used. For example, in the case of parallel shafts, a gear transmission may be used if their distance from each other is not too great. However, the farther the shafts are from each other, the heavier the gearbox becomes, leading to more demanding requirements with respect to alignment and lubrication. Belts and pulleys may be used for transmissions between shafts that are separated by a considerable length, but the problem of spatial arrangement remains. Moreover, the power to be transmitted becomes considerably limited due to the belt-to-pulley friction coefficient. Chain transmissions are noisy and require the shafts to be perfectly parallel with constant lubrication. Additionally, these types of transmissions – with the exception of some special arrangements of chains and belts – do not allow for continuous transmission ratios, as will be explained shortly.

Mechanical transmissions, in general, have the following limitations:

1. They require the driving and driven shafts to be relatively near one another.
2. They are usually not flexible with regard to the spatial arrangement of the components.
3. Typically, they do not provide a continuous transmission ratio, that is, the ratio between the angular speed of the driven shaft and the angular speed of the driving shaft assumes discrete values.

Hydrostatic transmissions transmit power through a hydraulic fluid that travels inside flexible hoses or other types of conduits. As a result, there is great spatial flexibility, and the input and output shafts can be placed almost anywhere in relation to one another. Continuous transmission ratios are an inherent property of hydrostatic transmissions, whereas in mechanical transmissions, continuous transmission ratios are usually attained through more complex mechanisms. Energy losses are typically higher in hydrostatic transmissions when compared to mechanical transmissions. The pros and cons of hydrostatic transmissions will become more clear throughout this book.

1.1.2 Definition

The transmission ratio, R_T, is commonly defined as the quotient between the angular speed of the input shaft and the angular speed of the output shaft:

$$R_T = \frac{\omega_i}{\omega_o}$$

However, the definition of transmission ratio given above is frequently inconvenient. For example, in cases where the angular speed of the output shaft (ω_o) becomes zero, the transmission ratio becomes infinite. In this case, it is better to define the transmission ratio as the quotient between the output and the input speeds [1]. We adopt this convention in this book and use the following expression for R_T:

$$R_T = \frac{\omega_o}{\omega_i} \tag{1.1}$$

It is important to note that the concept of transmission ratio only makes sense when the power transmission occurs between two rotating shafts, as shown in Figure 1.1. When there is a conversion from a rotary to a linear motion, as is the case for mechanical or hydrostatic actuators, it makes no sense to use the transmission ratio as a design parameter.

> The term *transmission* implies that the power transfer occurs between two rotating shafts. If, in the process of power transmission, there is a conversion from a rotary motion to a linear or a limited angular motion, the term *actuator* will be applied instead.

1.1.3 Classification

In what follows, let $c_1, c_2, \ldots, c_N$ be nonzero real constants. Concerning the transmission ratio, we can have the following types of power transmission:

- Fixed-ratio transmission ($R_T = c_1$): There is only one possible value for the transmission ratio.
- Discretely variable transmission ($R_T = [c_1, c_2, c_3, \ldots, c_N]$): There is a set of finite values for the transmission ratio.

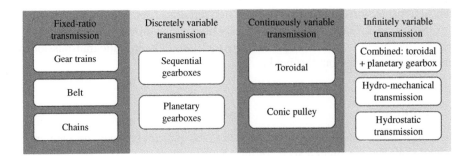

Figure 1.2   Classification of power transmissions according to the transmission ratio

- Continuously variable transmission (CVT) ($c_1 \leq R_T \leq c_2$ and $c_1c_2 > 0$): The transmission ratio varies continuously between two positive or negative[1] limits.
- Infinitely variable transmission (IVT) ($0 \leq |R_T| \leq |c_1|$): The absolute value of the transmission ratio varies continuously between zero and a positive value [2]. Strictly speaking, an IVT ratio would be such that $-\infty \leq R_T \leq +\infty$. However, due to engineering restrictions, this condition is usually relaxed[2] [1, 2].

Figure 1.2 shows a few selected types of power transmissions, organized according to their transmission ratio. The first two categories – fixed-ratio and discretely variable transmissions – cover most of the existing mechanical transmissions today. Continuously and infinitely variable mechanical transmissions are restricted to some special designs.

In the following sections, we briefly review each mechanical transmission type presented in Figure 1.2.

1.2   Mechanical Transmissions

There are many different types of mechanical transmissions, and it is not our goal to address all of them. Rather, we aim to give a succinct introduction to the theme in order to present an overview of the main transmission types. Therefore, only a small subset of the existing technologies will be covered here – enough to exemplify each type of mechanical transmission and give brief explanation about their characteristics.

1.2.1   Gear Trains

Gear trains are a typical example of fixed-ratio transmissions. They can be better understood if we first consider a disc train (Figure 1.3), where we assume that every disc is perfectly rigid

---

[1] A negative transmission ratio indicates a reversal of direction. Observe that the definition of a continuously variable transmission implies that the rotation of the output shaft is never reversed within the transmission range.

[2] Some authors have defined the term 'Infinitely Variable Transmission' based on the definition of the transmission ratio as being the ratio between the input and output speeds, ω_i/ω_o. In this aspect, a zero output speed would correspond to an infinite transmission ratio [2]. Using this argumentation, an infinitely variable transmission would be simply defined as a Continuously Variable Transmission for which $0 \leq |R_T| \leq |c_1|$.

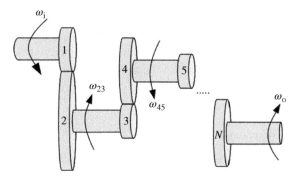

Figure 1.3 Train of friction discs

(non-deformable). This kind of transmission is not practical, but the equations obtained here are representative of the most general scenario. By assuming that no slip occurs between the discs, we can say that the tangential velocities on the surface of the two contacting discs are equal. This allows us to obtain the rotating speed of the shaft between discs 2 and 3 (represented by the subscript 23) through the following expression:

$$\omega_{23} = -\left(\frac{r_1}{r_2}\right)\omega_i \tag{1.2}$$

where r is the radius of the corresponding disc (disc 1 or disc 2).

If we proceed in this manner for every pair of discs, we obtain the following general relation for a train with $N/2$ pair of discs (N is an even number of discs, as shown in Figure 1.3):

$$\omega_o = (-1)^{\frac{N}{2}}\left(\frac{r_1 r_3 r_5 \cdots r_{N-1}}{r_2 r_4 r_6 \cdots r_N}\right)\omega_i \tag{1.3}$$

The transmission ratio, in this case, is given by:

$$R_T = \frac{\omega_o}{\omega_i} = (-1)^{\frac{N}{2}}\left(\frac{r_1 r_3 r_5 \cdots r_{N-1}}{r_2 r_4 r_6 \cdots r_N}\right) \tag{1.4}$$

We observe that friction discs lose their non-slip characteristic as soon as the resisting torque reaches a certain value. That is why gear trains are used instead of disc trains. Nevertheless, Eqs. (1.3) and (1.4) are still valid for gear trains if we replace r_i by the pitch radii of the gears, or, as it is usually the case, the corresponding number of teeth. An almost identical transmission ratio can be obtained for a chain of belts and pulleys under the non-slipping assumption.[3] It is also possible to show that chain drives behave similar to belts and pulleys.

---

[3] In fact, the only difference would be the absence of the reversal term, -1, in Eq. (1.4), given that the direction of rotation between each pair of pulleys is the same.

1.2.2   Gearboxes

Gearboxes are typical examples of discretely variable transmissions. We describe two types of gearboxes here: sequential and planetary. Within the automotive industry framework, sequential gearboxes are used in manual transmission cars, whereas planetary gearboxes are usually found in automatic transmission vehicles.

1.2.2.1   Sequential Gearboxes

The train shown in Figure 1.3 is limited because the whole mechanism gives only one constant transmission ratio, and that cannot be easily changed. In many applications, however, it is important for the transmission to offer the choice of more than one transmission ratio. The ideal case would be the one in which the transmission could be able to produce an infinite number of transmission ratios. Gearboxes allow for multiple transmission ratios and, in this sense, represent a step forward in relation to the fixed train represented by Figure 1.3.

Figure 1.4 shows a simple two-speed sequential gearbox. The collar C is mechanically engaged to the splined shaft B and is allowed to slide horizontally on the shaft ridges. Gears 2 and 4, on the other hand, are mounted on bearings, being mechanically decoupled from shaft B. Teeth are placed on the inner sides of gears 2 and 4 to allow for the coupling with the sliding collar and, as a result, with shaft B. Gears 1 and 3 are rigidly coupled to shaft A and rotate with it. The collar C may engage with either gear 2 or gear 4, with the aid of the fork F directly connected to a shifting lever, providing a choice of two transmission ratios as only one pair of gears is actually transferring power at a time: gears 1 and 2 or gears 3 and 4. We could proceed in this manner and add more gear couplings to the gearbox, creating a multiratio gearbox. Note, however, that still only a few discrete values for the transmission ratio would be available in the end, and the shifting between each ratio would

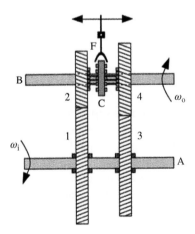

Figure 1.4   Schematic representation of a two-speed sequential gearbox

Figure 1.5 Sequential gearbox

require decoupling the driving shaft A from its prime mover by means of a clutch to avoid damaging the collar mechanism.

Figure 1.5 shows a typical sequential gearbox.

1.2.2.2 Planetary Gearboxes

In the gearboxes shown in Figures 1.4 and 1.5, the gears' teeth engage externally with one another. Figure 1.6 shows another arrangement called a planetary gearbox in which the engagements occur both externally and internally. Gears S and P (sun and planet) are held against each other by a rotating bar C (planet carrier) that is pivoted on each gear in a way that gear P can rotate around its centre and orbit around gear S simultaneously. Both gears P and S are placed inside an outer gear (ring R).

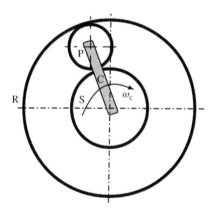

Figure 1.6 Schematic representation of a planetary gearbox

Planetary gearboxes can have different gear ratios depending on the way in which power is being transferred. For instance, consider the pitch radii of the sun and ring gears: r_S and r_R. It can be easily shown that the angular speeds of the ring, sun and carrier, ω_R, ω_S and ω_C, respectively, are related through the following equation[4]:

$$\frac{r_S}{r_R} = -\left(\frac{\omega_R - \omega_C}{\omega_S - \omega_C}\right) \tag{1.5}$$

Suppose, for example, that the input power source is connected to the carrier C. If we connect the output to the ring R and fix the sun S ($\omega_S = 0$), we get, from Eq. (1.5),

$$\frac{r_S}{r_R} = \frac{\omega_R - \omega_C}{\omega_C} \tag{1.6}$$

From Eqs. (1.6) and (1.1), we obtain the transmission ratio, in this case:

$$R_T = \frac{\omega_R}{\omega_C} = \frac{r_S}{r_R} + 1 \tag{1.7}$$

We may proceed in the same manner for other possible configurations. Note that when the carrier C is fixed ($\omega_C = 0$), a reversion of the rotation occurs (see Eq. (1.5)). As expected, Eqs. (1.5)–(1.7) can be written as a function of the gear teeth instead of the pitch radii by substituting r_S and r_R by the corresponding number of teeth z_S and z_R.

Planetary gearboxes are more complex to manufacture when compared to sequential gearboxes. However, changing gears only requires a set of simple devices to keep the right element stationary. For example, we may use a belt around the ring gear (brake band) to keep it stationary while connecting the output shaft to the sun gear with the aid of a clutch. A similar procedure can be applied to the other gears. In planetary gearboxes, the transmission ratio can be changed without the need to disengage the gears simply by holding the right elements; this makes this type of gearbox attractive in automatic automotive transmissions. Another important feature of planetary gearboxes is that they can be used as either mechanical power dividers or combiners. For example, by connecting the ring and the sun to different power sources, the total power input can be combined into an output shaft connected to the carrier. This feature will be better explained later on in this chapter when we introduce the theme of power-split transmissions. A typical planetary gearbox is illustrated in Figure 1.7.

1.2.3 Efficiency

The transmission ratio given by Eq. (1.4) was determined under the premise that the velocities on the surface of two contacting discs were equal. That is true as long as there is no slippage between the discs and no deformation in the point of contact so that the discs remain perfectly round. However, by transitioning to the usage of gears, the picture of two friction discs is replaced by two surfaces pressing against one another. For this reason we cannot use

---

[4] We leave this demonstration as an exercise for the student at the end of this chapter.

Figure 1.7 In-line planetary gear reducer 3:1-10:1 I AE series (courtesy of Apex Dynamics, USA)

either the external or the internal radii of the gears in relation (1.4); instead, we must use a virtual dimension called pitch radius. Fortunately, this is equivalent to using the number of gear teeth instead [3], which makes the determination of the transmission ratio much easier.

Figure 1.8 shows the moment when two gears are engaging. The instantaneous speed of a point in gear A, located at the intersection of the two contacting surfaces, is $v_A = r_A \omega_A$. Similarly, the instantaneous speed of a point in gear B, 'in touch' with point A, is $v_B = r_B \omega_B$. The projection of these two speeds onto line t, which is tangent to the contacting surfaces, will be different most of the time. For example, in the figure, we see that the projection of

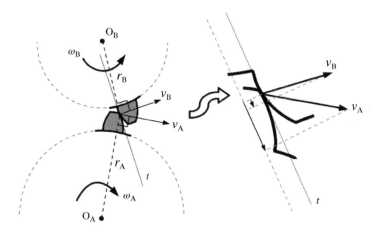

Figure 1.8 Two engaging gears (only one tooth profile per gear is shown)

v_A onto t is bigger than the projection of v_B onto t. Such difference causes the two gear teeth to slide against each other.

By multiplying the magnitude of the difference between the velocity projections by the friction force between the teeth, we obtain the instantaneous friction dissipative power that is lost in the form of heat. Friction dissipation will be greater when either the normal force or the friction coefficient between the teeth increases. The relative velocity also increases with the angular velocities ω_A and ω_B. Therefore, higher angular speeds favour energy losses.

From what has been examined so far, we observe that not all the input power is transferred to the output shaft; a certain amount of that power is lost. Moreover, the amount of energy dissipation is not constant, given that the losses depend on factors such as torque, angular speed, lubricant, surface roughness of the gear teeth and temperature. In the particular case of mechanical transmissions, the energy loss is translated into a smaller torque available at the output shaft.

It has been reported that gear transmissions are quite efficient, each pair being capable of transmitting as much as 99.1–99.9% of the input power [4]. However, it must be pointed out that depending on the size of the gearbox, its design and the choice of the lubricant, this efficiency could drop significantly. For example, miniature planetary gearboxes have been reported to have a maximum efficiency near 73% [5].

We finish this discussion about efficiency by giving it a formal definition that will be used throughout this book:

> Efficiency of a power transmission is a measure of the energy lost in the power transfer process and is defined as the ratio between the transmission power output, P_o, and the transmission power input, P_i:

$$\eta = \frac{P_o}{P_i} \tag{1.8}$$

If we recall that power is the product of the torque and the angular speed ($P = T\omega$), we may write Eq. (1.8) as,

$$\eta = \frac{T_o \omega_o}{T_i \omega_i} \tag{1.9}$$

Using the transmission ratio definition (Eq. (1.1)), we obtain,

$$T_o = \frac{\eta T_i}{R_T} \tag{1.10}$$

From Eq. (1.10), we see that, for a fixed transmission ratio, the relation between the input torque and the output torque in gearboxes is constant, as long as the efficiency remains constant. As said earlier, lower efficiencies reflect directly on the magnitude of the output torque.

In closing this section, it is very important to mention another aspect of gear transmissions: teeth geometry. Depending on the gear diameters and the number of teeth, the tooth of one of the gears may carve into the other's causing a serious problem known as *interference*. The way to avoid this is to limit the minimum number of teeth of the smallest gear; therefore, the size of the transmission must be inferiorly limited (the interested reader may consult [3] for a more detailed discussion on this topic). As the dimensions of the gear teeth must grow for higher torques, the size of the gearbox is also affected by the power to be transferred. Gearboxes, therefore, can become bulky and heavy, and this characteristic can be seen as a negative feature for some particular applications (such as in wind turbines [6]).

1.2.4   Continuously and Infinitely Variable Transmissions

1.2.4.1   Basic Design

The idea of a CVT is illustrated with the aid of three friction discs, as shown in Figure 1.9. The three discs A, B and C have radii r_A, r_B and r_C, respectively. In this figure, discs A and B rotate in opposite directions with vector angular speeds $\vec{\omega}_A$ and $\vec{\omega}_B$. The angular speed of the intermediate disc, C, is $\vec{\omega}_C$. Disc C touches both discs A and B in two points whose distance to the centre lines are given by r_1 and r_2, respectively. Although it is not shown in the figure, both r_1 and r_2 can have negative values. If r_1 and r_2 have opposite signs, the discs A and B rotate in the same direction. Observe that the distance between discs A and B varies with both r_1 and r_2, reaching a maximum for $r_1 = r_2$.

In order to transmit high powers, the device illustrated in Figure 1.9 needs a high friction between the discs. As a result, the contact pressure must be high in order to create sufficient friction torques and, since the contact between the discs occurs practically along a line segment, the superficial deformation becomes significant. This is the main reason why this kind of device is not used in practice. However, the toroidal transmission, which will be introduced next, constitutes a natural development of this basic scheme and has been commercially used.

From Figure 1.9, we obtain the transmission ratio between discs A and B:

$$R_T = -\frac{\omega_B}{\omega_A} = -\frac{r_1}{r_2} \tag{1.11}$$

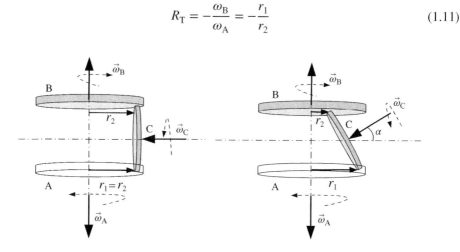

Figure 1.9   Basic continuously variable transmission

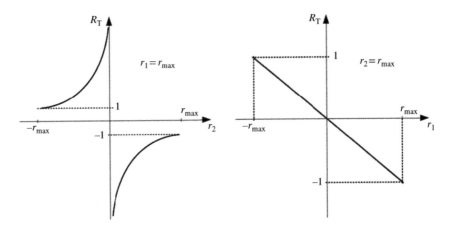

Figure 1.10   Transmission ratio in a mechanical IVT

Equation (1.11) shows that the transmission illustrated in Figure 1.9 can have an infinite transmission ratio since the values of r_1 and r_2 can range from a negative minimum to a positive maximum. Therefore, if we disregard any physical limitation, what we have is an IVT in the strictest sense of the expression. Figure 1.10 shows the transmission ratio in a graph format as given by Eq. (1.11). Two situations are shown: one in which we fix the radius r_1 at its maximum value, r_{max}, and another in which the radius r_2 is fixed at r_{max}. In both cases, we plot the variation of the transmission ratio for the other radius changing from $-r_{max}$ to $+r_{max}$. In the first case, we obtain a speed amplification at the output shaft as the transmission ratio grows infinitely when r_2 approaches zero. In the second case, the output speed is reduced in relation to the input speed, becoming zero at $r_1 = 0$ and smoothly reversing as disc C is gradually inclined.

1.2.4.2   Toroidal Continuously Variable Transmissions

Toroidal CVTs are based on the IVT model shown in Figure 1.9. However, because of the way in which they are built, they are not IVTs.

Figure 1.11 shows the working principle of a toroidal transmission. The input and output shafts are connected to the toroidal discs A and B, respectively. Discs C_1 and C_2 work in a synchronized way, moving symmetrically and playing the same role of disc C in Figure 1.9.[5] Discs A and B are free to slide over the vertical shaft that passes through them so that they may move towards or apart from one another as needed.

As can be seen in Figure 1.11, a reversal of the rotation is not possible in this configuration, and the transmission ratio is limited within a range that is determined by the geometry of the discs. Toroidal CVTs can be found in automobile transmissions, being regarded as 'smooth' and efficient [7].

Even though the toroidal design does not allow for the construction of an IVT or reverse the rotation of the driven shaft in relation to the driving shaft, it is possible to combine it with a planetary gearbox in a way that a reversal of the output speed can be continuously

---

[5] Although only one disc would be necessary in this case, two discs are used to keep the vertical shaft mechanically balanced.

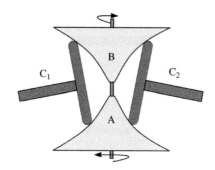

Figure 1.11   Schematic representation of a toroidal CVT

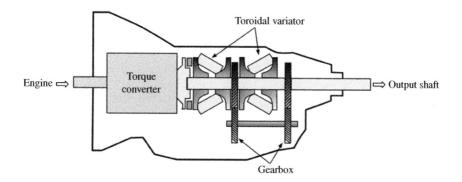

Figure 1.12   Toroidal power-split IVT [7]

attained [1]. The idea here is to 'split' the input power so that part of it goes through a CVT, also known as a *toroidal variator*,[6] and the other part goes through a discretely variable transmission (gearbox), as illustrated in Figure 1.12.[7]

The power-split concept shown in Figure 1.12 can be generalized to include other types of CVTs. For instance, as will be seen in Section 1.6, the same idea can be applied to a combined hydrostatic transmission and gearbox.

In order to understand the operation of a power-split transmission, consider the arrangement[8] shown in Figure 1.13. In this figure, we see that the prime mover shaft is connected to both the input shaft of the variator, A, and the carrier of a planetary gearbox, C. Furthermore, the output shaft of the variator, B, is connected to the sun gear of the gearbox, S. Finally, the output shaft of the combined transmission is connected to the planetary ring, R. Note that we have

$$\begin{cases} \omega_C = \omega_A = \omega_i \\ \omega_B = \omega_S = -\omega_A R_T^{CVT} = -\omega_i R_T^{CVT} \end{cases} \tag{1.12}$$

where R_T^{CVT} is the CVT ratio of the toroidal transmission.

---

[6] In power split transmissions, it is common to refer to the CVT component as *variator*.

[7] Torque converters will be introduced in Section 1.3.

[8] Not necessarily the same arrangement be used in the transmission shown in Figure 1.12.

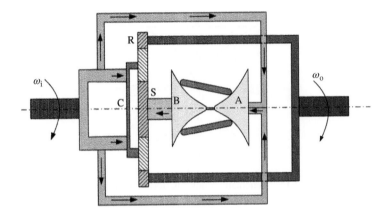

Figure 1.13 Combination of a toroidal CVT and a gearbox

From Eqs. (1.5) and (1.12), we have

$$\frac{r_S}{r_R} = -\left(\frac{\omega_o - \omega_i}{-R_T^{CVT}\omega_i - \omega_i}\right) \tag{1.13}$$

We can rewrite Eq. (1.13) in a more meaningful way:

$$\omega_o = \omega_i\left[1 + \frac{r_S}{r_R}\left(R_T^{CVT} + 1\right)\right] \tag{1.14}$$

The transmission ratio of the combined transmission, R_T, can now be obtained:

$$R_T = \frac{\omega_o}{\omega_i} = 1 + \frac{r_S}{r_R}(R_T^{CVT} + 1) \tag{1.15}$$

Table 1.1 shows the behaviour of the transmission ratio, R_T, in terms of the number of teeth of the sun and the ring gears, z_S and z_R, for different values of the toroidal transmission ratio, R_T^{CVT}. We observe that through the use of the combined toroidal and gearbox transmissions, we can obtain a transmission ratio that continuously ranges from a negative value to a positive value, that is, an **IVT**.

1.2.4.3 Variable Diameter Pulleys

Figure 1.14 shows a continuously variable mechanical transmission with conic pulleys and a steel belt [7] or a chain [8]. The pulley cones can slide over their shafts towards or apart from one another. In order to keep the belt tightly adjusted if one pair of pulley cones moves towards one another, the other must be set apart by the same length. For example, in Case 1, the pulley cones in the top are close to each other, whereas the pulley cones at the bottom are separated. When we consider pulley A as the input pulley, in this case, $R_T > 1$ ($\omega_A < \omega_B$),

Table 1.1 Transmission ratio range for the power-split transmission

CVT transmission ratio	Combined CVT and gearbox transmission ratio
$R_{\mathrm{T}}^{\mathrm{CVT}} < -\left(\dfrac{z_{\mathrm{R}}}{z_{\mathrm{S}}} + 1\right)$	$R_{\mathrm{T}} < 0$
$R_{\mathrm{T}}^{\mathrm{CVT}} = -\left(\dfrac{z_{\mathrm{R}}}{z_{\mathrm{S}}} + 1\right)$	$R_{\mathrm{T}} = 0$
$R_{\mathrm{T}}^{\mathrm{CVT}} > -\left(\dfrac{z_{\mathrm{R}}}{z_{\mathrm{S}}} + 1\right)$	$R_{\mathrm{T}} > 0$

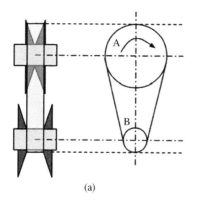 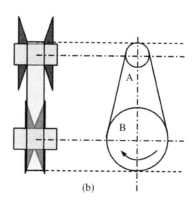

(a)   (b)

Figure 1.14 Continuously variable transmission with conic pulleys: (a) Case $1 - \omega_{\mathrm{A}} < \omega_{\mathrm{B}}$ and (b) Case $2 - \omega_{\mathrm{A}} > \omega_{\mathrm{B}}$

we have a speed amplification. In Case 2, the top pulley cones are separated, whereas the bottom pulley cones are brought together. The transmission ratio then changes to $R_{\mathrm{T}} < 1$ ($\omega_{\mathrm{A}} > \omega_{\mathrm{B}}$) and a reduction occurs. Note that it is not possible to reverse the direction of rotation in this transmission design.

Variable diameter transmissions have been widely used in the automobile industry. Figure 1.15 shows a typical CVT with a steel belt.

1.3 Hydraulic Transmissions

In hydraulic transmissions, power is transmitted by means of a fluid connecting a hydraulic pump to a hydraulic motor. The pump receives mechanical energy from a rotating shaft connected to the prime mover and transfers it to the fluid in the form of flow and pressure (hydraulic energy). The fluid then carries the hydraulic energy into the motor where it is transformed back into mechanical power at the output shaft connected to a mechanical

Figure 1.15   Variable diameter CVT (courtesy of Jatco Ltd)

device (Figure 1.16). The basic elements in a hydraulic transmission are, therefore, the
pump, the fluid and the motor, whose individual roles can be summarized as follows:

The pump converts mechanical energy into hydraulic energy.
The fluid transports hydraulic energy from the pump into the motor.
The motor converts hydraulic energy into mechanical energy.

Depending on the type of pump and motor used in the hydraulic transmission of
Figure 1.16, the result can be either a hydrodynamic transmission or a hydrostatic transmission. Despite this general definition, it is not unusual to find the terms 'hydraulic
transmission' and 'hydrostatic transmission' being used interchangeably, especially in
older references [9, 10]. In the automotive industry, the term 'hydrodynamic transmission'
itself is not very popular, being usually substituted by the expression 'torque converter', as
will be described shortly.

Hydrodynamic transmissions require the connection of a hydrodynamic pump to a
hydrodynamic motor. In a hydrodynamic motor, torque is created through a change in
the fluid velocity as it passes through the internal blades and channels. Similarly, the
torque input at a hydrodynamic pump causes the fluid velocity to change in intensity and
direction, producing a flow. This is better visualized in Figure 1.17, which shows a typical

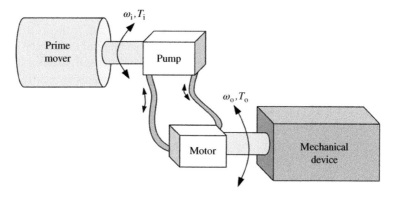

Figure 1.16   Schematic representation of a hydraulic transmission

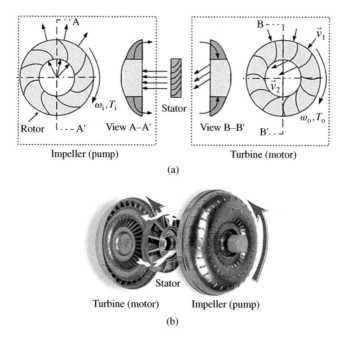

Figure 1.17   Torque converter: (a) schematic diagram and (b) exploded view (courtesy of BD Diesel Performance)

hydrodynamic transmission used in automatic cars, best known as a 'torque converter' as mentioned earlier.

In the schematic representation given in Figure 1.17(a), we identify the pump (impeller), the energy-carrying fluid and the motor (turbine). The pump receives power from the engine and rotates at an angular speed ω_i. As the rotor connected to the pump shaft revolves, the fluid is radially accelerated towards the case. Due to the case curvature, the flow changes direction and leaves the pump parallel and opposite to the incoming flow (see view A–A′). Likewise, as soon as the fluid reaches the turbine (motor), the case redirects the stream so that it now flows radially towards the centre. Due to the curvature of the turbine blades, the velocity of the fluid changes direction, as shown by their projections $\vec{v}_1$ and $\vec{v}_2$. The change in the velocity vector as the fluid travels from the turbine case towards the centre produces a tangential force and, as a consequence, an output torque, T_o, that depends on the magnitude of the incoming speed and the aerodynamic design of the blades.

Due to the geometry of the turbine blades, the flow returning from the turbine into the impeller is no longer horizontally oriented. If this flow, coming from the turbine, hit the impeller blades just as it were, that is, inclined in relation to the transmission axis, it would end up contributing to a deceleration of the impeller as an opposite torque would be created. Therefore, in order to redirect the flow from the turbine into the pump, a stator (a helical fluid redirector) is placed in between the impeller and the turbine [11]. Note that, even if the turbine stops, torque at the output shaft will still exist as long as fluid is pumped through the turbine blades. At this stage, the input power would be totally converted into torque and heat (due to the viscosity of the fluid). Because of this particular feature, it seems appropriate to use the term 'torque converter' instead of 'hydrodynamic transmission' when referring to the device illustrated in Figure 1.17, given that power is not always transmitted between the input and output shafts.[9]

When both pump and motor in a hydraulic transmission are 'hydrostatic', we have a hydrostatic transmission. In general terms, we may say that in hydrostatic pumps, fluid is literally 'pushed' into the circuit, while, on the motor side, it is the fluid pressure that causes the motor to move. A simple illustration of the operational principle of hydrostatic pumps and motors is given in Figure 1.18, where the crank and shaft mechanism on the left (pump) pushes the fluid with a force F_i, creating pressure p inside the hydraulic circuit. On the other hand, the piston on the right (motor) turns the crank by the action of this same pressure (p). Therefore, the input power is transmitted to the output shaft.

Depending on the relative diameters of the pump and motor pistons in Figure 1.18, we can obtain a speed amplification or reduction. Note, however, that differently from the torque converter shown in Figure 1.17, it is not possible to stop the motor shaft and still have the pump shaft rotating. In other words, if power (torque and speed) is input at the pump, we must have a power output at the motor[10] – the basic characteristic of any power

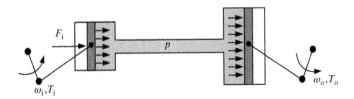

Figure 1.18 Operational principle of hydrostatic pumps and motors (left: pump and right: motor)

---

[9] Remember that power requires two components: torque and angular speed. Therefore, when the turbine shaft is stationary, although torque is being generated, no power is being transmitted.
[10] Here, we assume that there is no fluid leakage in the circuit.

Figure 1.19   External gear-type hydrostatic pump (courtesy of Parker-Hannifin Corp.)

transmission. Figure 1.19 shows a typical hydrostatic pump that can be used in hydrostatic transmissions.[11]

In what follows, we explore hydrostatic transmissions in more detail.

1.4   Hydrostatic Transmissions

Having introduced the main differences between hydrodynamic and hydrostatic transmissions, we now focus on hydrostatic transmissions, one of the theme subjects of this book. We begin by exploring the way in which hydrostatic transmissions operate, with emphasis on the circuit components. We then move on to give a formal definition of 'hydrostatic transmissions', followed by some considerations about efficiency and classification. Before we move on, however, we need to say something about the way in which hydraulic circuits are represented in this book.

As a general rule, the ISO 1219-1 hydraulic symbols (see Appendix A) are frequently used in this book to describe hydraulic circuits. However, a few non-standard symbols are also employed to enhance text comprehension or to represent a particular situation for which a standard symbol is not available.[12] Table 1.2 shows two non-standard symbols used throughout this book, representing a generic prime mover, such as an internal combustion engine or an electric motor.

1.4.1   Operational Principles

Figure 1.20 illustrates a simple situation where pump and motor are connected as a hydrostatic transmission. In Figure 1.20, the prime mover, PM, is connected to a reversible hydrostatic pump[13] whose input and output ports 1 and 2 link to the motor input and output

---

[11] There is no crank mechanism in the pump shown in Figure 1.19 (keep in mind that Figure 1.18 is merely illustrative). Details about hydrostatic pumps and motors, including the one shown in Figure 1.19, will be presented in Chapter 3.

[12] It must be noted here that the use of non-standard ISO symbols is a common practice in hydraulics, as can be confirmed by a number of references quoted in this book.

[13] A reversible hydrostatic pump is a pump that can output flow in both directions.

Table 1.2   Non-standard prime mover hydraulic symbols

Description	Symbol
Prime mover	PM
Variable speed prime mover	PM

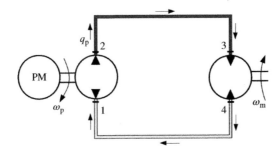

Figure 1.20   Pressure rise in the pump–motor conduit

ports 3 and 4. Imagine that the motor shaft is slowed down by a resistive load while the pump continues sending its flow, q_p, into port 3. As a consequence, we observe that the pressure in lines 2–3 will rise, as illustrated in the figure.[14]

One possible solution for lowering the pressure in the conduit 2–3 of the circuit shown in Figure 1.20 is to reduce the pump output flow. This can be done in two different ways in a hydrostatic pump:

1. By varying the prime mover speed.
2. By varying the pump displacement, that is, by adjusting the pump flow without changing its speed.[15] It is important to mention that, in this particular case, only variable-displacement pumps can alter their output flows for a constant speed of the shaft. For fixed-displacement pumps, this is not possible.

We will give a formal definition of displacement shortly. In the meantime, note that the adoption of the first or the second solution changes the hydraulic symbols of Figure 1.20 accordingly. Figure 1.21 shows the representation of a variable-speed prime mover and a variable-displacement pump, respectively.

---

[14] As seen in Section 1.3, this is one of the basic differences between a hydrostatic and a hydrodynamic transmission. In hydrodynamic transmissions, stopping the motor shaft will also elevate the internal pressure of the circuit, whereas in hydrostatic transmissions, the situation is much more complicated because stopping the motor would imply stopping the pump as well. Thus, the rise in pressure can be extremely high in hydrostatic transmissions, depending on the input power. A means of alleviating the pressure in the circuit must, therefore, be provided.

[15] *Displacement* is a concept unique to hydrostatic pumps and motors.

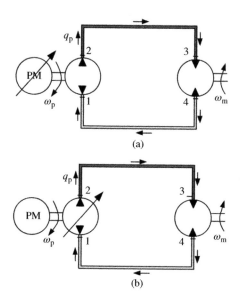

Figure 1.21   Two different ways of controlling the pump output flow: (a) variable-speed prime mover and (b) variable-displacement pump

1.4.1.1 Pump Displacement and Flow

We have mentioned that a hydrostatic pump can have a fixed or a variable displacement. A general definition for displacement in a pump is as follows:

> Displacement of a hydrostatic pump is a technical term given to the maximum theoretical fluid volume displaced by the pump when the rotor performs a complete revolution, considering the hypothetically perfect situation where no volumetric losses are present.

It is easy to see that a fixed-displacement pump, connected to a constant-speed prime mover, will output a constant flow.[16] For instance, if the pump displacement is $100\,\text{cm}^3/\text{rev}$ and the prime mover rotates at $1800\,\text{rpm}$, the pump will output a flow of $180{,}000\,\text{cm}^3/\text{min}$, or, in more common units, $180\,\text{l/min}$. We can, therefore, express the average pump output flow, q_p as a function of its displacement, D_p, and the angular speed of its shaft, ω_p:

$$q_p = D_p\omega_p \tag{1.16}$$

In a typical variable-displacement pump, D_p changes continuously between two limiting values. As a result, the output flow, q_p, can be altered even if the angular speed, ω_p,

---

[16] It will be shown in Chapter 3 that the pump flow is not, actually, constant in time. Therefore, q_p as defined by Eq. (1.16) must be seen as an average value [12], obtained for the hypothetical case where no volumetric losses are present. The same observation is valid for hydrostatic motors.

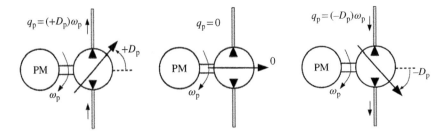

Figure 1.22   Pump output flow and displacement

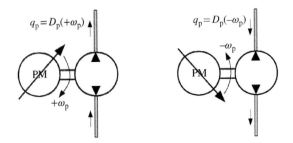

Figure 1.23   Pump output flow and variable prime mover speed

does not change. Moreover, in a variable-displacement pump, whenever the displacement changes sign, the pump flow changes its direction.[17] This is equivalent to say that the output port of the pump becomes the input port, and vice versa. Figure 1.22 illustrates the three possibilities.

From Eq. (1.16), we observe that we can also alter the output flow by varying the rotation of the pump shaft, ω_p. In the particular case, when ω_p is reversed, the pump flow changes its direction as illustrated in Figure 1.23.

1.4.1.2   Motor Displacement and Speed

We saw that the pump output flow can be controlled by changing the displacement, D_p. Similarly, the angular speed of hydrostatic motors, ω_m, is related to the displacement, D_m, and the input flow, q_m, through the following expression:

$$\omega_m = \frac{q_m}{D_m} \qquad (1.17)$$

The concept of displacement in a hydraulic motor[18] is a little bit different from the concept given for the pump. Note that, here, the fluid that flows into the motor causes its shaft to

---

[17] Note that in Eq. (1.16), the pump displacement, D_p, can be negative, positive or zero.

[18] It is a common practice to use the term 'hydraulic motor' instead of 'hydrostatic motor'. Likewise, the term 'hydro-dynamic motor' is usually replaced by some more specific term, such as 'turbine'. In this book, to keep with the usual convention, whenever we refer to a hydraulic motor, we are, in fact, referring to a hydrostatic motor.

rotate, whereas in the pump, it was the rotation of the pump shaft that caused the fluid to flow. We can, therefore, say that:

> Displacement of a hydraulic motor is a technical term given to the theoretical fluid volume that, when flowing through the motor, causes the shaft to perform a complete revolution, considering a perfect situation where no volumetric losses are present.

It is easy to see that the units of the motor displacement are the same as the units used for the pump displacement. Also, similar to pumps, motors can be of either fixed-displacement or variable-displacement types.

In Eq. (1.17), if the motor displacement approaches zero, the angular speed, ω_m, approaches infinity. Mathematically speaking, $D_m = 0$ is a singular point in Eq. (1.17). We know, by experience, that infinite angular speeds are not possible. Moreover, due to dynamic considerations, every rotary machine must have a maximum speed beyond which operation is not recommended. Thus, variable-displacement hydraulic motors should never operate near zero displacement when there is flow coming from the pump. In fact, the displacement of variable hydraulic motors is usually set between two limiting values so that there is no risk of reaching the maximum angular speed. Figure 1.24 illustrates what has been said so far. Notice that an inversion of the flow direction automatically results in a reversal of the motor rotation.

1.4.1.3 Motor Displacement and Pressure

Suppose that the motor input flow comes from a fixed-displacement pump, as shown in Figure 1.25. If we assume that there are no energy losses between the pump and the motor, we can write the following relation between the torque applied to the pump shaft and the torque output by the motor (see Eq. (1.10)):

$$T_m = \frac{T_p}{R_T} \qquad (1.18)$$

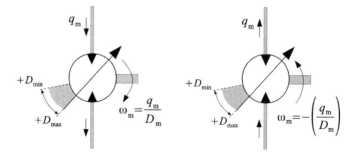

Figure 1.24   Motor output flow and displacement

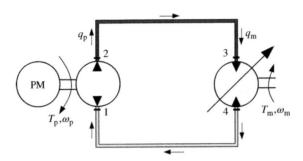

Figure 1.25 Pressure in a hydrostatic transmission with a variable-displacement motor

We observe from Eq. (1.18) that a high transmission ratio produces a low torque at the transmission output, and vice versa.

From Eqs. (1.1) and (1.18), the torque at the pump, T_p, necessary to produce the torque T_m, at the motor, is given by

$$T_p = T_m R_T = T_m \left(\frac{\omega_m}{\omega_p} \right)$$ (1.19)

In the hypothesis that the motor displacement becomes zero, we have seen that the angular speed of the motor shaft becomes infinite and, therefore, the torque at the pump also goes to infinity (Eq. (1.19)). We shall see, in Section 3.4.4, that the input torque for an ideal and completely efficient pump can be written as

$$T_p = D_p(p_2 - p_1)$$ (1.20)

where p_2 is the pressure at the pump output port (high-pressure line) and p_1 is the pressure at the pump input port (low-pressure line). Observe that in a 100% efficient transmission, $p_2 = p_3$ and $p_4 = p_1$, as shown in Figure 1.25.

By writing $q_p = q_m$ and using Eqs. (1.16) and (1.17), the transmission ratio in the absence of transmission losses is

$$R_T = \frac{\omega_m}{\omega_p} = \frac{D_p}{D_m}$$ (1.21)

After substituting T_p and R_T, given by Eqs. (1.20) and (1.21), into Eq. (1.19), we arrive at the following expression for the pressure at the pump output, p_2:

$$p_2 = \frac{T_m}{D_m} + p_1$$ (1.22)

Equation (1.22) shows that whenever the motor displacement becomes small, the pressure in the pump output, p_2, becomes high. Moreover, it tends to infinity when the motor displacement, D_m tends to zero. Therefore, we have another limitation for the minimum displacement of the motor. This limitation now describes the whole circuit, not just the motor itself (the high pressure in line 2–3 will affect everything that is connected to the output port of the pump).

1.4.1.4 Pressure Overshoot Attenuation

We have seen that the solution for the problem of the high pressures in the pump–motor line 2–3 (Figure 1.25) is to reduce the pump output flow. However, there may be momentary pressure elevations for which a faster relief is necessary (e.g. during unexpected high loads at the motor shaft). In those situations, there is usually no time to reduce the pump output flow so that the pressure is quickly attenuated. The usual solution is to use a *pressure relief valve* (Figure 1.26), which opens a passage between the two lines, 2–3 and 3–4, as soon as the pressure rises in one of the conduits.

Figure 1.27 illustrates the use of pressure relief valves in hydrostatic transmissions. In this figure, a variable-displacement pump is coupled with a fixed-displacement motor. Note that two valves are employed, and each one of them can become operational for a corresponding flow direction.

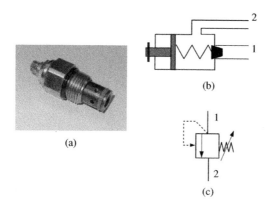

Figure 1.26 Pressure relief valve: (a) picture, (b) operational principle and (c) ISO representation

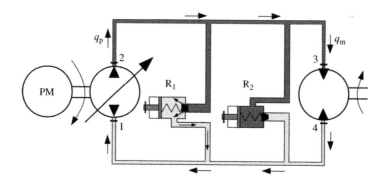

Figure 1.27 Pressure overshoot attenuation using relief valves

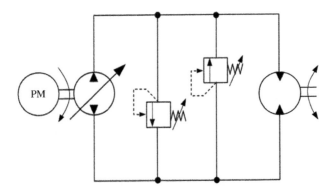

Figure 1.28 ISO Representation of the hydrostatic transmission shown in Figure 1.27

In a pressure relief valve, the pressure acts against the force of a spring. As the pressure rises beyond a certain value, the corresponding valve spring is compressed, opening a passage from the high-pressure conduit to the low-pressure conduit, as shown in the figure for the valve R_1. The pressure in the circuit can then be relieved, preventing an eventual damage of the parts connected to the high-pressure line.

The ISO representation of the circuit in Figure 1.27 is shown in Figure 1.28.

1.4.1.5 Leakages and Fluid Replenishment

Up until this point, we have been considering that all the flow that comes out of the pump makes its way through the conduits, passes through the motor and returns to the pump. However, this is not what happens in real life. Hydraulic components usually leak to some degree. To understand where the leakages happen in a hydrostatic transmission, consider the flow at point 1 immediately before the pump (Figure 1.27), and let us follow it in a clockwise direction. As we travel from point 1 to point 2, part of the flow leaks through the internal clearances of the pump into the pump case. From point 2 to point 3, there may be some flow through the relief valve R_1, as indicated in the figure, but this flow remains in the circuit, being diverted into line 4–1. From point 3 to 4, fluid leaks from the high-pressure chambers of the motor into the motor case in the same way it leaked in the pump. In the return line 4–1, no leakages are likely to happen. When we reach point 1 again, the circuit will be depleted of hydraulic fluid because of the external leakages at the pump and motor.

> In a hydrostatic transmission, there are two kinds of leakages: internal and external. Internal leakages remain in the main circuit (pump–motor), whereas external leakages happen between the main circuit and the outside. By 'the outside', we mean to describe the elements through which the main flow does not circulate, such as the pump and the motor cases.

If we consider the external leaking flows we have seen so far (between points 1 and 2 and points 3 and 4), it is clear that the flow reaching the pump input, q_1, will be smaller than

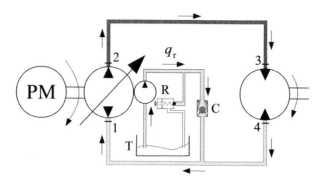

Figure 1.29   Charge circuit in a hydrostatic transmission

the flow leaving the pump, q_2. Because of that, a charge circuit is provided to replenish the fluid that has been lost through external leakages. Usually, this circuit consists of a small pump (driven by the same prime mover, PM, that is connected to the main pump), a pressure relief valve (to keep the charge pressure limited), an oil reservoir (tank) and a check valve (which allows for the passage of the fluid in only one direction). The check valve is important because we want the replenishing fluid to flow into the main circuit while preventing the fluid in the main circuit from flowing back to the reservoir. In many cases, the charge pump also powers the pump servo-control cylinder that is responsible for changing the pump displacement. Figure 1.29 illustrates how the charge circuit operates in a simple unidirectional transmission. Relief valves between lines 2–3 and 4–1 are not shown in the figure to keep things simplified.

In Figure 1.29, we see that the check valve, C, is simply represented by a sphere that is loosely placed inside a case. If the flow comes from the indicated direction, the sphere is pushed downwards until it stays held in the mid-position so that fluid can flow into the conduit 4–1. If the flow tries to go the other way around, the sphere will close the passage between the conduit 4–1 and the charge pump. The relief valve, R, keeps the pressure in the line 4–1 at a pre-determined level,[19] given by the degree of compression of its spring. As shown in the figure, the relief valve R usually possesses an external adjustment to the spring compression, which allows for the control of the pressure at the charge circuit. Figure 1.30 shows a typical check valve used in hydraulic circuits.[20]

Figure 1.31 shows the ISO representation of the complete circuit, with all the elements represented in Figures 1.27 and 1.29. Note that there are two independent check valves, C_1 and C_2, each one connecting to a different side of the circuit.

1.4.1.6   Energy Storage and Hydraulic Accumulators

One of the great advantages of using a fluid as an energy transportation medium is the capacity to store energy for later reuse. The way in which this is usually done is by either

---

[19] Typical values for the pressure in the charge circuit lie between 7.5 and 26 bar [13].

[20] Despite the ISO representations shown in Figure 1.30(c), commercially available check valves are usually built with an internal spring. The same observation is valid for the circuit shown in Figure 1.29. In this book, we adopt a more relaxed approach and use both symbols in Figure 1.30(c) to represent check valves in general.

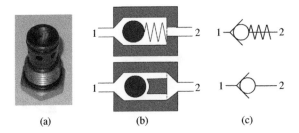

Figure 1.30 Check valve (spring loaded and standard): (a) picture, (b) operational principles and (c) ISO representations

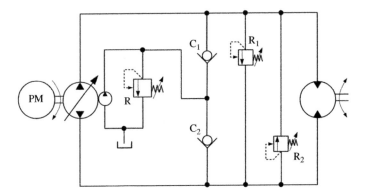

Figure 1.31 ISO representation of a hydrostatic transmission with charge circuit

having the hydraulic fluid press an elastic element (a spring or a gas encapsulated in a chamber) or lift a weight. In any case, the energy storage element is called *hydraulic accumulator.*

There are, basically, three types of hydraulic accumulators, namely, the spring-loaded accumulator, the gas-loaded accumulator and the weight-loaded (or gravity) accumulator. Figure 1.32 illustrates the operational principle of each type. In weight-loaded accumulators, a volume of fluid, V, is kept at a constant pressure, p, by the weight, W, acting on the piston. Spring-loaded accumulators, on the other hand, store elastic energy in the spring. Given that for the usual spring type, the elastic spring force, F_s, is proportional to the spring compression, the pressure inside the accumulator changes linearly between zero and a maximum value as the volume of stored hydraulic fluid increases. Gas-loaded accumulators also store energy elastically, though the pressure does not vary linearly with the stored fluid volume as in spring-loaded accumulators. The elastic medium is a gas held in an internal chamber by a flexible membrane, for example[21] (Figure 1.32(c)). The initial volume and pressure[22]

---

[21] Other configurations are possible that are not shown in Figure 1.32. For instance, the gas could have been kept inside a bladder or separated from the oil by a piston. Under restrictive conditions, it is even possible to have the gas mixed-up with the oil itself [12].

[22] The initial pressure at the gas chamber, p_{g0}, is called 'pre-charge pressure' and is usually 70–90% of the minimum operating pressure at the accumulator [12]. The volume V_{g0}, on the other hand, is a design parameter that defines the accumulator size.

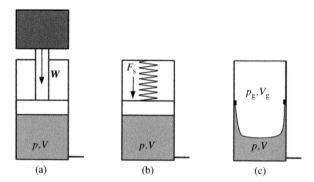

Figure 1.32   Schematics of typical accumulators: (a) weight-loaded accumulator, (b) spring-loaded accumulator and (c) gas-loaded membrane (diaphragm) accumulator

inside the gas chamber are V_{g0} and p_{g0}, respectively. As the accumulator discharges, the gas expands causing the pressure inside the gas chamber, p_g, to decrease.[23]

If we compare the three types of accumulators illustrated in Figure 1.32, we observe that the weight-loaded accumulator is the only one that supplies a constant pressure while discharging. However, some limitations are evident. For example, there is a need for the accumulator to be stationary and vertically placed. Another drawback, which is also shared with the spring-loaded accumulator, is the use of a piston separating the oil chamber from the energy storage element. Since there must be a clearance between the piston and the case, energy will inevitably be lost through leakages, especially at high pressures. Moreover, the fatigue of the spring, when subject to cyclic operation will, in the end, affect its elasticity and, as a result, the energy storage capacity.

Some types of gas-loaded accumulators, such as membrane and bladder accumulators, do not leak internally and are, by far, the most used in practical applications. Due to their mode of operation, these accumulators can store fluid at very high pressures, even for small gas volumes. As a result, the energy storage capacity is considerably high.[24] Moreover, since there is no need for a moving piston, these accumulators are much lighter and provide a faster response, being able to immediately compensate for pressure variations in the circuit. Figure 1.33 qualitatively compares the typical behaviour of the three types of accumulators studied so far. The figure shows the oil pressure versus time when the accumulator is being discharged considering a (hypothetical) quasistatic[25] process.

Membrane and bladder accumulators are pre-charged with a non-reacting gas (usually nitrogen) to prevent explosion during an eventual failure of the membrane/bladder at high

---

[23] If an idealized reversible process is considered, the pressure of the gas inside the gas chamber, p_g, can be related to its volume V_g through the state equation: $p_g = C/V_g^n$ where C is a constant and n is the polytropic index ($1 \leq n \leq 1.4$). In actual situations, however, the process of charging and discharging the accumulator is irreversible and the relation between p_g and V_g can be considerably complex [14].

[24] It is important to note here that thermal losses due to the irreversible charge and discharge process of gas-loaded accumulators are significant and may amount to 40% of the input energy [14]. The use of elastomeric foam inside the gas chamber in bladder accumulators, for example, has proved to minimize such losses by reducing the temperature gradients gas (see [15] for a thorough discussion about the theme).

[25] 'Quasistatic' means that the charge/discharge of the accumulator is so slow that the process can be considered as thermodynamically reversible.

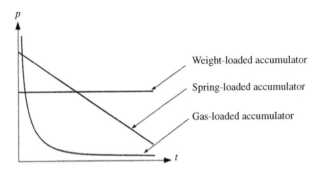

Figure 1.33 Oil pressure, p, as a function of time during the accumulator discharge

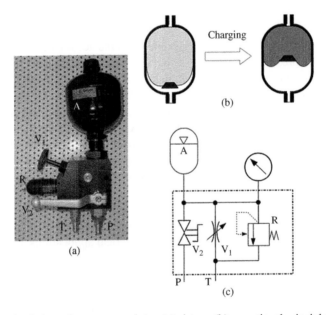

Figure 1.34 Gas-loaded membrane accumulator: (a) picture, (b) operational principle and (c) ISO representation of the corresponding hydraulic circuit

pressures.[26] Figure 1.34(a) shows a small membrane accumulator with the usual shut-off and relief valves required for safety. The corresponding hydraulic scheme is shown in Figure 1.34(b). The relief valve, R, limits the maximum pressure inside the accumulator. The shut-off valve, V_2, connects the accumulator to the main hydraulic line through port, P, and the throttle valve, V_1, is needed to discharge the accumulator to the tank through port T in case of an eventual emergency.

Figure 1.35 illustrates one of the ways in which energy can be stored in a hydrostatic transmission using accumulators. As the weight W descends (Figure 1.35(a)), the pulley

---

[26] We must remember that hydraulic fluids can combust when exposed to the air oxygen at certain temperature and pressure conditions.

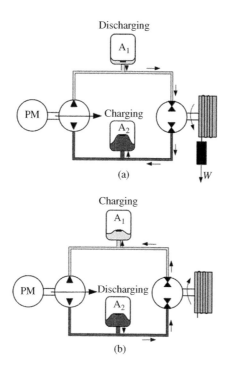

Figure 1.35   Energy storage (a) and regeneration (b) using hydraulic accumulators

turns the motor shaft, which, as a result, operates as a pump. The gravitational potential energy of the weight is, therefore, stored inside the accumulator A_2, which acts as a braking device. Note that the pump displacement has been set to zero during the energy recovery process.[27] At this moment, the prime mover, PM, can be disconnected from the pump or even turned off, as represented in the figure.

The energy stored in the accumulator A_2 can be reused at a posterior moment, as illustrated in Figure 1.35(b) where we have removed the weight, W, to eliminate the load at the pulley. At this time, the pressure differential between the accumulators A_2 and A_1 causes the fluid to flow back and rotate the motor shaft the other way around. The energy that had been stored in the previous moment is now recovered and turned into mechanical work again.

Before we move on to the next topic, it is important to mention another application of accumulators in hydraulic circuits. In addition to being used to store energy, accumulators can also act as shock absorbers, making operation smoother and reducing vibrations. Note, however, that the presence of accumulators might affect the response time of the transmission. By 'response time', we mean the time lapse the motor takes to respond to an angular rotation of the pump shaft. Accumulators reduce the 'stiffness' of the hydraulic medium, causing the motor to take longer to sense an input at the pump. All these things must be carefully considered when using accumulators in a hydraulic circuit.

---

[27] A more technical term is *energy regeneration*. We will discuss this theme more deeply at Chapter 6, when we study hydrostatic actuators.

1.4.2   Formal Definition of Hydrostatic Transmissions

It may seem strange that we have delayed a formal definition of hydrostatic transmissions
until now. However, the information provided so far will prove to be valuable at this stage.
Interestingly, in spite of the apparent simplicity of the matter, we notice that in the search
for a suitable definition of the term 'hydrostatic transmission', we must be cautious not to
be too general. For example, if our definition is too broad to cover all the circuits in which
power is transferred from a prime mover into a mechanically driven rotary device, we will
eventually conclude that any hydraulic circuit containing a pump on one end and a motor
on the other can be categorized as a hydrostatic transmission.

In this book, we adopt the following line of thought: whenever we speak about a 'hydrostatic transmission', we speak of a pump, connected to a motor, with the particularity that the
motor control is performed exclusively by the prime mover, the pump or the motor itself. As
an example, consider the scheme shown in Figure 1.25. The speed of the motor is directly
controlled by the pump flow or the motor displacement. If we had placed a flow control
valve between the pump and the motor in order to control the motor speed, or else if we had
made use of a directional valve to control the direction of the motor rotation, we would no
longer have a hydrostatic transmission, but a conventional hydraulic circuit. We, therefore,
give the following definition for hydrostatic transmissions:

> The term *hydrostatic transmission* encompasses all the hydrostatic power transmissions
> between two rotating shafts whereby the motor (output) is directly controlled by the
> prime mover, the pump, the motor, or by any combination of the three elements.

The definition given above has not been arbitrarily created but should be viewed as an
attempt to summarize the general concept found in the specialized literature. For instance,
in Ref. [16] we find that 'a hydrostatic transmission is simply a pump and motor connected
in a circuit'. On the other hand, in Ref. [17], it is said that 'hydrostatic transmissions are
hydraulic systems specifically designed to have a pump drive a hydraulic motor'. From these
statements, it is clear that a hydrostatic transmission should be composed of one pump and
one motor, and that the pump should control the motor. Merritt [18] speaks of a hydrostatic
transmission as being a type of pump controlled system, which he defines as consisting of
'… a variable delivery pump supplying fluid to an actuation device'. If we add the possibility of a variable-displacement motor to these descriptions, we naturally arrive at our
definition.

At this point, it is interesting to introduce the general classification of hydraulic circuits
proposed in Ref. [19]. In this classification, the term 'hydrostatic drive' is used as a synonym
to the hydraulic circuits where an actuator (cylinder or motor) is driven by a flow source
(pump) or a pressure source (accumulator). The hydrostatic drives have been divided into
displacement-controlled and resistance-controlled drives. Within each of these categories,
we also have the flow-supplied circuits and pressure-supplied circuits. Figure 1.36 shows
the details.

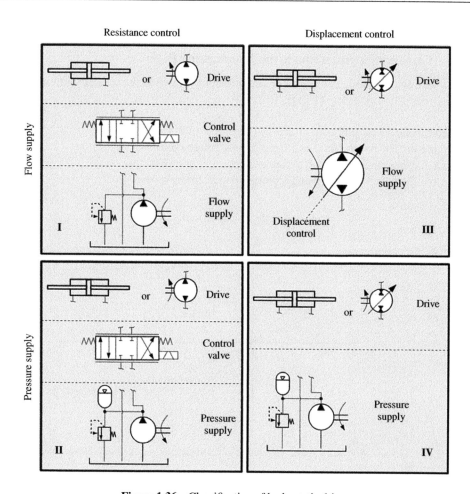

Figure 1.36   Classification of hydrostatic drives

If we carefully observe Figure 1.36, it is easy to see that the circuits represented in region III, that is, flow-supplied and displacement-controlled circuits, precisely fit into our definition of hydrostatic transmissions in the particular case where the actuator is a hydraulic motor.[28] Therefore, based on the classification of Figure 1.36, a 'hydrostatic transmission' could be regarded as a flow-supplied/displacement-controlled drive where the actuator is a hydraulic motor.

---

[28] Similarly, as will be seen later in this chapter, hydrostatic actuators can also be identified with the circuits in region III in the case where the actuator is a hydraulic cylinder. In addition, regions II and IV define the 'Common Pressure Rail' systems, which will be addressed in Chapter 6, and region I can be identified with valve-controlled actuators in general.

1.4.3   Classification of Hydrostatic Transmissions

1.4.3.1   Classification According to the Transmission Ratio

We can have four possible pump–motor combinations related to the transmission ratio of a hydrostatic transmission (Eq. (1.21)):

1. Fixed-displacement pump and fixed-displacement motor.
2. Variable-displacement pump and fixed-displacement motor.
3. Fixed-displacement pump and variable-displacement motor.
4. Variable-displacement pump and variable-displacement motor.

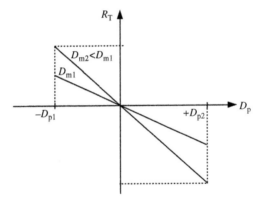

Figure 1.37   Transmission ratio for a variable-displacement pump and fixed-displacement motor transmission

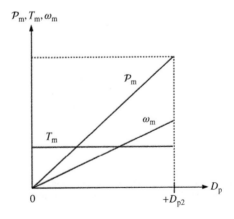

Figure 1.38   Power, speed and torque output for a variable-displacement pump and fixed-displacement motor transmission for a constant load at the motor

In what follows, we briefly explore the main characteristics of each of these transmission types:

1. *Fixed-displacement pump and fixed-displacement motor.* In this configuration, the transmission ratio, R_T, is constant given that both D_p and D_m in Eq. (1.21) have constant values. Therefore, the only means of obtaining different speeds at the motor is through altering the speed of the prime mover that is directly connected to the pump. This kind of hydrostatic transmission is the least flexible and constitutes a fixed-ratio transmission. The only advantage lies in the spatial flexibility of the whole set, which allows for both pump and motor to be placed mostly anywhere, in contrast to the strict requirements of mechanical transmissions.

2. *Variable-displacement pump and fixed-displacement motor.* This constitutes the most popular configuration and can be regarded as an IVT. By keeping the motor displacement, D_m, constant while changing the pump displacement from negative $(-D_{p1})$ to positive $(+D_{p2})$, the transmission ratio can have any value within the interval $[-D_{p1}/D_m, +D_{p2}/D_m]$. Note that a reversal in the motor shaft rotation happens smoothly and continuously in this configuration. Figure 1.37 shows the transmission ratio for two different values of the motor displacements D_{m1} and D_{m2} ($D_{m2} < D_{m1}$).

   In situations where the motor needs to drive a constant torque, the pressure differential between the input and output ports of the motor remains constant (Eq. (1.22)). On the other hand, since the motor speed can only be altered through changing the pump flow, the power output at the motor will not be constant in this situation (remember that power is the product between torque and speed). Figure 1.38 shows the speed (ω_m), torque (T_m) and power (P_m) for a fixed-displacement motor driving a constant load, with the pump displacement varying between 0 and $+D_{p2}$.

3. *Fixed-displacement pump and variable-displacement motor.* This configuration is not very practical due to the limitations associated with small motor displacements (see comments after Eq. (1.22)). Figure 1.39 shows a typical transmission ratio curve for a fixed value of the pump displacement and the motor displacement changing between $-D_{m1}$

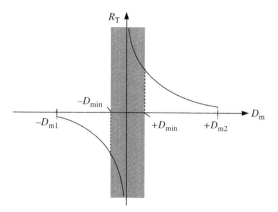

Figure 1.39 Transmission ratio for a fixed-displacement pump and variable-displacement motor transmission

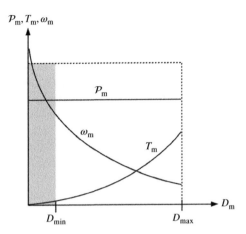

Figure 1.40   Power, speed and torque output for a fixed-displacement pump and variable-displacement motor transmission

and $+D_{m2}$. We observe that the magnitude of the motor displacement should not become smaller than a minimum established value, D_{min}, to keep the circuit pressure and angular speed within acceptable limits (Eqs. (1.21) and (1.22)).

Typically, the fixed-displacement pump and variable-displacement motor configuration produces a constant power transmission, as can be seen in Figure 1.40, where the torque, speed and power output at the motor are shown for D_m varying between D_{min} and D_{max}.

4. *Variable-displacement pump and variable-displacement motor.* This configuration is the most flexible one. In theory, we may say that this is an example of an IVT in the strictest sense of the term, that is, $-\infty < R_T < +\infty$. The graphical representation of a transmission of this kind is similar to the mechanical IVT shown in Figure 1.10, and can be seen in Figure 1.41 for symmetrical values of the pump and motor displacements (both changing between $-D$ and $+D$). The graph on (a) shows the situation where the pump displacement is kept constant while the motor displacement varies between $-D_{max}$ and $+D_{max}$. In (b), we invert the situation and keep the motor displacement constant while changing the pump displacement. We see that, by varying D_p and D_m, the transmission ratio can, in theory, have any value within the interval $]-\infty, +\infty[$. However, that when this configuration is used, we are not free from the high pressures that occur when the motor displacement approaches zero. This demands a limiting value for R_T in such a way that, in the real case scenario, the transmission ratio should stay within the finite interval $[-R_{T\,min}, +R_{T\,max}]$.

1.4.3.2   Classification According to the Spatial Arrangement

It is common to classify compact (or integral) hydrostatic transmission units according to the geometrical placement of pump and motor. It is possible to find expressions such as 'in-line',

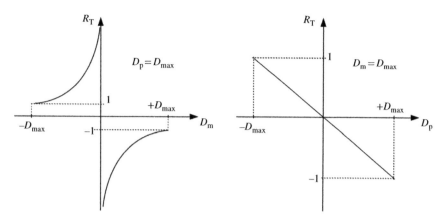

Figure 1.41 Transmission ratio for a variable-displacement motor and variable-displacement pump transmission

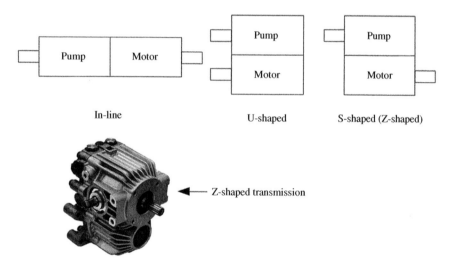

Figure 1.42 Compact unit transmission configurations and picture of a Z-shaped hydrostatic transmission (courtesy of Danfoss Power Solutions)

'U-shaped', 'S-shaped' (also called 'Z-shaped') and 'split' associated with each particular hydrostatic transmission type [16]. Figure 1.42 shows a picture of an actual Z-shaped hydrostatic transmission (bottom) and the schematic representation of the three configurations mentioned earlier. The 'split' configuration corresponds to hydrostatic transmissions where pump and motor are placed separately, as in the generic scheme shown in Figure 1.16.

1.4.3.3 Classification According to the Circuit Construction

Hydrostatic transmissions can also be divided according to the way their circuits are designed. In this context, we can have either a closed-circuit or an open-circuit transmission.[29] Technically speaking, a closed-circuit transmission is one in which the flow that leaves the motor is immediately directed to the pump input, whereas in an open-circuit transmission, the flow that comes out of the motor returns to a reservoir from where it is pumped again to the circuit. An example of a closed-circuit transmission is given in Figure 1.43, which is similar to the circuit shown in Figure 1.31 with the addition of an in-line oil filter F, a heat exchanger H, and the pump and motor drain lines connecting the cases to the tank.

With reference to Figure 1.43, a filter F is placed between the charge pump and the tank to filter out any eventual fluid impurity. Note that only the replenishing fluid is filtered.[30] The heat exchanger H may or may not be present, depending on the amount of heat generated by the viscous flows inside the transmission. Sometimes, the fluid temperature drops down naturally as heat is exchanged with the environment through conduits, pump and motor surfaces. However, there are occasions when a heat exchanger is necessary to bring the oil back to the temperature recommended by the manufacturer. In those circumstances, the usual procedure is to overcool the replenishment fluid so that, as it mixes with

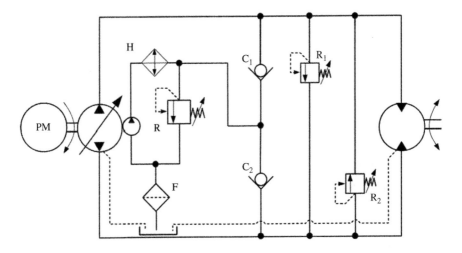

Figure 1.43   Closed-circuit transmission

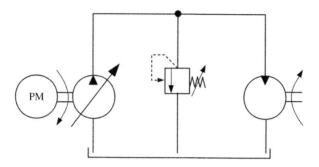

Figure 1.44   Open-circuit hydrostatic transmission

the main flow, the temperature inside the transmission can be brought down to an acceptable level. An additional technique consists of sporadically flushing fluid from the lower pressure conduit into the tank so that a larger fraction of oil can be filtered and cooled, as will be seen later on in Section 4.6.2, when we address the theme of loop-flushing in hydrostatic transmissions.

Figure 1.44 shows an open-circuit configuration. Note that this time, the hydraulic fluid flows from the oil reservoir into the pump and then from the pump into the motor. The fluid that leaves the motor returns to the tank instead of going back to the pump. We observe that, in this particular circuit, there is no possibility of reversing the motor rotation by changing the direction of the pump flow unless the gauge pressure in the line between the pump output and the motor input becomes negative (an undesirable situation). There is also no need for a charge circuit, as the tank now plays the role of the low-pressure conduit, and the relief valve only exists to alleviate any pressure spikes in the line connecting the pump to the motor.

The open-circuit configuration illustrated in Figure 1.44 is relatively simple and, in typical situations, some new elements should be added for a sound operation. For example, consider the case where the load accelerates the motor, creating a low-pressure zone between the pump output and the motor input. Given that the motor output connects to the tank, we must add a counterbalance valve to the circuit, as illustrated in Figure 1.45, to prevent the risk of fluid evaporation in the pump–motor line.

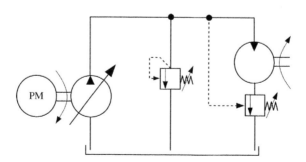

Figure 1.45   Counterbalance valve as a means of decelerating the motor

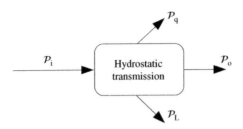

Figure 1.46   Power flow in a hydrostatic transmission

1.4.4   Efficiency Considerations

Figure 1.46 shows the energy balance in a hydrostatic transmission. The input and the output powers are represented by P_i and P_o. Losses in the transmission are divided into heat losses,[31] P_q and fluid leakage losses, P_L. Here, we see one of the reasons why mechanical transmissions are, in general, superior to hydrostatic transmissions in terms of efficiency. There is no fluidic system to transfer the power from the input to the output, and therefore, losses due to fluid leakages do not exist in mechanical transmissions. Therefore, one of the challenges of hydrostatic transmission technology is to reduce fluid leakages as much as possible.

External leakages are not the only factors detrimental to transmission efficiency; internal leakages also play an important part since the fluid that backflows from the high-pressure conduit into the low-pressure conduit will not transfer its energy to the motor in the end, but will return the energy to the pump input instead.

Considering the definition of efficiency given by Eq. (1.8), the efficiency of a hydrostatic transmission is given as

$$\eta = \frac{P_i - (P_q + P_L)}{P_i} \tag{1.23}$$

We will come back to Eq. (1.23) later on, in Chapter 4, when we study the steady-state operation of hydrostatic transmissions.

1.5   Hydromechanical Power-Split Transmissions

In order to take advantage of the high efficiency of gearboxes and the flexible transmission ratios of hydrostatic transmissions, a power-split scheme, similar to the one shown in Figure 1.13, can be used. The resulting scheme is denominated 'hydromechanical (power-split) transmission' (HMT) [23]. Technically, an HMT can be seen as 'an energy translation device in which mechanical energy at the input is converted into mechanical and hydrostatic energy and then is reconverted into mechanical energy before leaving at the output' [24]. The idea is illustrated in Figure 1.47, where we can visualize the power flow

---

[31] Heat loss is a direct result of the friction and viscous forces and, therefore, can also be thought of as mechanical losses.

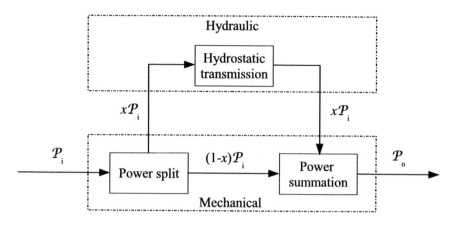

Figure 1.47   Power flow in a typical HMT

in a typical HMT. The input and output powers are represented by P_i and P_o, respectively, and x represents the fraction of the input power diverted into the hydraulic leg of the circuit. For simplicity, transmission losses have not been represented in the figure.

1.5.1   General Classification

HMTs are usually classified as 'input-coupled', 'output-coupled' and 'compound' [24]. This classification has to do with the way in which power is input to the hydrostatic transmission (or 'variator'). In an input-coupled HMT, there is a direct connection between the input shaft and the variator, whereas in an output-coupled HMT, the variator is connected to the output shaft. In the compound configuration neither the input nor the output shafts are directly connected to the variator. Figure 1.48 illustrates these three types[32] [23–25].

As an example, consider the particular case of the input-coupled architecture, shown in Figure 1.48(a) and presented in more details in Figure 1.49. In this configuration, the hydraulic unit M at the output side usually has a fixed displacement [23]. From the figure, we see that the input power $T_1 \omega_1$ is split between the shaft connected to the sun gear S of the planetary train and the variable-displacement pump[33] P through gears 1 and 2. Gear 3, on the other hand, receives the power $T_3 \omega_3$ coming from the hydrostatic transmission output. Finally, the powers entering the sun and the ring gears of the planetary gear train, S and R, are added into the output shaft, connected to the carrier C.

In what follows, we study the input-coupled HMT shown in Figure 1.49 in more detail.

---

[32] We are not going to study the three configurations shown in Figure 1.48 in detail in this book. Rather, we focus only on the input-coupled type. The interested student can consult Ref. [24] for further information on double- and compound-type HMTs.

[33] Observe that it is the input torque T_1 and not the input speed ω_1, which is responsible for splitting the input power between the hydrostatic transmission and the planetary train. In fact, given that $\omega_S = \omega_1$, the power entering the sun S can only be smaller than the input power $T_1 \omega_1$ because of the torque division at gear 1. Therefore, it is also common to refer to input-coupled HMTs as 'split-torque' HMTs [24].

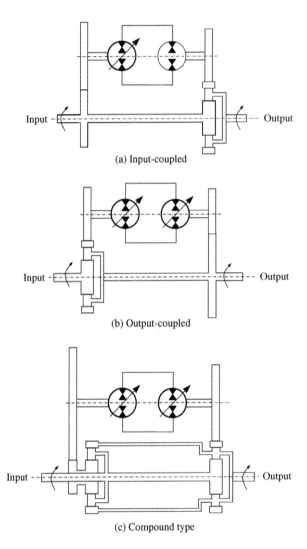

(a) Input-coupled

(b) Output-coupled

(c) Compound type

Figure 1.48 Typical architectures for an (a) input-coupled HMT, (b) output-coupled HMT and (c) compound-type HMT

1.5.2 Transmission Ratio

The relation between the output speed ω_C and the ring and sun speeds ω_R and ω_S in the planetary gear train can be obtained through substituting the pitch radii r_S and r_R by the corresponding number of teeth z_S and z_R in Eq. (1.5):

$$\frac{z_S}{z_R} = -\left(\frac{\omega_R - \omega_C}{\omega_S - \omega_C}\right) \tag{1.24}$$

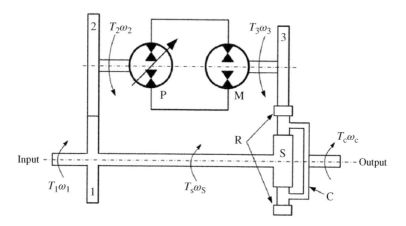

Figure 1.49 Input-coupled transmission

Considering the hypothetical situation of a 100% efficient transmission, we have (see Eq. (1.21)):

$$\frac{\omega_3}{\omega_2} = \frac{D_p}{D_m} \tag{1.25}$$

Observing Figure 1.49, the following additional relations can be written

$$\frac{\omega_2}{\omega_1} = \frac{\omega_2}{\omega_S} = -\frac{z_1}{z_2} \quad \text{and} \quad \frac{\omega_R}{\omega_3} = -\frac{z_3}{z_R} \tag{1.26}$$

From Eqs. (1.25) and (1.26), we obtain

$$\omega_R = -\frac{z_3}{z_R}\omega_3 = -\frac{z_3 D_p}{z_R D_m}\omega_2 = \frac{D_p z_1 z_3}{D_m z_2 z_R}\omega_S \tag{1.27}$$

Substituting ω_R, given by Eq. (1.27), into Eq. (1.24), we arrive at the following expression:

$$\frac{z_S}{z_R} = \frac{\omega_C - \dfrac{D_p z_1 z_3}{D_m z_2 z_R}\omega_S}{\omega_S - \omega_C} = \frac{\dfrac{\omega_C}{\omega_S} - \dfrac{D_p z_1 z_3}{D_m z_2 z_R}}{1 - \dfrac{\omega_C}{\omega_S}} = \frac{R_T - \dfrac{D_p z_1 z_3}{D_m z_2 z_R}}{1 - R_T} \tag{1.28}$$

where R_T is the transmission ratio ($R_T = \omega_C/\omega_1 = \omega_C/\omega_S$).

Equation (1.28) can be rearranged to give the transmission ratio explicitly as follows:

$$R_T = \frac{\dfrac{z_S}{z_R} + \dfrac{D_p}{D_m}\left(\dfrac{z_1 z_3}{z_2 z_R}\right)}{1 + \dfrac{z_S}{z_R}} \tag{1.29}$$

Observing Eq. (1.29), we see that it is possible to have the transmission ratio changing from $-R_{T\,min}$ to $+R_{T\,max}$, by varying the pump displacement from $-D_{p\,min}$ to $+D_{p\,max}$. In practice, as will be seen in Section 1.5.4, input-coupled HMTs are not efficient at negative transmission ratios [24], and a reversal of rotation at the output shaft is usually carried out mechanically [23].

1.5.3 Lockup Point

If we use a variable-displacement pump/fixed-displacement motor configuration in the hydrostatic transmission, as shown in Figure 1.49, the energy flow will be dictated by the pump displacement, D_p. For instance, it is easy to see that the choice of $D_p = 0$ corresponds to $x = 0$ in Figure 1.47. In this case, the ring of the planetary gearbox stops moving, and all the power is mechanically transmitted from the input to the output shaft. At this stage, the transmission reaches the point of highest efficiency, also known as 'lockup point' [24]. The corresponding transmission ratio R_{TL} can be easily obtained by making $D_p = 0$ in Eq. (1.29):

$$R_{TL} = \frac{\dfrac{z_S}{z_R}}{1 + \dfrac{z_S}{z_R}} \tag{1.30}$$

As will be seen in the following section, the lockup point is very significant to HMTs.

1.5.4 Power Relations

It is possible to obtain a mathematical expression for the proportion, x, of the input power that goes into the hydraulic branch of the input-coupled HMT, as shown in Figure 1.49.

Consider the forces acting on the shaft connecting gear 1 to the sun gear, S, of the planetary train, as shown in Figure 1.50. The nomenclature used in the figure is the following: F_{XY} indicates the tangential component of the force F, coming from element X onto element Y,

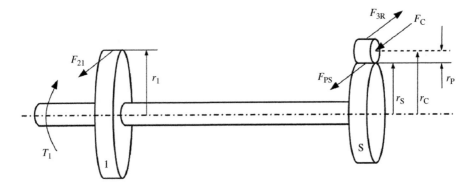

Figure 1.50 Forces acting on the sun gear shaft and the planet gear

where X and Y are generic representations of the transmission parts. Likewise, r_X is the radius corresponding to element X. For instance, F_{21} is the tangential component[34] of the force that gear 2 exerts on gear 1 at their point of contact, r_p is the pitch radius of the planet gear and so on. A momentum balance over the sun gear shaft and the planet gear[35] gives us the following equations:

$$\begin{cases} T_1 - F_{21}r_1 - F_{PS}r_S = 0 \\ F_{3R}r_P + F_{PS}r_P = 0 \end{cases} \tag{1.31}$$

Similar to what was assumed in Figure 1.50, let r_2 and r_3 be the pitch radii of gears 2 and 3 in Figure 1.49. The magnitude of the contact forces between gears 2 and 1 and between gear 3 and the ring of the planetary train can then be written as

$$F_{21} = \frac{T_2}{r_2} \quad \text{and} \quad F_{3R} = \frac{T_3}{r_3} \tag{1.32}$$

Using Eqs. (1.31) and (1.32), it is possible to obtain the following expression for the input torque T_1:

$$T_1 = T_2 \left(\frac{r_1}{r_2} \right) + T_3 \left(\frac{r_S}{r_3} \right) \tag{1.33}$$

The torques T_2 and T_3, on the other hand, can be related to the pump and motor displacements, D_p and D_m, and the pressure differential between the conduits, Δp (see Eq. (1.20)). If we also observe that $r_1/r_2 = \omega_2/\omega_1$, Eq. (1.33) can be rewritten as

$$T_1 = D_p \Delta p \left(\frac{\omega_2}{\omega_1} \right) + D_m \Delta p \left(\frac{r_S}{r_3} \right) \tag{1.34}$$

Equation (1.34) can be further modified into the following expression:

$$\frac{D_p \Delta p \omega_2}{T_1 \omega_1} = 1 - \frac{D_m \omega_1 \Delta p}{T_1 \omega_1} \left(\frac{r_S}{r_3} \right) \tag{1.35}$$

Note that

$$\frac{D_p \Delta p \omega_2}{T_1 \omega_1} = \frac{T_2 \omega_2}{T_1 \omega_1} = x \tag{1.36}$$

We can then write Eq. (1.35) as

$$x = 1 - \frac{D_m \Delta p}{T_1} \left(\frac{r_S}{r_3} \right) \tag{1.37}$$

---

[34] Tangential component, that is, the component of the contact force between the two gears that is tangent to the pitch circle (see Ref. [3] for a more complete discussion).

[35] Without loss of generality, we are treating the planetary gear train as if it had only one planet when we know that the torques exerted by the ring and the sun are equally divided between all the planets in the train. In our representation, all these forces are summed up in F_{3R} and F_{PS}.

Since there are no energy losses in the transmission, $T_1\omega_1 = T_C\omega_C$ in Figure 1.49. We can also perform another momentum balance over the planet gear having the pitch point of contact between the planet and the sun as a reference (Figure 1.50):

$$F_C = \frac{2F_{3R}r_p}{r_p} = 2F_{3R} \tag{1.38}$$

If we make $T_C = F_C r_C$, the following chain of relations can be developed:

$$T_1 = \frac{T_C\omega_C}{\omega_1} = \frac{F_C r_C \omega_C}{\omega_1} = \frac{2F_{3R}r_C\omega_C}{\omega_1} = \frac{2T_3 r_C\omega_C}{r_3\omega_1} = \frac{2D_m\Delta p r_C\omega_C}{r_3\omega_1} \tag{1.39}$$

Substituting T_1, given by Eq. (1.39) into Eq. (1.37), it is possible to arrive at the following expression for x (we leave the demonstration as an exercise for the student):

$$x = 1 - \frac{r_S}{2r_C}\left(\frac{\omega_1}{\omega_C}\right) = 1 - \frac{r_S}{2r_C}\left(\frac{1}{R_T}\right) \tag{1.40}$$

Equation (1.40) can be written in a more interesting way. First, we note that we can write the lockup transmission ratio, R_{TL}, in Eq. (1.30) as a function of the pitch radii of the gears, r_S and r_R, instead of the number of teeth, z_S and z_R:

$$R_{TL} = \frac{\dfrac{r_S}{r_R}}{1 + \dfrac{r_S}{r_R}} = \frac{\dfrac{r_S}{r_R}}{\dfrac{r_R + r_S}{r_R}} = \frac{r_S}{r_R + r_S} \tag{1.41}$$

From Figure 1.50, we obtain the following geometric relations:

$$\begin{cases} r_C + r_p = r_R \\ r_C - r_p = r_S \end{cases} \tag{1.42}$$

Adding each side of the two equations in (1.42), we get

$$r_R + r_S = 2r_C \tag{1.43}$$

Finally, using Eqs. (1.40), (1.41) and (1.43), we arrive at the following expression for x:

$$x = 1 - \frac{R_{TL}}{R_T} \tag{1.44}$$

Remarks

1. Given that the value of the lockup transmission ratio, R_{TL}, is constant, the proportion of hydraulic power, x, as a function of the transmission ratio, R_T, has the form of a hyperbole.
2. For $R_T > R_{TL}$, $0 < x < 1$, meaning that a fraction of the input power, P_i, is diverted through the hydrostatic variator to be later added to the input power through the ring gear of the planetary set. If $R_T = R_{TL}$, $x = 0$ and the HMT behaves like a mechanical transmission. On the other hand, if $R_T < R_{TL}$, x becomes negative, meaning that power coming from the variator is joined to the input power, P_i, before entering the planetary gear train through the sun gear. In the limit, when $R_T \to 0$, there will be infinitely more 'hydraulic power' than the input power, P_i, itself. This may be confusing or even ambiguous now, but we will make it clear shortly.

Figure 1.51 shows a graphical representation of Eq. (1.44). A brief analysis of the curves will be given in the sequence.

To better understand the curves in Figure 1.51, we have divided the picture into four regions, A through D, read from right to left. Observe that $\omega_C > \omega_1$ at region A and $\omega_C < \omega_1$ at regions B and C (remember that $\omega_C = R_T\omega_1$). In region D, a reversal of the output speed occurs as the transmission ratio becomes negative.

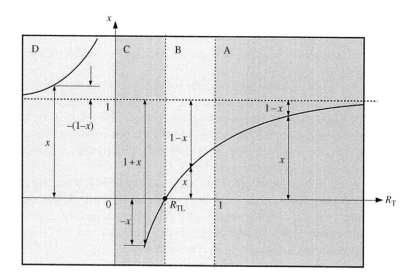

Figure 1.51   Power relation in a 100% efficient input-coupled HMT

1.5.4.1   Operation at Regions A and B

Observing the curve at the right of Figure 1.51, we see that for both regions A and B, $R_T >$ R_{TL}, and the proportion of input power going into the hydraulic leg of the circuit, x, increases from zero ($R_T = R_{TL}$), converging asymptotically to 1 as $R_T \to \infty$. Clearly this is merely theoretical because the pump displacement in Figure 1.49 is limited to a maximum value, $D_{p\,max}$, while the motor displacement is constant. Therefore, in practice, there is a maximum positive value of x that can be obtained in this HMT.

Given that the proportion of input power going into the hydrostatic transmission increases with x, the input-coupled HMT becomes less efficient as we move further to the right of the lockup point. As mentioned earlier, the highest efficiency is achieved when $R_T = R_{TL}$. At this point, all the input energy is mechanically transmitted to the output shaft. Figure 1.52 shows the HMT operation in regions A and B. Note that the output torque, T_C, becomes smaller than the input torque, T_1, in region A, when the output speed, ω_C, is higher than the input speed, ω_1 ($\omega_1 = \omega_S$). As expected, the opposite happens in region B, when $\omega_C < \omega_S$.

1.5.4.2   Operation at Region C

We have seen that the transmission ratio, R_T, can be reduced by reducing the pump displacement, D_p (see Eq. (1.29)). Therefore, if we want to decrease the transmission ratio down to values below the lockup ratio, R_{TL}, we need to make the pump displacement negative. This is what happens in region C, where we have $0 < R_T < R_{TL}$. However, if we make the pump displacement negative, the oil flow in the hydrostatic transmission also changes direction. As a result, gear 3, which is connected to the output shaft of the motor M, rotates the other way around, as does the ring gear of the planetary set, as shown in Figure 1.53(a). Note that the motor M now operates as a pump, recirculating the unused power back into the transmission again. In this case, the HMT is said to be operating in *regenerative* mode [24].

As $R_T \to 0^+$ (i.e. when R_T tends to zero from the positive side of the abscissa axis), more and more power recirculates through the hydrostatic transmission into gear 1. In addition, given that $\omega_C \to 0^+$ when $R_T \to 0^+$, the output power, $T_C\omega_C$, is also reduced. We know that in a 100% efficient transmission, the power input equals the power output. Therefore, the input power, $T_1\omega_1$, and, consequently, the input torque, T_1, also tend to zero as $R_T \to 0^+$, which results in $x \to -\infty$, as shown in Figure 1.51.

Figure 1.53(b) graphically illustrates the situation in an actual HMT, where the efficiency is smaller than 100%. Because of the losses in the hydrostatic transmission, the power entering gear 2 is smaller than the power leaving gear 3. Therefore, a minimal input power is required to cover the energy losses. In other words, in real-life situations, x will have a finite negative value when the transmission ratio, R_T, becomes zero.

1.5.4.3   Operation at Region D

We have seen that by changing the pump displacement to negative values, the ring gear rotates in the opposite direction of the sun gear. As a result, the output speed ω_C is reduced,

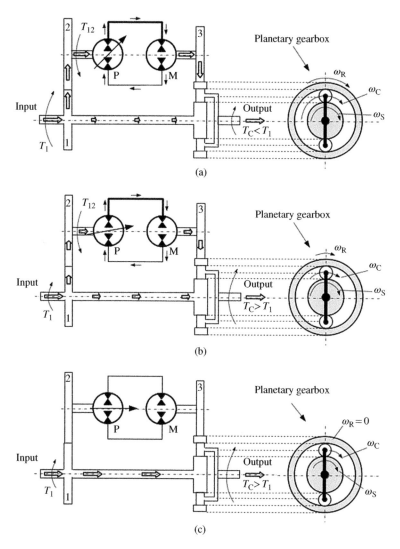

Figure 1.52   Energy flow (regions A and B in Figure 1.51): (a) region A, (b) region B and (c) lockup point

and eventually becomes zero.[36] It is, therefore, natural to infer that if we keep on increasing the magnitude of the pump displacement to the negative side, the ring gear will, eventually, drive the carrier to the opposite direction and the transmission ratio will become negative. This is precisely what happens, and is illustrated in Figure 1.54, where we represent the particular case when $-1 < R_T < 0$.

---

[36] We must observe here that it is usually difficult to precisely match the angular speeds of the ring and the sun gears in the planetary train in such a way that the neutral gear ($\omega_C = 0$) is obtained [26].

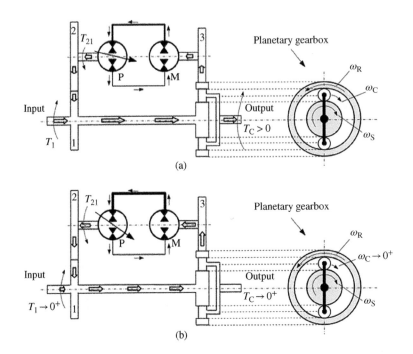

Figure 1.53 Energy flow (region C in Figure 1.51): (a) $0 < R_T < R_{TL}$ and (b) $R_T \to 0^+$

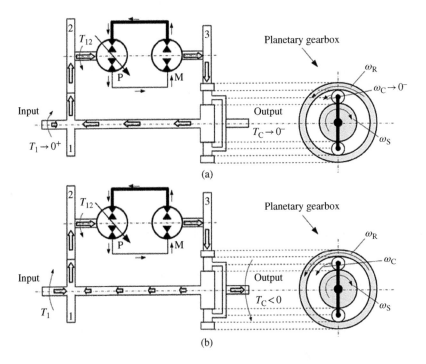

Figure 1.54 Energy flow (region D in Figure 1.51): (a) $R_T \to 0^-$ and (b) $R_T < 0$

Figures 1.54(a) and (b) can be understood as follows. The ring gear rotates in the opposite direction in relation to the sun gear, dragging the carrier counter-clockwise at the same time that it creates a positive torque at the sun gear, adding to the torque T_1.[37] The power flow must, therefore, be towards gear 1. The difference between the power entering and leaving the planetary train is then output to the planetary carrier. In the particular situation shown in Figure 1.54(a), the carrier speed approaches zero from the negative side (i.e. $\omega_C \to 0^-$). Therefore, almost all the power circulates back to the sun gear, and the output torque approaches zero. Because of the energy conservation, the input power is also very small, which causes the ratio between the hydraulic power and the input power, x, to go to plus infinity (see Figure 1.51). As the ring gear rotates faster and faster, more power is transmitted to the carrier and, consequently, the value of x is reduced. Note that x will always be higher than 1, meaning that in the negative range of transmission ratios, the power transfer mechanism will be mostly hydraulic. As a result, the input-coupled HMT becomes considerably less efficient when operating at negative transmission ratios [24].

The subject of hydromechanical power-split transmissions is vast and will not be addressed in more details in this book. What we have done here, however, can be used as a basis for the study of other configurations. For instance, it is possible to show that the ratio x, in an output-coupled configuration (Figure 1.48(b)), is given by [24]:

$$x = 1 - \frac{R_T}{R_{TL}} \qquad (1.45)$$

If we compare Eq. (1.44) with Eq. (1.45), we observe that the transmission ratio now appears in the numerator of the second term. As a result, the curve $x(R_T)$ will be a straight line and not a hyperbole as in Figure 1.51. We leave the demonstration of Eq. (1.45), and the subsequent analysis of the power flows as an exercise to the student.

1.6 Mechanical and Hydrostatic Actuators

In mechanical and hydrostatic transmissions, we saw that power was transferred between two rotating shafts. The term actuator is commonly used in the context where the input is a rotary device (e.g. an electric motor), and the output is either a semi-rotary device or a linear actuator. The way in which the power is transferred between the input and the output defines the type of actuator.

1.6.1 Mechanical Actuators

Consider the rack and pinion mechanism shown in Figure 1.55. As the pinion performs a rotary movement, clockwise and counter-clockwise, the rack moves to the left and to the right, respectively. This is an example of a mechanical actuator, where the prime mover, PM, connects to the gear (pinion) and the rack A is the linear actuator.

---

[37] Note that the planets reverse the direction of the torque coming from the ring onto the sun. Therefore, when the ring is being pushed in the opposite direction, it actually acts towards accelerating the sun gear.

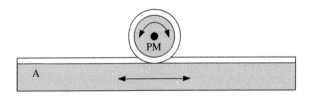

Figure 1.55   Simple mechanical actuator

Figure 1.56   A screw-type mechanical actuator

The mechanical actuator shown in Figure 1.55 has the spatial rigidity of mechanical transmissions. In addition, the prime mover must be near the actuator and is responsible for the actuator control. These characteristics are common to every mechanical actuator and make them a poor option for situations where the actuator must be placed away from the prime mover. However, it must be said that in many applications, such as the opening and closing of a gate, mechanical actuators like the one shown in Figure 1.55 can be the best choice.

Figure 1.56 shows another type of mechanical actuator, where a platform, P, is driven by an endless screw, S, mechanically connected to a stepper-motor,[38] M. As the motor shaft is turned to a certain angle, the platform moves accordingly. A linear potentiometer, R, is attached to the platform to detect its position. This type of device provides an excellent positioning and is widely used in machine tools.

1.6.2   Hydrostatic Actuators

Hydrostatic actuators are composed of a prime mover, a pump and a hydraulic cylinder. Figure 1.57 shows the basic configuration. The cylinder is connected to the pump through a hydraulic circuit. Depending on which element is responsible for the cylinder control – the variable-displacement pump or the variable-speed electric motor – the hydrostatic actuator is called *displacement-controlled actuator* or *electrohydrostatic actuator*. In both cases, the control of the hydraulic cylinder is entirely carried out by either the variable-displacement

---

[38] A *stepper-motor* is an electric motor whose shaft can rotate at discrete angles. This type of motor is very common in positioning devices.

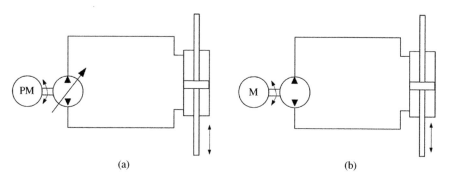

Figure 1.57 Displacement-controlled and electrohydrostatic actuators: (a) displacement-controlled actuator and (b) electrohydrostatic actuator

pump (displacement-controlled actuator) or by the electric motor (electrohydrostatic actuator).[39] We may therefore say that

> The term *hydrostatic actuator* encompasses all the hydrostatic power transmissions between a rotating shaft and a linear or rotary hydraulic actuator,[40] where the actuator control is carried out solely by the pump or by the prime mover.

1.6.3 Hydrostatic Actuation Versus Valve Control

As an example of a valve-controlled actuator, consider the situation illustrated in Figure 1.58, where the cylinder is controlled through a four-way, three-position directional valve. The number of ways (4) represents the number of external connections and the positions (3) correspond to the possible spool stops (we only see one stop in the figure, where the spool is totally displaced to the left; other stops are the spool at the centre and at the right position). A relief valve limits the pressure inside the cylinder. With the valve positioned as it is, the cylinder moves to the right until it reaches the end of its stroke. When that happens, the flow into the left chamber is blocked and the pressure rises in the conduit connected to the pump output. In the simplest case scenario, the pump flow is diverted into the tank through the relief valve.[41] In order to bring the cylinder back to the left position, we must move the spool of the directional valve to the right. If the spool is centred, the cylinder is held in place.

Figure 1.59 shows the ISO representation of the circuit illustrated in Figure 1.58.

---

[39] Note the change from the generic label PM into the more specific label M (electric motor) in the electrohydrostatic actuator scheme (Figure 1.57).

[40] We make the distinction here between a rotary actuator and a motor, in that a rotary actuator moves within a limited angle and a motor rotates continuously.

[41] Here we are assuming that the pump has a fixed displacement and that the prime mover has a constant speed. A better solution is to use a variable-displacement pump so that the flow can be reduced when the output line is blocked (pressure-compensated circuit).

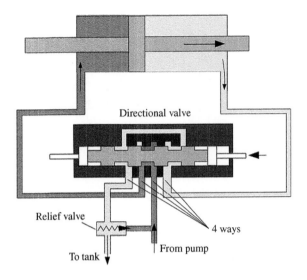

Figure 1.58   Example of a valve-controlled actuator

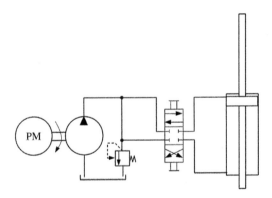

Figure 1.59   ISO representation of the hydraulic circuit shown in Figure 1.58

An example of a hydrostatic actuator is given in Figure 1.60. The circuit is identical to the circuit of the hydrostatic transmission shown in Figure 1.31 and is only repeated here for convenience. Again, the charge pump exists only to replenish the external leakage[42] and to keep the low-pressure side of the cylinder at a minimal level, regulated by the relief valve, R. The connection between the charge circuit and the main lines is made by two symmetrically disposed check valves, C_1 and C_2, in the same way it was done in

---

[42] It must be observed that the use of a charge pump is not the only way to replenish the external volumetric losses of the actuator. Alternative pressure sources like accumulators or even direct connections to the oil reservoir can be used as well, given that the need for fluid replenishment is usually much smaller when compared to hydrostatic transmissions.

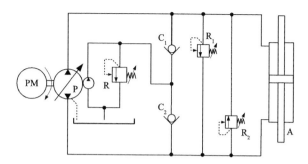

Figure 1.60   Double-rod hydrostatic actuator

the hydrostatic transmission of Figure 1.31. Since pressure overshoots can also exist in hydrostatic actuators, a couple of relief valves, R_1 and R_2, must be present in the circuit for safety reasons.

1.6.4   Multiple Cylinder Actuators

The hydrostatic actuator concept can be easily extended to applications where multiple cylinders are involved. Figure 1.61 illustrates how this can be done for two displacement-controlled, double-rod actuators sharing one single-charge circuit and the same prime mover, PM (pressure overshoot relief valves are not shown for simplicity reasons). The two cylinders A_1 and A_2 are independently controlled by the variable-displacement pumps P_1 and P_2, which are driven by the same prime mover. The prime mover also connects to the charge pump, which together with the check valves C_1–C_4 and the relief valve R is responsible for replenishing the external volumetric losses in the circuit. A considerable drawback lies in the fact that multiple cylinder actuators require multiple pumps, which increases the complexity and the weight of the system.

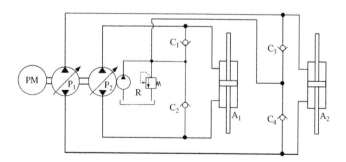

Figure 1.61   Two-cylinder actuator system

Exercises

(1) For two gears to engage perfectly, their module, m, defined as the ratio between the pitch diameter and the number of teeth ($m = D_p/z$), must be the same. Use this fact to prove that Eq. (1.3) can be written as

$$\omega_i = (-1)^{\frac{N}{2}} \left(\frac{z_2 z_4 z_6 \cdots z_N}{z_1 z_3 z_5 \cdots z_{N-1}} \right) \omega_o$$

(2) Based on the planetary gearbox illustration (Figure 1.6), give a formal demonstration of Eq. (1.5).

(3) Suppose that the diameter of the ring gear of the planetary gearbox represented by Figure 1.6 is 300 mm. Consider that the minimum number of teeth allowed for the smallest gear in the gearbox is 18 (to avoid interference), and that the module of the gears is 2 mm. Assuming that one of the elements – sun, ring or carrier – remains fixed while the others are allowed to turn, obtain the number of teeth of the sun and the planet gear that gives the highest transmission ratio possible. What is the element that needs to be fixed in that case?

(4) Show that the slippage between the two gear teeth shown in Figure 1.8, given by the difference between the linear speed projections onto the line, t, increases with the angular speeds of the gears.

(5) The gear train shown in Figure 1.62 must be placed inside a metal case, and is therefore vertically limited by the dimension $L_m = 300$ mm. To avoid interference, the number of teeth of each gear must be equal to or greater than 18. Considering that the distance between the pitch circle and the external circle in each gear is equal to the module $m = 2$ mm (see problem 1 for the definition of *module*), what is the maximum transmission ratio that can be obtained and what is the corresponding number of teeth of each gear?

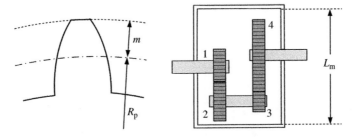

Figure 1.62   Two-speed gearbox

(6) Consider the dual-motor transmission shown in Figure 1.63 and prove that the following relation holds (disregard the circuit losses):

$$\frac{D_p}{D_{m2}} = R_{T1} \left(\frac{D_{m1}}{D_{m2}} \right) + R_{T2}$$

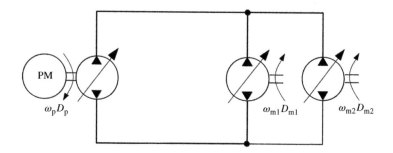

Figure 1.63   Dual-motor circuit

(7)   Consider the hydrostatic actuator shown in Figure 1.60 and show that the (constant) linear speed of the piston rod, in the absence of leakages, is given by

$$v = \frac{4\omega_p D_p}{\pi(d_p - d_r)^2}$$

where ω_p is the pump speed, D_p is the pump displacement and d_p and d_r are the diameters of the piston and the rod of the cylinder, respectively.

References

[1] Brockbank C (2010) Multi regime full-toroidal IVT for high torque and commercial vehicle applications. 9th International CTI-Symposium, Detroit.

[2] Beachley NH, Frank AA (1979) Continuously variable transmissions: theory and practice. Lawrence Livermore Laboratory, University of California, Report UCRL-15037.

[3] Mabie HH, Ocvirk FW (1957) Mechanisms and dynamics of machinery. John Wiley & Sons, USA.

[4] Petry-Johnson TT, Kahraman A, Anderson NE, Chase DR (2007) Experimental investigation of spur gear efficiency. Proceedings of the ASME 2007 International Design Engineering Technical Conferences & Computers and Information in Engineering Conference – IDETC/CIE – 35045.

[5] Faulhaber F (2002) A second look at Gearbox efficiencies. Machine Design: for engineers by engineers. http://machinedesign.com/article/a-second-look-at-gearbox-efficiencies-0620. Accessed 18 January 2011.

[6] Whitby RD (2010) Hydraulic fluids in wind turbines. A hydraulic transmission system makes wind turbines lighter, cheaper and more robust. Tribology and lubrication technology, 66 (3): 72.

[7] NISSAN (2011) Extroid CVT For application to rear-wheel-drive cars powered by large engines. NISSAN. http://www.nissan-global.com/PDF/tcvt_e.pdf. Accessed 26 January 2011.

[8] SUBARU (2011) Lineartronic continuously variable transmission (video). Subaru. http://www.subaru.com/engineering/transmission.html. Accessed 10 March 2011

[9] Nekrasov B (1969) Hydraulics for aeronautical engineers. Peace Publishers, Moscow

[10] Newton Jr. GC (1947). Hydraulic variable-speed transmissions as servomotors. Journal of the Franklin Institute, 243(6): 439–469.

[11] Stone R, Ball JK (2004) Automotive engineering fundamentals. SAE International, USA.

[12] Rabie MG (2009) Fluid power engineering. McGraw-Hill, USA.

[13] Sauer-Danfoss (2010) Series 40, axial piston pumps: technical information, 520L0635-Rev EJ, USA

[14] Pourmovahed A, Beachley NH, Fronczak FJ (1992) Modeling of a hydraulic energy regeneration system – Part 1: analytical treatment. Transactions of the ASME 114: 155–159.

[15] Stroganov A, Sheshin L (2011) Improvement of heat-regenerative hydraulic accumulators. Revlia za fluidno tehniko, avtomatizacijo in mehatroniko-Ventil 17(4): 322–332.

[16] Cundiff JS (2002) Fluid power circuits and controls: fundamentals and applications. CRC Press, Boca Raton, USA.

[17] Doddannavar R, Barnard A (2005) Practical hydraulic systems: Operation and troubleshooting for engineers and technicians. Elsevier Science & Technology Books

[18] Merritt HE (1967) Hydraulic control systems. John Wiley & Sons, USA.

[19] Murrenhoff H (1999) Systematic approach to the control of hydrostatic drives. Proceedings of the Institution of Mechanical Engineers, 213(Part I): 333–347.

[20] Stringer J (1976) Hydraulic systems analysis. Halsted Press, USA.

[21] Klocke C (2011) Control terminology for hydrostatic transmissions. Proceedings of the 52nd National Conference on Fluid Power. 23.3: 629–636, Las Vegas, USA.

[22] Johnson JL (2010) Understanding hydrostatic transmissions. Hydraulics & Pneumatics. http://www .hydraulicspneumatics.com/200/GlobalSearch/Article/False/86140/Understanding+hydrostatic+transm. Accessed 19 January 2011.

[23] Jacobson E, Wright J, Kohmäscher T (2011). Hydro-mechanical power split transmissions (HMT) – Superior technology to solve the Conflict: Tier 4 vs. machine performance. Proceedings of the National Conference on Fluid Power (paper 5.3), Las Vegas, USA.

[24] Kress JH (1968). Hydrostatic power-splitting transmissions for wheeled vehicles – Classification and theory of operation. SAE Paper No. 680549, pp. 2282–2306.

[25] Cheong KL, Li PY, Sedler SP, Chase TR (2011). Comparison between input coupled and output coupled power-split configurations in hybrid vehicles. Proceedings of the 52nd National Conference on Fluid Power (Paper 10.2), Las Vegas, USA.

[26] Hall W (2014) Novel Use of a U-Style hydrostatic transmission to develop a low-power dual-mode transmission. Proceedings of the International Fluid Power Exposition (IFPE), LA, USA, paper 24.1.

2

Fundamentals of Fluid Flows in Hydrostatic Transmissions

The energy-carrying element in a hydrostatic transmission is the hydraulic fluid. This chapter focuses on two related subjects: hydraulic fluids and internal flows in hydrostatic transmissions. Because of the frequent use of fluid dynamics equations along the chapter, it may be beneficial for the reader to have a look at Appendix C, which gives a brief exposition on the theme.

Although our discussion will focus primarily on hydrostatic transmissions in this chapter, the concepts explored here equally apply to hydrostatic actuators. We will use the terms *hydraulic fluid* and *hydraulic oil* interchangeably to mean the fluid that is used in hydraulic circuits.

2.1 Fluid Properties

We begin this section by exploring two important characteristics of hydraulic fluids: viscosity and compressibility.

2.1.1 Viscosity

2.1.1.1 Newton's Law of Viscosity

Viscosity can be viewed as the internal resistance of the fluid to tangential stresses. Take, for example, the situation displayed in Figure 2.1, where two parallel plates slide relatively to one another. In this example, the plates are kept apart by a layer of a Newtonian fluid[1] and the slide happens by moving the upper plate to the right while keeping the lower plate

---

[1] Simply stated, a Newtonian fluid is one for which the viscosity depends only on temperature and pressure, not on the forces acting upon it. Hydraulic oils, in general, can be regarded as Newtonian fluids [1].

Hydrostatic Transmissions and Actuators: Operation, Modelling and Applications, First Edition.
Gustavo Koury Costa and Nariman Sepehri.
© 2015 John Wiley & Sons, Ltd. Published 2015 by John Wiley & Sons, Ltd.
Companion Website: www.wiley.com/go/costa/hydrostatic

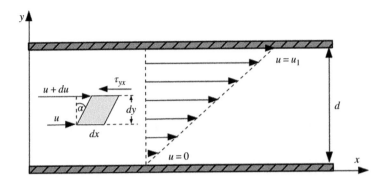

Figure 2.1   Viscous flow between two plates caused by fluid dragging

stationary. The fluid molecules adjacent to the upper plate move with it to the right with a velocity u_1, while the molecules adjacent to lower plate remain stationary. Because of the continuity of the flow,[2] a velocity profile is formed between the upper and lower surfaces and the velocity gradient generates a resisting tension, which opposes the movement of the upper plate. This tension exists because the lower molecules try to decelerate the faster ones, responding similar to friction between two sliding bodies.

We now consider the distortion of an infinitesimal fluid element $dxdy$, placed between the upper and lower plates, in Figure 2.1. The Newton's law of viscosity, when applied to this particular situation, states that the magnitude of the viscous tension in the x-direction, on a plane perpendicular to axis y, τ_{xy}, is given by

$$\tau_{yx} = \mu \frac{\partial u}{\partial y} \tag{2.1}$$

where μ is called the *absolute viscosity* of the fluid.[3] Since the velocity component along the x-direction, u, may vary with time as well as with x, a partial derivative is taken for the variation of u in relation to y.

Remark

It is important to mention here that the Newton's law of viscosity is only a simplification of what actually happens in a fluid flow. In fact, fluid flows are a 3D phenomena, and the viscous tension in Eq. (2.1) is only part of a more complex matrix (see Appendix C for a more detailed study of viscous forces in a fluid).

It can be shown that if the absolute viscosity, μ, is constant, the velocity profile shown in Figure 2.1 is linear. The resulting flow is known as *Couette flow*.[4]

---

[2] We can understand continuity in the following way: a continuous fluid that does not have 'holes' in it, and its molecules constantly interact with one another – that is, the motion of one molecule is always perceived by its neighbours.

[3] Sometimes, the absolute viscosity is also called *dynamic viscosity*.

[4] In honour of the French scientist Maurice Marie Alfred Couette (1858–1943).

2.1.1.2 Kinematic Viscosity

For Newtonian fluids [2], the kinematic viscosity, v, is defined as the ratio between the absolute viscosity, μ, and the fluid density, ρ:

$$v = \frac{\mu}{\rho} \tag{2.2}$$

The unit of the absolute viscosity in the centimetre-gram-second (CGS) system of units is the *Poise*[5] (P). Given that the viscosity values of typical fluids are much smaller than 1 P, it is common to use centipoises (cP) instead (1 P = 100 cP). Table 2.1 gives some usual units for absolute and kinematic viscosities.

A commonly used unit for the kinematic viscosity is the *Stoke*[6] (St), given by

$$1 \text{ St} = 10^{-4} \text{ m}^2/\text{s} \tag{2.3}$$

Even adding the 10^{-4} factor to m²/s, 1 St is still a very large viscosity measurement. Therefore, it is common to use centistokes (cSt) instead. The relation between stokes and centistokes is straightforward: 1 St = 100 cSt.

Hydraulic fluids are usually characterized by their kinematic viscosities in industrial catalogues. It is also common to refer to the kinematic viscosity when classifying hydraulic fluids, as will be seen shortly. Therefore, if no explicit mention is made to the viscosity type, whenever we speak of viscosity in this book, we will be referring to the kinematic viscosity.

2.1.1.3 Viscosity and Temperature

Fluid viscosity varies with temperature, and this variation is usually significant enough to be considered. In addition, the temperature of the hydraulic fluid may undergo severe fluctuations, especially in outdoor applications, at places where the temperature changes considerably during the four seasons. The degree to which the oil viscosity varies with changes in temperature is so important that a viscosity index (VI) has been created to classify the hydraulic oils according to their viscosity range (DIN ISO 2909). The higher the viscosity index, the less variable the oil viscosity is with temperature changes. Most hydraulic fluids have a VI value of 90–110. However, higher VI oils can also be found [3]. Some hydraulic oils are chemically designed to compensate for viscosity changes. These so-called

Table 2.1 Absolute and kinematic viscosity units

	SI[a]	CGS
Absolute (μ)	N s/m²	dyn s/cm²
Kinematic (v)	m²/s	cm²/s

[a]SI is an abbreviation of the French 'Système International d'Unités' (International System of Units).

---

[5] In honour of the French physician and physicist Jean-Louis-Marie Poiseuille (1799–1869).
[6] In honour of the Irish physicist and mathematician George Gabriel Stokes (1819–1903).

multigrade oils are capable of raising their own viscosity at higher temperatures and lowering it down at lower temperatures. Such a characteristic can be highly desirable given that at the first moments of system operation, the oil temperature is always lower. Lower oil temperatures cause the viscosity to become higher, leading to cavitation[7] and pump starvation[8] risks [4].

2.1.1.4 Viscosity Grades

The 'ISO 3448 Viscosity Classification for Industrial Liquid Lubricants' has been commonly used for hydraulic fluids. In this system, the oils are classified according to viscosity grades, VG. The ISO VG and their corresponding viscosities are given in Table 2.2.[9]

Table 2.2   ISO viscosity classification of hydraulic fluids

ISO VG	Minimum v(cSt), at 40 °C	Maximum v(cSt), at 40 °C
2	1.98	2.42
3	2.88	3.52
5	4.14	5.06
7	6.12	7.48
10	9.00	11.0
15	13.5	16.5
22	19.8	24.2
32	28.8	35.2
46	41.4	50.6
68	61.2	74.8
100	90.0	110
150	135	165
220	198	242
320	288	353
460	414	506
680	612	748
1000	900	1100
1500	1350	1650

---

[7] *Cavitation* is the formation of oil vapour bubbles at the pump inlet that results from the low suction pressure caused by difficulty in suctioning high viscous oil. These vapour bubbles eventually collapse (implode) due to the quick change in the oil pressure as they travel from the input to the output port, causing a high wear in the internal parts of the pump. Cavitation is frequently confused with *aeration*, which refers to the presence of air bubbles in the circuit. Although the effect of aeration is just as bad, cavitation and aeration are two physically different phenomena.

[8] *Pump starvation* is a phenomenon caused when high-viscosity oils do not penetrate into the small pump clearances, causing solid friction to develop, favouring a rapid wear of the mechanical parts.

[9] The viscosity grades in Table 2.2 refer to their kinematic viscosity at 40 °C, chosen as a representative temperature for most of the usual hydraulic applications. For example, an ISO VG 100 is expected to have a viscosity around 100 cSt at 40 °C. Note that the ISO standards allow the viscosity to fall within a range in such a way that the viscosity grade is always the rounded average between one maximum and one minimum viscosity value. We must be aware that the ISO 3448 classification only takes the viscosity range into consideration, leaving out the viscosity index and eventual chemical additives, which must be provided by the oil manufacturer.

2.1.1.5   Viscosity Choice

As a rule, the higher the viscosity is, the slower the fluid flows. If we apply this principle to a hydrostatic pump, where a mechanical device pushes the hydraulic fluid through conduits and narrow internal passages, we immediately observe that high-viscosity oils are more difficult to be pumped. In fact, viscosity may be viewed as the cause of internal 'friction' in the fluid and is therefore directly associated with mechanical losses.[10]

> A high viscosity increases the resistance to fluid flow and, as a consequence, mechanical losses also increase.

On the other hand, a low viscosity increases the chance of fluid leakage as that the fluid can flow through the circuit clearances more easily. Additionally, since the hydraulic oil also works as a lubricant, a low viscosity may also favour eventual solid friction by the thinning of the lubricant film.

> Low-viscosity fluids are more likely to leak and therefore contribute to draining out the hydraulic power. Low-viscosity fluids may also contribute to excessive wear of the moving components.

Figure 2.2 shows the effects of the oil viscosity on the performance of a hydrostatic pump [4]. There are two types of efficiency curves in the graph: mechanical and volumetric.

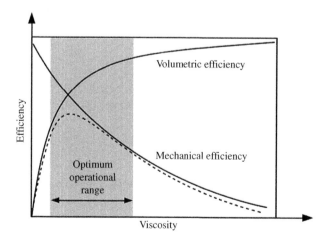

Figure 2.2   Effect of the oil viscosity on the pump performance

---

[10] Also known as *mechanical–hydraulic* losses (see, e.g. Ref. [5]). These losses are responsible for the heat generation inside hydraulic circuits and are directly related to the fluidic friction between the moving parts of the hydraulic machine and within the fluid itself.

The mechanical efficiency is related to the mechanical losses due to friction inside the pump. This efficiency drops when the fluid viscosity increases. The volumetric efficiency is related to the leakages in the circuit. This efficiency rapidly decreases at lower viscosities but shows a less accentuated variation for higher viscosities. This last behaviour can be explained by the fact that, after a certain viscosity value, the oil has already become so thick that leakages no longer quickly increase. The same effect is not observed for the mechanical efficiency, because fluid friction will invariably become higher for higher oil viscosities. The dashed curve represents the product between the mechanical and volumetric efficiencies. It will be seen later on in Chapter 3 when we formally define efficiency that the energetic efficiency of a pump/motor can be measured by the product of the volumetric and mechanical efficiencies. We can therefore conclude from Figure 2.2 that there is a certain viscosity range for which we obtain the best energetic performance of the pump (shaded area).

From our previous discussion, we conclude that neither a high viscosity nor a low viscosity is desirable for hydraulic fluids. In fact, hydraulic components usually have an optimal operational range for the oil viscosity, which is determined by the manufacturers. Since viscosity varies with temperature, a cooling system may be needed to keep the circuit temperature within established limits.[11]

2.1.2 Compressibility

Strictly speaking, liquids are not incompressible; however, in some situations, the assumption that the hydraulic fluid is incompressible can be acceptable. In fact, fluid compressibility plays a major role during transients where dynamic effects become important, as will be seen in Chapters 5 and 7. On the other hand, if we are only interested in the steady-state analysis of a hydraulic circuit, we usually consider the fluid as incompressible. In this section, we explore the theme of compressibility in hydraulic circuits. We begin by considering the fluid alone and then move on to study the role of other circuit components.

2.1.2.1 Bulk Modulus

One way to measure the compressibility of fluids is through the bulk modulus β. The isothermal bulk modulus[12] can be defined through the following expression [6]:

$$\delta p|_T = -\beta \left(\frac{\delta v}{v} \right) \bigg|_T \tag{2.4}$$

where the subscript T indicates that we are keeping the temperature constant during the process, p is the pressure and v is the specific volume of the fluid.[13]

---

[11] When choosing the right oil for a hydrostatic transmission, we must consider the recommended viscosity range for the pump and the motor given by the manufacturers. Ideally, the viscosity requirements are similar for both motor and pump so that the optimum operational viscosity can be adequate for the transmission as a whole.

[12] The fluid bulk modulus can be either determined isothermally (constant temperature) or adiabatically (constant entropy). However, it can be shown that in the particular case of liquids, there is no significant difference between their values [1].

[13] The specific volume, v, is the inverse of the density, ρ ($\rho = $ mass/volume).

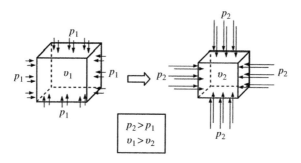

Figure 2.3   Mass of fluid under compression

Figure 2.3 describes Eq. (2.4). Suppose that a certain mass of fluid with a specific volume v_1 is exposed to a hydrostatic pressure p_1 and is isothermally compressed until its specific volume becomes equal to v_2 ($v_2 < v_1$). The pressure differential, $\delta p = p_2 - p_1$, that must be applied to compress the fluid will be proportional to the relative variation in the specific volume of the fluid, $\delta v / v$. The proportionality constant is the bulk modulus β. We can understand the bulk modulus as the reciprocal of the fluid compressibility (the higher the value of β, the less compressible the fluid). An average value of the bulk modulus,[14] β, for hydraulic fluids is 17×10^8 N/m² [9].

As an example, consider that a pressure differential of 500 bar (500×10^5 N/m²) is applied to a certain oil volume. Using the value of 17×10^8 N/m² for the bulk modulus, the following volume reduction is obtained:

$$\frac{\delta v}{v} = -\frac{500 \times 10^5}{17 \times 10^8} = -2.94\%$$

Given that the mass, m, of the compressed portion of the fluid is not altered during compression, we may write Eq. (2.4) as a function of the volume V, where the fluid is contained:

$$\delta p = -\beta \left[\frac{\delta (V/m)}{V/m} \right] = \delta p = -\beta \left(\frac{\delta V}{V} \right) \tag{2.5}$$

2.1.2.2   Effective Bulk Modulus

In practical applications, the elasticity of hydraulic conduits and the eventual presence of air in the fluid must be considered in the determination of the bulk modulus. For instance, Figure 2.4 shows a schematic view of the pressurized line in a typical hydrostatic transmission 'contaminated' with undissolved (or entrapped) air, represented by the white circles. Here, we assume that all the conduits have elastically expanded (the conduit volumes 1–5

---

[14] Strictly speaking, the bulk modulus is not constant but depends on the temperature, the pressure and the specific volume of the fluid itself. In this aspect, we must understand the value of 17×10^8 N/m² as an average value, which serves the purposes of this book. Much discussion about the bulk modulus value has been carried out and the student should be aware of the complexity of the theme (see, e.g. Ref. [7] and references therein). The situation is even more complex when other circuit components are considered alongside with the fluid [8].

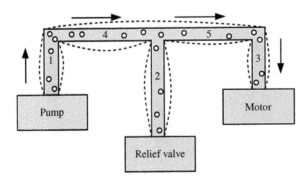

Figure 2.4 Presence of air and volumetric deformation in hydraulic conduits

appear exaggeratedly distorted due to the high pressure). There will also be some deformation in the pressurized chambers of the pump and motor, which is not shown in the figure.

We begin by considering only the presence of air inside the liquid in Figure 2.4. Let m_L be the mass of liquid and m_G the mass of entrapped gas (air) inside the conduits 1–5. The inner volume of the conduits is V. Thus, we can write the specific volume of the liquid–gas mixture as

$$v = \frac{V}{m_G + m_L} \tag{2.6}$$

In the same way that we defined the bulk modulus, β, in Eq. (2.4), we can define the effective bulk modulus, β_E, in this case. In fact, an isothermal compression within the circuit, δp, can be related to the reduction of the specific volume of the mixture liquid–gas through the following equation:

$$\delta p = -\beta_E \frac{\delta v}{v} = -\beta_E \left[\frac{\delta \left(\frac{V}{m_G + m_L} \right)}{\frac{V}{m_G + m_L}} \right] \tag{2.7}$$

In the absence of leakages in the circuit, the total mass $m_G + m_L$ remains constant after compression. We may then apply the same considerations we used for Eq. (2.5) and rewrite Eq. (2.7) as

$$\delta p = -\beta_E \frac{\delta V}{V} \tag{2.8}$$

We now split the total volume, V, into the sum of the volume occupied by the liquid, V_L, and the volume occupied by the air, V_G, to obtain

$$\delta p = -\beta_E \left(\frac{\delta V_L + \delta V_G}{V_L + V_G} \right) \tag{2.9}$$

After rearranging the terms in Eq. (2.9), we arrive at the following expression:

$$\frac{1}{\beta_E} = -\left(\frac{1}{\delta p} \frac{\delta V_L}{V_L} \right) \frac{V_L}{V} - \left(\frac{1}{\delta p} \frac{\delta V_G}{V_G} \right) \frac{V_G}{V} \tag{2.10}$$

Equation (2.10) can be rewritten as

$$\frac{1}{\beta_E} = \frac{1}{\beta_L}\frac{V_L}{V} + \frac{1}{\beta_G}\frac{V_G}{V} \tag{2.11}$$

where β_L and β_G are the bulk moduli of the liquid and the air, respectively.

By making $V_L = V - V_G$ in Eq. (2.11), we obtain

$$\frac{1}{\beta_E} = \frac{1}{\beta_L} - \frac{1}{\beta_L}\frac{V_G}{V} + \frac{1}{\beta_G}\frac{V_G}{V} \tag{2.12}$$

The volume occupied by the entrapped air, V_G, is relatively small when compared to V. On the other hand, the fluid bulk modulus, β_L, is considerably high; therefore, we can make $1/\beta_L \cong 0$. With these considerations, Eq. (2.12) simplifies to

$$\frac{1}{\beta_E} = \frac{1}{\beta_L} + \frac{1}{\beta_G}\frac{V_G}{V} \tag{2.13}$$

Using the perfect gas law, we obtain for the isothermal bulk modulus[15] of the air, β_G:

$$\beta_G = -v\frac{\delta p}{\delta v} = -v\frac{\delta\left(\dfrac{RT}{v}\right)}{\delta v} = \frac{RT}{v} = \rho RT = p \tag{2.14}$$

where R is the gas constant; T and p are the absolute temperature and pressure, respectively.

Equation (2.13) can then be rewritten as

$$\frac{1}{\beta_E} = \frac{1}{\beta_L} + \frac{V_G}{pV} \tag{2.15}$$

Assuming that $\beta_L = 17 \times 10^8$ N/m^2, we can plot the effective bulk modulus as a function of the gauge pressure[16] for different values of V_G/V. Figure 2.5 shows the results.

We conclude from Figure 2.5 that the effective bulk modulus grows with pressure and is highly influenced by the presence of undissolved air in the fluid. For instance, even at a relatively high pressure (200 bar), 1% of entrapped air is already capable of reducing the bulk modulus in more than 50%.

Equation (2.13) has been obtained by considering the compression of the mixture fluid/air alone. However, as seen in Figure 2.4, the conduits (and the other elements subject to high pressures, not shown in the figure) also undergo deformation so that the total variation in the volume, V, occupied by the fluid is given by

$$\delta V = \delta V_L + \delta V_G - \sum_i \delta V_i \tag{2.16}$$

where δV_i is the inner volume increment of every elastic element of the circuit.

---

[15] If we consider the air compression as an isotropic process, the equation $pV^{1.4} = $ constant should be used instead of the perfect gas equation [1, 10]. This would result in $\beta_G = 1.4p$ instead of $\beta_G = p$ as given by Eq. (2.14).

[16] Note that p in Eq. (2.15) is the absolute and not the gauge pressure.

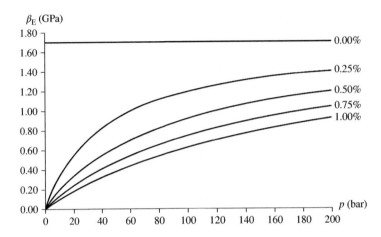

Figure 2.5 Effective bulk modulus as a function of the gauge pressure in the presence of entrapped air for different percentages of V_G/V

The negative sign at the summation in Eq. (2.16) is necessary because of the fact that, while δV_L and δV_G are negative (the liquid and gas volumes are reduced under pressure), the inner volumes of conduits and other parts (δV_i) increase. From Eqs. (2.8) and (2.16), we obtain

$$\delta p = -\beta_E \left(\frac{\delta V_L + \delta V_G - \sum_i \delta V_i}{V_L + V_G} \right) \tag{2.17}$$

We can rewrite Eq. (2.17) as follows:

$$\frac{1}{\beta_E} = -\frac{1}{\delta p}\frac{\delta V_L}{V_L}\left(\frac{V - V_G}{V} \right) - \frac{1}{\delta p}\frac{\delta V_G}{V_G}\frac{V_G}{V} + \frac{1}{\delta p}\frac{\sum_i \delta V_i}{V} \tag{2.18}$$

After reorganizing the terms and considering the fact that $\beta_G \ll \beta_L$, Eq. (2.18) becomes

$$\frac{1}{\beta_E} = \frac{1}{\beta_L} + \frac{1}{\beta_G}\frac{V_G}{V} + \frac{1}{\beta_C} \tag{2.19}$$

where β_C is the equivalent bulk modulus for the conduits and the other elastic parts of the circuit:

$$\beta_C = \left(\frac{1}{\delta p}\frac{\sum_i \delta V_i}{V} \right)^{-1} \tag{2.20}$$

In order to calculate β_C, we must know the elastic behaviour of the circuit elements so that we can determine the individual volumetric increments, δV_i. This is not an easy task and one generally has to use numerical methods and experiments to calculate shaped volumes (e.g. the interior of pumps and motors). However, it is possible to estimate the conduit bulk modulus, as will be seen in the sequence. Two cases will be presented here: straight metal conduits and hydraulic hoses.

2.1.2.3 Straight Metal Conduits

Here we consider that the conduit is a straight, thick-walled tube.[17] Thick-walled tube theory is relatively complex when compared to thin-walled tubes, and therefore we will limit ourselves to show only some resulting equations in this book. The interested reader can find a thorough explanation of the theme in Ref. [11].

A thick-walled tube filled with a fluid at a pressure p undergoes a radial stress, σ_r, a circumferential or hoop stress, σ_H, and a longitudinal stress, σ_L, at the radius, r, as indicated in Figure 2.6. These three stresses generate three corresponding strains along the radial, circumferential and longitudinal directions, respectively, which are responsible for the change in the internal volume as infinitesimally taken as dV. Thus, if the volume V of the cylinder taken at radius r (dashed lines in Figure 2.6) is $L\pi r^2$, the volumetric strain, dV/V, will be

$$\frac{dV}{V} = \left[\left(\frac{1}{L\pi r^2} \right) \frac{d}{dr} (\pi L r^2) \right] dr = \frac{dL}{L} + \frac{2dr}{r} = \varepsilon_L + 2\varepsilon_r \tag{2.21}$$

It is usual to disregard the longitudinal deformation in the volumetric strain (i.e. $\varepsilon_L \ll 2\varepsilon_r$) [12], in which case, Eq. (2.21) simplifies to

$$\frac{dV}{V} = 2\varepsilon_r \tag{2.22}$$

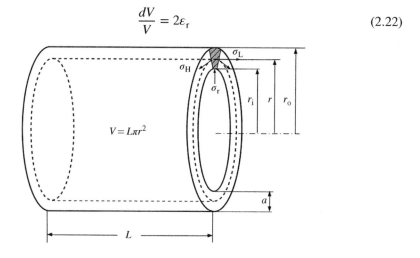

Figure 2.6 Thick-walled tube under internal pressure

---

[17] A thick-walled tube is a tube for which the wall thickness, a, is larger than 10% of the inner radius, r_i, that is, $a/r_i > 0.1$ [11]. Due to the high pressures involved, standard hydraulic tubes are generally thick walled (see Table 2.3).

Considering the module of elasticity of the metal, E, and the Poisson ratio,[18] v_p, the following relation holds for the circumferential and radial strains, ε_H and ε_r:

$$\varepsilon_H = \frac{\sigma_H - v_p\sigma_r}{E} = \frac{d(2\pi r)}{2\pi r} = \frac{dr}{r} = \varepsilon_r \tag{2.23}$$

Substituting ε_r from Eq. (2.23) into Eq. (2.22), we have

$$\frac{dV}{V} = 2\left[\frac{\sigma_H - v_p\sigma_r}{E}\right] \tag{2.24}$$

Normally, the pressure outside the tube can be disregarded when compared to the inner pressure, p. Under these circumstances, it is possible to show that the stresses σ_r and σ_H are given by [11]

$$\begin{cases} \sigma_r = \frac{pr_i^2}{r_o^2 - r_i^2}\left(1 - \frac{r_o^2}{r^2}\right) \\[3mm] \sigma_H = \frac{pr_i^2}{r_o^2 - r_i^2}\left(1 + \frac{r_o^2}{r^2}\right) \end{cases} \tag{2.25}$$

From Eqs. (2.25) and (2.24), we obtain

$$\frac{dV}{V} = \frac{2}{E}\left[\frac{pr_i^2}{r_o^2 - r_i^2}\left(1 + \frac{r_o^2}{r^2}\right) - \frac{v_p pr_i^2}{r_o^2 - r_i^2}\left(1 - \frac{r_o^2}{r^2}\right)\right] \tag{2.26}$$

Note that Eq. (2.26) does not give the total volumetric strain for the tube but the strain as a function of the radius r. In order to calculate the equivalent bulk modulus of the cylinder, it suffices to make $r = r_i$ in Eq. (2.26), in which case the volumetric deformation, dV/V, becomes

$$\frac{dV}{V} = \frac{2p}{E}\left[\frac{r_i^2 + r_o^2 + v_p\left(r_o^2 - r_i^2\right)}{r_o^2 - r_i^2}\right] \tag{2.27}$$

Using the definition of the equivalent bulk modulus, β_T, given by Eq. (2.20), we obtain from Eq. (2.27)[19]:

$$\beta_T = \frac{E(r_o^2 - r_i^2)}{2[r_i^2 + r_o^2 + v_p(r_o^2 - r_i^2)]} \tag{2.28}$$

---

[18] Named in honour of the French mathematician Siméon Denis Poisson (1781–1840), the Poisson ratio measures the secondary strains that appear in directions perpendicular to the stress direction.

[19] Note that we have assumed that the initial pressure inside the tube is zero, so that $\delta p = p$ in Eqs. (2.20) and (2.28). We have also substituted the general subscript 'C' in β_C for the more specific subscript 'T', meaning that we are applying the definition in Eq. (2.20) to 'tubes' specifically. The same reasoning will be used when we talk about hoses, where the subscript 'H' will be applied.

The Poisson ratio, v_p, for isotropic materials[20] is $\frac{1}{4}$ [12]. Thus, substituting $v_p = 1/4$ into Eq. (2.28), we, finally, obtain

$$\beta_T = \frac{E(r_o^2 - r_i^2)}{2[r_i^2 + r_o^2 + 0.25(r_o^2 - r_i^2)]} \tag{2.29}$$

Table 2.3 gives a list of standard tube dimensions used in hydraulic systems [13]. Note that the dimensions of the wall thicknesses, a, and the inner diameters, $d = 2r_i$, characterize every tube as thick walled. The table also shows the values of β_T obtained for steel tubes ($E = 2.069 \times 10^{11}$ N/m^2 [14]) based on Eq. (2.29).

The two shaded lines in Table 2.3 show the 'most rigid' and 'most flexible' tubes in the list. The effective bulk modulus for both cases, β_{E_1} and β_{E_2}, can be obtained from Eq. (2.19), assuming that no air is entrapped in the fluid ($V_G = 0$):

$$\begin{cases} \beta_{E_1} = \left(\dfrac{1}{17 \times 10^8} + \dfrac{1}{74.78 \times 10^9} \right)^{-1} = 16.62 \times 10^8 \text{ N/m}^2 \\[4mm] \beta_{E_2} = \left(\dfrac{1}{17 \times 10^8} + \dfrac{1}{10.07 \times 10^9} \right)^{-1} = 14.54 \times 10^8 \text{ N/m}^2 \end{cases} \tag{2.30}$$

We observe that the use of steel tubes does not affect the bulk modulus substantially when compared to the effect of undissolved air (Figure 2.5). For instance, in the presence of 1% of entrapped air, the effective bulk modulus reduces from 1.7 to 0.63 GPa at 100 bar. A total of 63% is a considerable reduction when compared to the 15% reduction (1.7–1.45 GPa) obtained for the more flexible steel tube in Table 2.3. The influence of the conduits on the effective bulk modulus factors into the tube material as well, and we must always check these values through Eq. (2.29). As an example, the use of copper, whose modulus of elasticity E is 1.103×10^{11} N/m^2 [14], as the tube material would result in an effective bulk modulus of 1.291×10^9 N/m^2 for the most flexible tube in Table 2.3. This value represents 76% of the bulk modulus of the hydraulic fluid (a 24% reduction).

Although it is possible to arrive at an analytical expression for the effective bulk moduli of straight metal tubes, a different situation occurs when the conduit is a flexible hose. In this case, it is generally very difficult to apply the elasticity theory to calculate β_C; the best way is to recur to experimental values, as will be seen in the following section.

2.1.2.4 Hydraulic Hoses

Given that hoses are, by nature, more flexible than metal pipes, we would expect a higher influence on the effective bulk modulus in this case. Figure 2.7, based on the data published

---

[20] Isotropic materials are those whose elastic properties are the same in all directions [12]. In general, metals can be treated as isotropic.

Table 2.3 Standard dimensions and bulk moduli for steel metal tubes

d_o(mm)	a(mm)	d(mm)	a/r_i	β_T(GPa)	d_o(mm)	a(mm)	d(mm)	a/r_i	β_T(GPa)
4	0.5	3.0	0.3	27.07	18	3.0	12.0	0.5	36.30
4	0.8	2.5	0.6	40.86	20	1.5	17.0	0.2	16.02
4	1.0	2.0	1.0	53.97	20	2.0	16.0	0.3	21.53
5	1.0	3.0	0.7	43.56	20	2.5	15.0	0.3	27.07
6	0.8	4.5	0.3	27.07	20	3.0	14.0	0.4	32.62
6	1.0	4.0	0.5	36.30	20	3.5	13.0	0.5	38.13
6	1.5	3.0	1.0	53.97	20	4.0	12.0	0.7	43.56
6	2.0	2.0	2.0	68.97	22	1.5	19.0	0.2	14.53
6	2.3	1.5	3.0	74.78	22	2.0	18.0	0.2	19.52
8	1.0	6.0	0.3	27.07	22	2.5	17.0	0.3	24.55
8	1.5	5.0	0.6	40.86	22	3.0	16.0	0.4	29.59
8	2.0	4.0	1.0	53.97	25	2.0	21.0	0.2	17.12
8	2.5	3.0	1.7	65.59	25	2.5	20.0	0.3	21.53
10	1.0	8.0	0.3	21.53	25	3.0	19.0	0.3	25.96
10	1.5	7.0	0.4	32.62	25	4.0	17.0	0.5	34.83
10	2.0	6.0	0.7	43.56	25	4.5	16.0	0.6	39.22
10	2.5	5.0	1.0	53.97	25	5.0	15.0	0.7	43.56
10	3.0	4.0	1.5	63.43	28	1.5	25.0	0.1	11.35
12	1.0	10.0	0.2	17.85	28	2.0	24.0	0.2	15.24
12	1.5	9.0	0.3	27.07	28	2.5	23.0	0.2	19.16
12	2.0	8.0	0.5	36.30	28	3.0	22.0	0.3	23.11
12	2.5	7.0	0.7	45.34	30	2.0	26.0	0.2	14.20
12	3.0	6.0	1.0	53.97	30	2.5	25.0	0.2	17.85
12	3.5	5.0	1.4	61.94	30	3.0	24.0	0.3	21.53
14	1.5	11.0	0.3	23.11	30	4.0	22.0	0.4	28.92
14	2.0	10.0	0.4	31.04	30	5.0	20.0	0.5	36.30
14	2.5	9.0	0.6	38.91	35	2.0	31.0	0.1	12.13
14	3.0	8.0	0.8	46.61	35	2.5	30.0	0.2	15.24
14	3.5	7.0	1.0	53.97	35	3.0	29.0	0.2	18.37
14	4.0	6.0	1.3	60.85	35	4.0	27.0	0.3	24.69
15	1.0	13.0	0.2	14.20	38	2.5	33.0	0.2	14.01
15	1.5	12.0	0.3	21.53	38	3.0	32.0	0.2	16.89
15	2.0	11.0	0.4	28.92	38	4.0	30.0	0.3	22.69
15	3.0	9.0	0.7	43.56	38	5.0	28.0	0.4	28.53
16	1.5	13.0	0.2	20.15	38	6.0	26.0	0.5	34.36
16	2.0	12.0	0.3	27.07	38	7.0	24.0	0.6	40.14
16	2.5	11.0	0.5	34.00	42	2.0	38.0	0.1	10.07
16	3.0	10.0	0.6	40.86	42	3.0	36.0	0.2	15.24
18	1.0	16.0	0.1	11.78	42	4.0	34.0	0.2	20.47
18	1.5	15.0	0.2	17.85	50	6.0	38.0	0.3	25.96
18	2.0	14.0	0.3	23.99	50	9.0	32.0	0.6	39.22
18	2.5	13.0	0.4	30.15	65	8.0	49.0	0.3	26.64

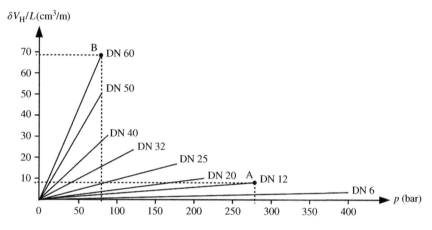

Figure 2.7 Volumetric dilatation as a function of the inner pressure for different nominal diameters (DNs) of hydraulic hoses (reproduced courtesy of Contitech AG)

by a hydraulic hose manufacturer (Contitech AG [15]), shows an example of how the internal hose volume increases with pressure for different nominal diameters.[21] Observing the figure, we conclude that it is possible to write the volumetric dilatation of the hose, $\delta V_H(cm^3)$, as a function of the pressure variation, δp(bar), and the conduit length, L(m):

$$\frac{\delta V_H}{L} = k_H \delta p \tag{2.31}$$

where k_H is a constant that depends on the nominal hose diameter.

The hose equivalent bulk modulus, β_H, is given by

$$\frac{1}{\beta_H} = \frac{1}{\delta p}\left(\frac{\delta V_H}{V}\right) \tag{2.32}$$

where V is the inner hose volume, which can be expressed as a function of the length, L, and the inner radius, r_i, as $V = L\pi r_i^2$. Note that we have used the subscript 'H' instead of the more generalist subscript 'C' in Eq. (2.20), following the same reasoning used for Eq. (2.28).

Substituting δV_H, obtained from Eq. (2.31) into Eq. (2.32), we have

$$\frac{1}{\beta_H} = \frac{1}{\delta p}\left(\frac{Lk_H \delta p}{L\pi r_i^2}\right) = \frac{k_H}{\pi r_i^2} \tag{2.33}$$

As an example, for a hose with a nominal diameter of 12 mm ($r_i = 6$ mm), the constant k_H can be obtained from the line corresponding to DN = 12 mm in Figure 2.7 and Eq. (2.31).

---

[21] The nominal diameter of a hose can be approximately defined as the inner diameter [15].

Considering the volumetric dilatation of $7\,cm^3/m$ caused by the pressure increase from 0 to 280 bar (point A in Figure 2.7), we have

$$k_H = \frac{7(cm^3/m)}{(280-0)\,bar} = \left(\frac{1}{40}\right)\frac{cm^3}{bar.m} \cong 2.5 \times 10^{-13}\,m^4/N \tag{2.34}$$

Substituting k_H given by Eq. (2.34) into Eq. (2.33), we obtain for β_H:

$$\beta_H = \left(\frac{k_H}{\pi r_i^2}\right)^{-1} = \frac{3.14 \times (0.006)^2}{2.5 \times 10^{-13}} = 4.52 \times 10^8\,N/m^2 \tag{2.35}$$

The effective bulk modulus in the absence of entrapped air can then be calculated as

$$\beta_E = \left(\frac{1}{17 \times 10^8} + \frac{1}{4.52 \times 10^8}\right)^{-1} = 3.57 \times 10^8\,N/m^2 \tag{2.36}$$

We clearly see that there is a much higher reduction of the effective bulk modulus when comparing hoses to steel tubes. In fact, the use of hoses as conduits has reduced the effective bulk modulus in 79% (compare 0.357–1.7 GPa). A somewhat greater reduction would occur for bigger hose diameters. For instance, it could be easily verified that if a $DN = 60\,mm$ hose was used, we would have obtained $k_H = 0.8625\,cm^3/bar.m$, based on the coordinates of point B in Figure 2.7 (80 bar, $69\,cm^3/m$). In this case, the effective bulk modulus would be $2.36 \times 10^8\,N/m^2$ (a reduction of 86% in relation to the fluid bulk modulus).

2.1.2.5 Hydraulic Accumulators

We briefly commented about the use of accumulators in hydrostatic transmissions in Chapter 1 (Section 1.4.1) and how they affect the transmission response. In fact, accumulators have a considerable impact on the effective bulk modulus of a hydraulic circuit. To see that, consider the situation illustrated in Figure 2.8, where a branch of a hydraulic circuit is connected to a bladder accumulator whose initial gas volume is V_G. The volume of liquid inside the conduits and accumulator is V_L so that the total volume is $V = V_G + V_L$. Note that the accumulator bladder resembles a huge air bubble trapped inside the fluid mass.

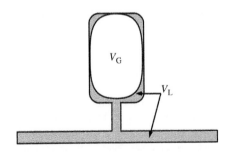

Figure 2.8 Accumulator in the circuit

We can have an idea of the influence of the accumulator in the circuit by assuming an isothermal compression, to which Eq. (2.14) applies.[22] Therefore, from Eqs. (2.14) and (2.12), we obtain

$$\frac{1}{\beta_E} = \frac{1}{\beta_L} - \frac{1}{\beta_L} \frac{V_G}{V} + \frac{1}{p} \frac{V_G}{V} \tag{2.37}$$

As an example, suppose that the initial gas volume, V_G, is 1 litre (l) at a given pre-charge pressure. If we assume that the bladder initially extends to the limits of the accumulator walls, as indicated in Figure 2.8, the liquid volume, V_L, will correspond to the inner volume of the conduit alone. Consider also that the inner diameter of the conduit is 12.7 mm and the total length is 2 m, corresponding to an internal volume of approximately 0.26 litres. In this case, the effective bulk modulus becomes

$$\beta_E = \left[\frac{1}{17 \times 10^8} - \frac{1}{17 \times 10^8} \left(\frac{1}{1 + 0.26} \right) + \frac{1}{p} \left(\frac{1}{1 + 0.26} \right) \right]^{-1} \tag{2.38}$$

Figure 2.9 is a plot of Eq. (2.38) and shows how the effective bulk modulus changes with pressure. As expected, the presence of an accumulator has greatly reduced β_E (note that the scale has changed from GPa to MPa, when Figure 2.9 is compared to Figure 2.5). This should not come as a surprise, since one of the functions of the accumulator in the circuit is to absorb shocks. One of the ways in which this can be done is by making the circuit 'less stiff', which is practically accomplished by reducing the effective bulk modulus.

Smaller bulk moduli ultimately mean that more energy coming from the pump gets trapped in the accumulator before reaching the motor (or actuator). This has a direct affect on the response time of the transmission, which, on the other hand, is directly related to the propagation of the pressure waves inside the circuit, as will be seen shortly.

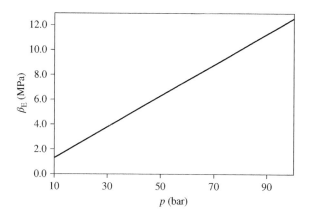

Figure 2.9   Effective bulk modulus in the presence of a hydraulic accumulator

---

[22] This is quite a substantial approximation, given the thermodynamic complexity of the charge and discharge process (see Section 1.4.1.6).

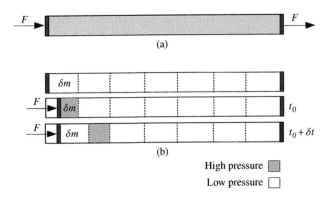

Figure 2.10   Force transmission in a conduit for an incompressible fluid (a) and a compressible fluid (b)

2.1.2.6   Pressure Waves

Figure 2.10(a) shows a pipe filled with hydraulic fluid and limited at the left and right sides by two pistons. If the fluid were incompressible and in the absence of friction between the piston and the case, the force, F, applied on the left piston, would be integrally transmitted to the right piston. Thus, if the force, F, were suddenly applied on the left piston, an instant reaction would appear on an obstacle placed in front of the right piston.

The situation becomes different when we consider fluid compressibility, as seen in Figure 2.10(b), where the force F is applied on the left piston at time t_0. Because of the fluid elasticity, the element, δm, adjacent to the left piston, is compressed and then subsequently expands at time $t_0 + \delta t$, compressing the next layer of fluid. The process is then repeated for the next fluid elements and the pressure is 'transferred' from layer to layer.[23] Although it is not shown in the figure, a careful reader will conclude that the pressure wave generated by the sudden application of the force F on the left piston will eventually bounce back when it gets to the right piston. Since these waves involve localized motion of fluid particles, the presence of viscous forces within the fluid will, in the end, disperse the pressure wave, which eventually ceases.

The pressure, p, inside the conduit in Figure 2.10(b) is a function of both time and space, that is, $p = p(x, t)$, where the spatial coordinate, x, is measured along the conduit. The determination of $p(x, t)$ requires the solution of a partial differential equation (or set of equations, depending on the complexity of the physical model adopted). For instance, if viscous losses are disregarded, the unidimensional wave equation can be used:

$$\frac{\partial^2 p}{\partial t^2} = c^2 \frac{\partial^2 p}{\partial x^2} \qquad (2.39)$$

---

[23] The compression of the fluid happens continuously and not discretely as shown in the figure. The basic principles illustrated in Figure 2.10 remain valid.

where c is the wave propagation speed, which is related to the fluid density, ρ, and the effective bulk modulus, β_E, through the following expression [16]:

$$c = \sqrt{\frac{\beta_E}{\rho}} \qquad (2.40)$$

The analytical solution of Eq. (2.39), together with the proper boundary and initial conditions, is lengthy and will not be given here (see Ref. [17] for a detailed analysis).

In Eq. (2.40), we observe that the speed, c, at which the pressure waves travel inside the conduit, depends directly on the effective bulk modulus, β_E. For a homogeneous fluid, whose bulk modulus and density are β and ρ, c will correspond to the speed of sound in the fluid. For instance, considering $\rho = 870$ kg/m^3 (usual density for hydraulic fluids [18]) and $\beta = 17 \times 10^8$ N/m^2, we obtain

$$c = \sqrt{\frac{17 \times 10^8}{870}} = 1397.86 \text{ m/s} \qquad (2.41)$$

We have seen that the effective bulk modulus can be considerably reduced in relation to the fluid bulk modulus because of the elasticity of the circuit components. Such reduction can have a great impact on the wave speed, c. For example, suppose that 1% of undissolved air at 10 bar is present in the hydraulic fluid. Moreover, assume that the conduit is a hose with an equivalent bulk modulus of 4.52×10^8 N/m^2 (Eq. (2.35)). Making $\beta_G = p$ in Eq. (2.19), we obtain for the effective bulk modulus, in this case:

$$\beta_E = \left(\frac{1}{17 \times 10^8} + \frac{0.01}{10 \times 10^5} + \frac{1}{4.52 \times 10^8} \right)^{-1} = 0.78 \times 10^8 \text{ N/m}^2 \qquad (2.42)$$

Interestingly, while the effective bulk modulus is significantly affected by the presence of entrapped air, the same does not happen with the density, which remains practically unaltered. For instance, considering the air as a perfect gas, the density is given by

$$\rho = \frac{p}{RT} \qquad (2.43)$$

where R is the 'gas constant' for the air ($R = 287.058$ J/kg K).

Now, assuming that the oil density is 870 kg/m^3, and using ρ as given by Eq. (2.43), we obtain the density of the mixture air–oil at room temperature (20 °C or 293 K):

$$\rho = 870 \times 0.99 + \left[\frac{(10 + 1) \times 10^5}{287.058 \times 293} \right] \times 0.01 = 861.43 \text{ kg/m}^3 \qquad (2.44)$$

which is very close to the density of the oil itself.

With the values of the mixture density (Eq. (2.44)) and the effective bulk modulus (Eq. (2.42)), we can determine the pressure propagation speed from Eq. (2.40) as

$$c = \sqrt{\frac{2.08 \times 10^8}{861.43}} = 491.38 \text{ m/s} \tag{2.45}$$

Equation (2.45) shows that the pressure waves travel inside the conduits at a speed almost three times slower than the speed of sound in the fluid (Eq. (2.41)). However, it is still a high speed and, in most of the situations, it can be safely assumed that pressure rises homogeneously along the hydraulic lines.[24] Such assumption makes mathematical modelling much easier, as will be seen later on in Chapters 5 and 7. On the other hand, if an accumulator was connected to the pressurized line, the situation would be quite different. To see this, let us consider a pressure of 40 bar for the simplified accumulator illustrated in Figure 2.8. In this situation, we can estimate the effective bulk modulus as $\beta_E = 5 \times 10^6$ N/m^2 (see Figure 2.9). Since we are not mixing air and oil together, the density of the fluid is not altered, that is, $\rho = 870$ kg/m^3. The pressure wave speed, c, in this case becomes

$$c = \sqrt{\frac{5 \times 10^6}{870}} = 75.81 \text{ m/s} \tag{2.46}$$

Note that the assumption that pressure develops uniformly along the conduit cannot be held, especially in long lines. As an example, if the line was 10 m long, it would take 0.13 s for a pressure wave to propagate from one end to another. Therefore, we conclude that the presence of accumulators in the circuit can significantly delay the response of the output device (e.g. motor) to an input signal coming from the pump.

2.1.2.7 Compressibility Flowrates

Consider the illustration in Figure 2.11, where two hypothetical non-leaking cylinders are connected to each other. The cylinder R on the right has its rod fixed so that it cannot move. The circuit is filled up with unpressurized oil at the start (Figure 2.11(a)) and then, by the action of an external force, F, the cylinder L on the left compresses the oil inside the upper conduit as it moves upward, reaching an equilibrium position as soon as the elastic forces balance the external force.

Because of the oil elasticity, the dislocation of cylinder L causes a reduction of volume at the upper branch of the circuit, $-\delta V$, and a corresponding increase in the volume at the lower branch, $+\delta V$. Since the compression/expansion process happens in a period of time, δt, it is natural to attribute a 'flowrate' to both sides of the circuit. The term *compressibility flowrate* [9] denotes the ratio $\delta V/\delta t$, which can be written as

$$\frac{\delta V}{\delta t} = -\left(\frac{V}{\beta_E}\right)\frac{\delta p}{\delta t} \tag{2.47}$$

[24] In other words, we are assuming that the pressure is a function of time alone, not a function of the length x, as in Eq. (2.39).

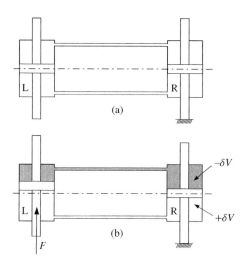

Figure 2.11 Compressibility flowrate

Closed circuits, where the fluid circulates between the pump and the motor (or actuator) without an intermediary tank, are to a certain extent similar to the circuit shown in Figure 2.11 and will develop compressibility flowrates whenever there is a pressure surge in the conduits during transients. As a result, an external flow must be supplied to the low-pressure conduit to avoid fluid evaporation. This can be accomplished through a charge circuit, as explained in Chapter 1.

The flow demands caused by a pressure surge may be considerably high. For instance, suppose the force F elevates the pressure in the upper conduit from 5 to 100 bar in 1/20 s, the compressibility flowrate for an effective bulk modulus $\beta_E = 3.57 \times 10^8$ N/m^2 (Eq. (2.36)) and an inner volume $V = 1$ litre will be

$$\frac{\delta V}{\delta t} = -\left(\frac{1 \times 0.001}{3.57 \times 10^8}\right)\frac{(95 \times 10^5)}{1/20} = 5.322 \times 10^{-4} \ \text{m}^3/\text{s} = 31.93 \ \text{l/min} \qquad (2.48)$$

2.2 Fluid Flow in Hydraulic Circuits

As mentioned earlier, there are situations in which the assumption of an incompressible fluid is necessary. One particular case is the study of hydraulic-mechanical losses in the circuit, where we are interested in the overall energy exchanges within the internal flows. This section deals with the details of internal incompressible flows in hydraulic circuits with emphasis on energy losses. We begin with some general considerations about typical flow patterns (or regimes) found in hydraulic circuits.

2.2.1 Flow Regimes

Generally speaking, there are two patterns for fluid flows: laminar and turbulent. Laminar flows resemble an organized pile of fluid 'layers', just like a stack of papers being gently

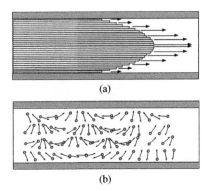

Figure 2.12   Laminar (a) and turbulent (b) flows between two parallel plates

sheared apart (Figure 2.12(a)). Each fluid layer slides on top of the other as the flow progresses. Turbulent flows, on the other hand, are made of chaotic movements of fluid particles (Figure 2.12(b)). Most of the existing relations for turbulent flows come from empirical experiments, and no elementary theory exists that can be precisely applied to every situation. Turbulent flows are not desirable in hydraulic circuits because the random motion of the fluid particles causes a considerable loss of energy. Fortunately, there is a means of predicting when the flow will become turbulent: the *Reynolds number*,[25] *Re*. Considering the flow between parallel plates shown in Figure 2.12, the Reynolds number is given by

$$Re = \frac{\rho \bar{u} d}{\mu} = \frac{\bar{u} d}{\nu} \tag{2.49}$$

where $\bar{u}$ is the average velocity of the fluid, d is the vertical distance between plates, ρ is the fluid density μ and ν are the absolute and kinematic viscosities of the fluid, respectively.

In other geometries, the distance d in Eq. (2.49) should be substituted by another representative dimension. For instance, for internal pipe flows, d becomes the inner diameter of the pipe. In more complex flow configurations, as in external flows over wings, the choice of d is not straightforward. In this book, however, we focus on internal flows, and the Reynolds number always relates to the geometric dimension, which characterizes the space where the flow is confined (e.g. the distance between the plates, the pipe diameter or the annular space between a piston and its case).

The parabolic profile of the laminar flow in Figure 2.12(a) is a result of the viscous action of each layer of fluid onto the other. Such a profile is known as the 'boundary layer' and is only completely developed after a certain entrance length, L_{min}, measured from the point where the flow first touches the solid surface (the plate edges in Figure 2.12). It has been experimentally determined that the entrance length in straight pipes can be estimated by [19]

$$L_{min} = (0.06d)Re \tag{2.50}$$

where d is the inner diameter of the pipe.

---

[25] In honour of the Irish scientist Osborne Reynolds (1842–1912).

2.2.2 Internal Flow in Conduits

In order to study internal flows in hydraulic conduits, we must first determine the flow regime. In this aspect, it is useful to write the Reynolds number as a function of the volumetric flow, q_c, the kinematic viscosity, v and the cross-sectional area, $\pi d^2/4$. Thus, from Eq. (2.49), we have

$$Re = \left(\frac{4q_c}{\pi d^2}\right)\left(\frac{d}{v}\right) = \frac{4q_c}{\pi d v} \tag{2.51}$$

It is generally accepted that usual pipe flows experience a transition from laminar to turbulent at $Re \approx 2300$ [19]. Therefore, we can use Eq. (2.51) to predict the type of flow regime in the conduit. For example, considering an ISO VG 32 hydraulic fluid ($v = 32 \times 10^{-6}$ m^2/s), the flow q_c (l/min), at the turbulence limit, can be written as a function of the inner diameter, d (mm), as

$$q_c = \left(\frac{2300\pi}{4}\right)\left(32 \times 10^{-6} \times \frac{100\ \text{dm}^2}{\text{min}} \times 60\right) d \times \left(\frac{1}{100}\right) \cong 3.47d \tag{2.52}$$

From Eq. (2.52), we conclude that a hose with an inner diameter of 10 mm should not carry a flow greater than 34.7 l/min if we expect the flow to be laminar.

2.2.2.1 Laminar Flow Theory for Pipes

Consider the straight pipe shown in Figure 2.13, through which a fully developed laminar flow is passing.[26] The pipe has a total length L and an inner radius r_i ($r_i = d/2$). The pressures

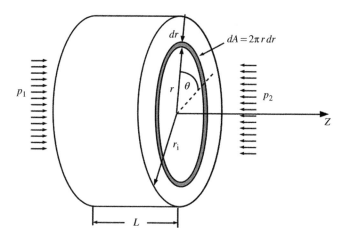

Figure 2.13 Laminar flow inside a cylindrical conduit

[26] We consider here that the boundary layer is fully developed, as shown in Figure 2.12(a). The student must be aware that, even for laminar flows, this assumption can only give us an estimate of the actual pressure losses in hydraulic conduits (see, e.g. Refs. [1, 19] for a more detailed analysis).

at the left and the right extremities are p_1 and p_2, respectively. Because of the cylindrical symmetry, velocity and pressure are expressed in cylindrical coordinates (r, θ, z). From the pattern observed in the laminar flow between parallel plates in Figure 2.12(a), we conclude that the following boundary conditions apply in this case:

$$\begin{cases} u\left(r_i, \theta, z\right) = 0 \\ p(r, \theta, 0) = p_1 \\ p(r, \theta, L) = p_2 \\ \dfrac{\partial u}{\partial r}(0, \theta, z) = 0 \end{cases} \tag{2.53}$$

The flow inside the pipe represented in Figure 2.13 is unidimensional, meaning that the velocity at every point of the flow is solely a function of the radius r. If we consider the steady-state regime, when the flow pattern no longer varies in time, the Navier–Stokes equations reduce to (see Eq. (C.29) in Appendix C):

$$\frac{\mu}{r}\frac{\partial}{\partial r}\left(r\frac{\partial u}{\partial r}\right) = \frac{\partial p}{\partial z} \tag{2.54}$$

where μ is the absolute viscosity.

In Eq. (2.54), we observe that $u = u(r)$ and $p = p(z)$. Therefore, the equality (2.54) can only exist if both members are equal to the same constant, k, that is[27]:

$$\begin{cases} \dfrac{\mu}{r}\dfrac{d}{dr}\left(r\dfrac{du}{dr}\right) = k \\ \dfrac{dp}{dz} = k \end{cases} \tag{2.55}$$

Equations (2.55) can be individually integrated, with the subsequent substitution of the boundary conditions (2.53). Thus, in what follows, let c_0, c_1 and c_2 be real constants. Integrating the first equation once with respect to r, we obtain

$$\mu\int \frac{d}{dr}\left(r\frac{du}{dr}\right)dr = k\int r\,dr \Rightarrow r\frac{du}{dr} = \left(\frac{k}{2\mu}\right)r^2 + c_0 \tag{2.56}$$

From the last boundary condition in (2.53), we obtain for the constant c_0:

$$\frac{du}{dr}\bigg|_{r=0} = 0 \Rightarrow r\cdot 0 = \left(\frac{k}{2\mu}\right)\cdot 0 + c_0 \Rightarrow c_0 = 0 \tag{2.57}$$

---

[27] Suppose that the left-hand side of Eq. (2.54) is a function of r and the right-hand side is a function of z; then, the equality (2.54) cannot be verified. This leaves us with the only possible solution, that is, both sides being equal to the same constant value. Notice also the shift in notation from partial to ordinary differentials between Eqs. (2.54) and (2.55).

A new integration with respect to r gives

$$\int \left(\frac{du}{dr}\right) dr = \left(\frac{k}{2\mu}\right) \int r\, dr \Rightarrow u = \left(\frac{kr^2}{4\mu}\right) + c_1 \tag{2.58}$$

We obtain c_1 from the first boundary condition in (2.53):

$$u(r_i) = 0 \Rightarrow 0 = \left(\frac{kr_i^2}{4\mu}\right) + c_1 \Rightarrow c_1 = -\frac{kr_i^2}{4\mu} \tag{2.59}$$

Integration of the second equation in (2.55), between $z = 0$ and $z = L$, gives

$$\int_0^L \left(\frac{dp}{dz}\right) dz = k \int_0^L dz \Rightarrow p_2 - p_1 = kL \Rightarrow k = -\frac{\Delta p}{L} \tag{2.60}$$

where

$$\Delta p = p_1 - p_2 \tag{2.61}$$

Combining Eqs. (2.58)–(2.60), we obtain

$$u = \frac{\Delta p r_i^2}{4\mu L}\left[1 - \left(\frac{r}{r_i}\right)^2\right] \tag{2.62}$$

The flow through the pipe, q_c, is given by

$$q_c = \int_0^{r_i} u 2\pi r\, dr = 2\pi \int_0^{r_i} r\left(\frac{\Delta p r_i^2}{4\mu L}\right)\left[1 - \left(\frac{r}{r_i}\right)^2\right] dr \tag{2.63}$$

If we develop the integral in Eq. (2.63), we obtain

$$q_c = \left(\frac{\pi \Delta p r_i^2}{2\mu L}\right) \int_0^{r_i} \left(r - \frac{r^3}{r_i^2}\right) dr = \left(\frac{\pi \Delta p r_i^2}{2\mu L}\right)\left(\frac{r_i^2}{2} - \frac{1}{r_i^2}\frac{r_i^4}{4}\right) = \left(\frac{\pi r_i^4}{8\mu L}\right)\Delta p \tag{2.64}$$

Equation (2.64) can be rewritten in a more convenient form:

$$\Delta p = \varepsilon_c q_c \tag{2.65}$$

where ε_c is constant for a constant viscosity:

$$\varepsilon_c = \frac{8\mu L}{\pi r_i^4} = \frac{128\mu L}{\pi d^4} = \frac{128 v \rho L}{\pi d^4} \tag{2.66}$$

Note that the coefficient ε_c has units of (pressure)/(volumetric flow):

$$\varepsilon_c = \frac{\text{N/m}^2}{\text{m}^3/\text{s}} = \text{kg/m}^4\,\text{s} \tag{2.67}$$

2.2.2.2 Generalized Pressure Loss Equation and Turbulence

Equation (2.65) requires the knowledge of the pipe flow, q_c, to be solved for the pressure losses, Δp. We can always write q_c as a function of the inner diameter, d, and the average fluid speed, $\bar{u}$, obtained at a cross section of the conduit as follows:

$$q_c = \bar{u}\left(\frac{\pi d^2}{4}\right) \tag{2.68}$$

Substituting ε_c, given by Eq. (2.66), and q_c, given by Eq. (2.68), into Eq. (2.65), we obtain

$$\Delta p = \left(\frac{128\mu L}{\pi d^4}\right)\left[\frac{\bar{u}\left(\pi d^2\right)}{4}\right] = \frac{32\mu L\bar{u}}{d^2} = f\left(\frac{L}{d}\right)\left(\frac{\rho\bar{u}^2}{2}\right) \tag{2.69}$$

where f is the *friction factor* [19] and is given by

$$f = \frac{64\mu}{u\rho d} = \frac{64}{Re} \tag{2.70}$$

The expression for the pressure drop (2.69) is very convenient. If we observe that pressure has units of energy per volume, Eq. (2.69) ultimately states that the energy losses are proportional to the kinetic energy per unit volume of the fluid and the dimensionless ratio between length and diameter of the conduit. The fact that we have used the average fluid speed to obtain the flow, q_c, characterizes Eq. (2.69) as an overall energy balance, despite the fact that the coefficient f itself has been obtained from a differential analysis using the Navier–Stokes equations. By a similar reasoning, we can infer that the pressure drop will always be proportional to the average kinetic energy per unit volume and the ratio L/d, independent of the flow regime in view. This is, in fact, one practical way of dealing with turbulent flows in pipes. Therefore, Eq. (2.69) can be generalized to represent the pressure losses in turbulent flows as well. All that needs to be changed is the friction factor, f, which must now be experimentally determined, given that Eq. (2.54) cannot be applied for turbulent flows.[28]

Many correlations have been suggested for the friction factor, which provide a reasonable approximation for Δp at turbulent regimes. Perhaps the simplest one has been proposed by Blasius[29] for flows in which $2300 < Re < 10^5$:

$$f = \frac{0.316}{Re^{0.25}} \tag{2.71}$$

Equations (2.69)–(2.71) have been used in practical applications and are currently adopted by some industrial manufacturers to calculate pressured drops in straight metal tubes (see, e.g. Ref. [13]). It is true to say, however, that turbulent flows should still be avoided because of the higher energy losses [20]. Therefore, in the majority of the practical applications, the friction factor will be calculated by Eq. (2.70).

---

[28] For instance, we note that the last boundary condition in Eq. (2.53) cannot be used if the flow is turbulent.
[29] Paul Richard Heinrich Blasius, German physicist (1883–1970).

As an example of a pressure drop calculation, suppose 20 l/min (3.33×10^{-4} m^3/s) of an ISO VG 32 hydraulic fluid ($v = 32 \times 10^{-6}$ m^2/s) is pumped through a pipe whose inner diameter is ½ in. (0.0127 m). The flow will be laminar, as confirmed by Eq. (2.51):

$$Re = \frac{4 \times 3.33 \times 10^{-4}}{3.14 \times 0.0127 \times 32 \times 10^{-6}} \cong 1044 < 2300 \qquad (2.72)$$

From Eqs. (2.65) and (2.66), we obtain the pressure drop per unit length, $\Delta p / L$:

$$\frac{\Delta p}{L} = \frac{128 v \rho q_c}{\pi d^4} \qquad (2.73)$$

Substituting the numerical values into Eq. (2.73), we obtain for a typical hydraulic fluid with density $\rho = 870$ kg/m^3:

$$\frac{\Delta p}{L} = \frac{128 \times 32 \times 10^{-6} \times 870 \times 33 \times 10^{-4}}{3.14 \times (0.0127)^4} = 143962.24 \text{ Pa/m} \cong 1.44 \text{ bar/m} \qquad (2.74)$$

Although the theory developed in this section focused on straight pipes alone, it is generally accepted that the resulting equations equally apply to flexible hoses as well[30] [10]. The situation is not as easy when more complex geometries such as bends, junctions and other types of hose and pipe connections are involved. Pressure losses can become considerably high in those elements and must be accounted for in an actual circuit design. However, since no specific theory exists to model pressure losses in such cases, the designer should rely on experimental data from the manufacturers.[31]

2.2.3 Flow Through Orifices

The flow regime through small passages such as orifices (Figure 2.14) is difficult to characterize and depends on many factors, such as the sharpness of the orifice edges and geometry.

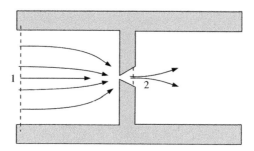

Figure 2.14   Orifice flow

---

[30] Here, we consider that the hose is not excessively bent.
[31] Usually, companies give the pressure drops for their manufactured connections in a graphical manner, as a function of the flow rate, for example. As for pressure losses in geometries other than straight conduits, it is common to recur to classical experimental data, which assigns an equivalent length in Eq. (2.69) to different geometries [21, 19]. We will give a practical example of how these concepts can be applied later in Chapter 4.

However, we can always approach this type of flow macroscopically through an energy balance equation, as will be seen in the sequence.

In Figure 2.14, the average pressure and speed at sections 1 and 2 are p_1, u_1 and p_2, u_2, respectively. If we disregard the energy losses in the passage from sections 1 to 2, we can write that (see Eq. (C.57), Appendix C)

$$p_1 + \frac{\rho u_1^2}{2} = p_2 + \frac{\rho u_2^2}{2} \tag{2.75}$$

In Eq. (2.75), the speeds u_1 and u_2 can be written as a function of the flow through the orifice, q, and the cross-sectional areas at sections 1 and 2, A_1 and A_2, respectively (i.e. $u_1 = q/A_1$ and $u_2 = q/A_2$). Thus, we have

$$p_1 + \frac{\rho q^2}{2A_1^2} = p_2 + \frac{\rho q^2}{2A_2^2} \tag{2.76}$$

Rearranging Eq. (2.76), we obtain

$$q = \left[A_2 \sqrt{\frac{1}{1 - (A_2/A_1)^2}} \right] \sqrt{\frac{2(p_1 - p_2)}{\rho}} = C_d A_2 \sqrt{\frac{2(p_1 - p_2)}{\rho}} \tag{2.77}$$

where C_d is given by

$$C_d = \sqrt{\frac{1}{1 - (A_2/A_1)^2}} \tag{2.78}$$

Equation (2.77) can be rewritten as

$$q = C \sqrt{\frac{2\Delta p}{\rho}} \tag{2.79}$$

where $C = C_d A_2$ and $\Delta p = p_1 - p_2$.

In an actual flow, there are pressure losses that have not been accounted for in Eq. (2.76). Therefore, in practice, the coefficient, C, should be experimentally determined rather than calculated through geometric parameters.

Equation (2.79) is commonly referred to as the 'orifice flow equation'. We can also write it in a slightly different form, which will be useful later on in this chapter:

$$q = C\sqrt{2}\sqrt{\frac{\Delta p}{\rho}} = k_T \sqrt{\frac{\Delta p}{\rho}} \tag{2.80}$$

where $k_T = C\sqrt{2}$.

2.2.4 Leakage Flow in Pumps and Motors

In every moving part of a hydraulic circuit, there will exist a relative motion between two solid surfaces. Such is the case, for example, of pumps, motors and cylinders. Because of the pressure differentials across the clearances formed between these surfaces, leakages are likely to happen.

Gaps between moving parts of hydraulic circuits are usually very small. A typical leakage can be compared to the flow between two parallel plates separated by a very small distance, with a pressure differential applied between the extremities. In these circumstances and considering the Reynolds number given by Eq. (2.49), it is easy to conclude that leaking flows are likely to be laminar. For example, consider the flow of a very low-viscosity fluid ($v = 10 \times 10^{-6} \, \text{m/s}^2$) through a $50 \, \mu\text{m}$ clearance (a relatively large clearance size for actual hydraulic devices [5]). It has been reported that, in the case of flows through tiny clearances between two stationary plates, turbulence can happen for Reynolds numbers bigger than 1400 [19]. In our example, this would require an average fluid speed of $280 \, \text{m/s}$. Such speed magnitude leads to the conclusion that for turbulence to develop, there must be an extremely fast oil stream being literally pumped through the clearance. Therefore, it seems reasonable to assume that leakage flows must be laminar. However, it must be said that when dealing with leakages inside pumps and motors, the limiting surfaces of the clearances are generally in motion in relation to one another. In addition, there will always be some degree of mechanical vibration involved, making it difficult to determine which flow regime we are dealing with. This fact appears to be one of the reasons why some authors have favoured the assumption that the flows inside these small clearances might become turbulent in some parts of the pump or motor, and have included a turbulence term in their leakage equations [22].

In this book, we consider leakage flow as laminar (see, e.g. Refs. [23–25] for a similar approach). The reason is twofold: (a) it is a fairly reasonable assumption given the small clearance dimensions [23], and (b) to this date, there is no exact model for pump and motor leakages in general. In fact, any proposed theoretical equation still relies on experimental data to be properly adjusted to a particular situation. Therefore, in what follows, we use the laminar leakage assumption to obtain a generalized expression for the volumetric and mechanical losses in a hydrostatic pump and motor. We have chosen to follow a pioneering paper written by Wilson [26], which has been considered a classical work for many years [24].

2.2.4.1 Volumetric Losses

In order to obtain a suitable model for leakages, consider the vane pump illustrated in Figure 2.15. In this pump, the shaft has radial slots, which contain the vanes. As the shaft rotates counter-clockwise, the vanes extend radially and press against the pump case, causing

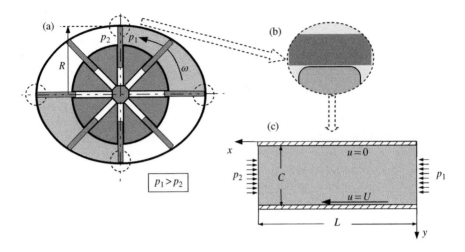

Figure 2.15   Vane pump model

the volume between two consecutive vanes to increase and decrease in every quarter of a revolution. As the volume increases in one quadrant, hydraulic fluid is sucked into the space between the two vanes. In the next quadrant, the volume between the vanes is reduced and the fluid is expelled.[32]

We observe in Figure 2.15(b) that each vane is separated from the case by a thin layer of oil. Here, we assume that the geometry of the vane/case clearance can be approximated by two parallel plates, whose dimensions extend infinitely in a plane perpendicular to the plane of the paper (z), as shown in Figure 2.15(c). The situation is, therefore, similar to the Couette flow, shown in Figure 2.1. The basic idea is that leakage flows inside pumps and motors can be compared with the flow between two plates separated by a tiny layer of fluid, in which one of the plates is moving, just as in Figure 2.1.

In Figure 2.15(c), p_1 and p_2 are the pressures at the output and input ports of the pump, respectively, u is the fluid speed, L is the vane width and C is the clearance between the vane and the case. The linear speed at a point located on the extremity of the vane is $U = \omega R$, where R is the distance between the centre of the pump and the vane tip, and ω is the angular speed of the pump shaft.

Since we are assuming that the flow between the vane and the case is unidimensional and laminar, we can model it with the help of Eq. (C.27):

$$\mu \left(\frac{\partial^2 u}{\partial y^2} \right) = \frac{\partial p}{\partial x} \tag{2.81}$$

---

[32] See Section 3.3.4 for a more detailed description of vane pumps and motors.

From Appendix C (Eqs. (C.25)–(C.27)), we have that $p = p(x)$ and $u = u(y)$ in Eq. (2.81). Therefore, we may write that[33]

$$\begin{cases} \mu \left(\dfrac{d^2u}{dy^2} \right) = k \\ \dfrac{dp}{dx} = k \end{cases} \tag{2.82}$$

where k is a constant.

To solve Eqs. (2.82), the following boundary conditions can be applied:

$$\begin{cases} u(x, 0) = 0 \\ u(x, C) = U \\ p(0, y) = p_1 \\ p(L, y) = p_2 \end{cases} \tag{2.83}$$

We can, then, integrate the first equation in (2.82) twice, with respect to y:

$$\iint \left(\frac{d^2u}{dy^2} \right) dy = \frac{k}{\mu} \iint dy \;\Rightarrow\; u = \left(\frac{k}{\mu} \right) \frac{y^2}{2} + c_1 y + c_0 \tag{2.84}$$

where c_0 and c_1 are constants.

The first and the second boundary conditions in (2.83) give

$$\begin{cases} u(y = 0) = 0 \;\Rightarrow\; c_0 = 0 \\ u(y = C) = U \;\Rightarrow\; c_1 = \dfrac{U}{C} - \left(\dfrac{k}{\mu} \right) \dfrac{C}{2} \end{cases} \tag{2.85}$$

If we substitute c_0 and c_1, given by Eqs. (2.85) into Eq. (2.84), we arrive at

$$u(y) = \frac{Uy}{C} + \frac{k}{\mu} \left(\frac{y^2}{2} - \frac{Cy}{2} \right) \tag{2.86}$$

In Eq. (2.86), k can be obtained from the integration of the second equation in (2.82) and the last two boundary conditions in (2.83):

$$k = \int_0^L \left(\frac{dp}{dx} \right) dx = \frac{p_2 - p_1}{L} = -\frac{\Delta p}{L} \tag{2.87}$$

---

[33] See Eq. (2.55) for a similar treatment.

where $\Delta p = p_1 - p_2$. Substituting k, given by Eq. (2.87), into Eq. (2.86), we finally obtain

$$u(y) = \frac{Uy}{C} - \frac{\Delta p}{\mu L}\left(\frac{y^2}{2} - \frac{Cy}{2}\right)$$

(2.88)

Equation (2.88) shows that if the pump is stationary ($U = 0$), the pressure differential between the oil chambers separated by the vane and the case originates a *Poiseuille flow*, with a parabolic velocity profile.[34] On the other hand, if there is no pressure differential between the chambers, the velocity profile will be linear and, therefore, the flow will be a Couette flow similar to the one illustrated in Figure 2.1.

Now that we know the velocity profile, we can obtain an expression for the leakage through one individual vane of the pump. Note that, in order to calculate the leakage between the vane and the case, we must leave the 'infinitely thick' model behind (Figure 2.15) and consider the vane thickness, B, in the z-direction. Strictly speaking, this is a step backwards, since by assuming that the dimension perpendicular to the paper plane is limited in Figure 2.15(c), we are implicitly assuming a variation of the velocity, u, along the z-axis. This is best seen in Figure 2.16 where two distinct situations are represented. Figure 2.16(a) shows the flow between two infinite parallel plates. Because there is no limiting surface in the z-direction, the velocity of the fluid, u, does not change with z. In Figure 2.16(b), a vertical plate is positioned at the back. Due to the viscosity, the fluid velocity on the plate surfaces must be zero. As a result, a speed gradient must appear along the z-axis. The situation of an actual vane pump would be similar to the one in Figure 2.16(b), with the addition of another vertical plate at the front.

Despite the actual fluid speed variation, it is usual to adopt a simpler approach and assume that u is constant along z. Such simplification was adopted in Ref. [26] and will be followed here. Therefore, considering $u(y)$ given by Eq. (2.88) and the cross-sectional area BC, the magnitude of the leakage, $|\delta q_L|$, will be given by

$$|\delta q_L| = B\left|\int_0^C u\,(y)\,dy\right| = B\left|\int_0^C \left[\frac{Uy}{C} - \frac{\Delta p}{\mu L}\left(\frac{y^2}{2} - \frac{Cy}{2}\right)\right]dy\right|$$

(2.89)

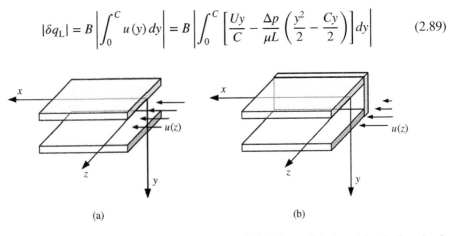

(a)   (b)

Figure 2.16 Difference between a flow developed between infinitely parallel plates (a) and a flow developed between limited parallel plates (b)

---

[34] In honour of the French scientist: Jean Louis Marie Poiseuille (1797–1869). Poiseuille flows are generally related to laminar flows between concentric tubes; however, it is also common to generalize them to a larger class of laminar unidimensional flows, driven solely by the pressure differential similar to the one discussed here.

After developing the integral in Eq. (2.89), we obtain

$$|\delta q_{L}| = \left| \left(\frac{RBC}{2} \right) \omega + \frac{B\Delta p C^3}{12\mu L} \right| \tag{2.90}$$

To make the notation simpler, we will always refer to the magnitude of the leakages instead of considering its signal (direction). We, therefore, drop the module notation in Eq. (2.90) and write the flow loss, δq_{L}, in a slightly different manner:

$$\delta q_{L} = k_u \omega + k_s \left(\frac{\Delta p}{\mu} \right) \tag{2.91}$$

where k_u and k_s are constants that depend on the geometric characteristics of the clearance and are given by the following equations:

$$k_u = \frac{RBC}{2} \tag{2.92}$$

$$k_s = \frac{BC^3}{12L} \tag{2.93}$$

Wilson [26] did not regard the term containing k_u in Eq. (2.91) as being part of the volumetric losses, given that it resulted from the natural dragging of the fluid by the moving surface (Couette flow). Thus, only the term containing the pressure differential (Poiseuille flow) was considered as leakage. In this book, we follow the same reasoning used by Wilson and write δq_{L} as

$$\delta q_{L} = k_s \left(\frac{\Delta p}{\mu} \right) \tag{2.94}$$

If we observe the vane pump in Figure 2.15(a), we note that there are four shaded volumes. These volumes are the ones from where fluid is expelled into the pressurized line of the circuit, that is, every shaded region in Figure 2.15(a) is at the pressure, p_1. Similarly, the non-shaded volumes are connected to the suction line and are exposed to the lower pressure, p_2. In the end, we observe that the same situation that led to Eq. (2.91) occurs four times, at the circled regions shown in Figure 2.15(a). If we assume that similar leakages are present in other parts of the pump, we can write the total volumetric losses at the pump/motor, q_{L}, as:

$$q_{L} = \sum (\delta q_{L}) = \left(\sum k_s \right) \left(\frac{\Delta p}{\mu} \right) = K_s \left(\frac{\Delta p}{\mu} \right) \tag{2.95}$$

where the volumetric loss coefficient, $K_s = \sum k_s$, includes every leakage of the pump/motor.

2.2.4.2 Torque Losses

Referring once again to Figure 2.15, we observe that the viscous forces act opposite to the motion of the rotating vane. In fact, the viscous tension adjacent to the vane tip ($y = C$)

produces a resisting torque, T_ν, whose magnitude is given by

$$T_\nu = (\tau_{yx} RBL)_{y=C} = \mu \left(\frac{du}{dy} RBL \right)_{y=C} \tag{2.96}$$

Substituting the expression for $u(y)$ given by Eq. (2.88) into Eq. (2.96), we obtain

$$T_\nu = RBL \left(\frac{\mu \omega R}{C} - \frac{C \Delta p}{2L} \right) \tag{2.97}$$

We can rewrite Eq. (2.97) in the following manner:

$$T_\nu = k_\omega (\mu \omega) - k_u \Delta p \tag{2.98}$$

where

$$k_\omega = \frac{BLR^2}{C} \tag{2.99}$$

Wilson disregarded the term with the coefficient k_u in Eq. (2.98). On the other hand, he added a constant term, T_f, to account for the mechanical friction between the moving parts of the pump/motor.[35] We follow his idea and write the total resisting torque at the pump/motor, T_L, as:

$$T_L = \left(\sum k_\omega \right) (\mu \omega) + T_f = K_\omega (\mu \omega) + T_f \tag{2.100}$$

In Eq. (2.100), the mechanical loss coefficient, $K_\omega = \sum k_\omega$, and the mechanical resistance, T_f, must be obtained experimentally.

Remarks

1. In Eqs. (2.95) and (2.100), the pressure differentials were taken between the high-pressure and low-pressure chambers of the pump. Leakages can also occur between the high-pressure chamber and the case, which is usually at a pressure different from p_2.
2. As will be seen in Section 2.2.5, some authors have included a term to account for the fluid compressibility in Eq. (2.95) (see, e.g. Ref. [24]). Others have assumed that, in some parts of the pump/motor, turbulence could also occur and combine laminar and turbulent equations into one expression (see Refs. [5, 22]). Several other models for pump and motor losses have been proposed over the years (see, for instance, Refs. [23, 27]). We will have the chance to see some of these models in the following section.

---

[35] Wilson also considered a pressure dependent friction term, which has not been included in Eq (2.98) for simplicity.

3. The coefficients K_s and K_ω in Eqs. (2.95) and (2.100) depend on the clearance dimensions (Figure 2.15). The clearance dimensions, on the other hand, are directly affected by the contact forces within the pump and motor, which, in turn, depend on the pressure differential. As a result, these coefficients will not usually be constant, and their determination must rely heavily on experimental data obtained from the pump and motor manufacturers, as will be seen in the next chapter.

4. The mechanical loss term, T_f, in Eq. (2.100), is actually dependent on the external load, acting on the pump and motor shaft. This is because internal friction depends on the forces between the contacting surfaces. This term has been designated as a Coulomb term, in a direct reference to the classic Coulomb friction model [23, 24].

2.2.5   Other Loss Models

Equations (2.95) and (2.100) can be seen as the simplest volumetric and torque loss models available for pumps and motors. Over the years, several attempts have been made to find one equation that can best fit the extensive collection of experimental data referent to the equally large number of pump and motor designs and sizes. The task is not easy at all, and to date we do not have one model that can be considered perfectly suited for every situation.

We finish this chapter by listing some volumetric and torque losses equations that have been developed through the years. No attempt will be made to explain these equations in details (the interested student can consult [23] and the references therein). Despite the fact that we are not using these additional models in the chapters to come, it is important for the student to have an overview of some alternative equations at this stage. We start with an addition to Wilson's model that has not been considered in this chapter:

- *Turbulence leakage models:* Wilson [26] presented the following equations (as compared to Eqs. (2.95) and (2.100)), based on an average energy balance between pump and motor leaks, which consider a turbulent leaking flow:

$$\begin{cases} q_L = K_{st}\sqrt{\dfrac{\Delta p}{\rho}} - q_{LV} \\ T_L = K_{\omega t}\rho\omega^2 + T_f \end{cases} \tag{2.101}$$

where K_{st} and $K_{\omega t}$ are turbulence coefficients and q_{LV} accounts for eventual flow losses due to fluid vaporization and/or liberation of entrained gases.

- *Compressibility volumetric loss model:* Dorey [23] considered the volumetric losses due to fluid compressibility and the effects of the pump and motor displacements with the following expression for q_L:

$$q_L = D_s\left(\frac{\Delta p}{\mu}\right) + B_s\left(\frac{\Delta p\omega}{\beta}\right)\left[V_r + \left(\frac{1+D^*}{2}\right)\right] \tag{2.102}$$

where D^* is the ratio between the actual displacement and the maximum displacement, D_{max} ($D^* = D/D_{max}$), β is the fluid bulk modulus and V_r is the ratio between the total clearance volume inside the pump/motor at maximum displacement and the sweep volume, that is, the variable volume swept by the moving parts of the pump and motor during operation. Similar to the other coefficients we have seen so far, the coefficients D_s and B_s depend on the pump/motor size and design.

- *Speed-dependent volumetric loss model:* Dorey [23] and Tessmann [28] proposed the following expression for the volumetric losses, where q_L was considered to vary with the angular speed as well as with pressure:

$$q_L = K_1 \left(\frac{\Delta p}{\mu} \right) - K_2 \omega \tag{2.103}$$

where the coefficients K_1 and K_2 depend on the internal geometry of the pump and motor.

- *Pressure-dependent torque loss model:* In a second paper by Wilson [29], mentioned later by Dorey [23], an expression similar to Eq. (2.104) was proposed in which a pressure-dependent term, K_p, appears, together with the pump/motor displacement, D:

$$T_v = DK_\omega(\mu\omega) + DK_p\Delta p + T_f \tag{2.104}$$

Exercises

(1) A thin and homogeneous layer of ISO VG 68 hydraulic oil with an average kinematic viscosity of 68 cSt is placed between two parallel steel plates, as shown in Figure 2.17. Assuming that the oil density is 870 kg/m^3, what is the traction, T, necessary to drag the upper plate to the right at 1 m/s?

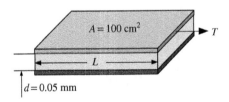

$A = 100$ cm^2

T

L

$d = 0.05$ mm

Figure 2.17   Fluidic friction

(2) Consider that the fluid viscosity in Figure 2.17 varies continuously between 2 and 150 cSt. Using the problem data given in Figure 2.17, plot the traction force, T, as a function of the kinematic viscosity v (consider the oil density as constant and equal to 870 kg/m^3). Compare the results with the mechanical efficiency curve shown in Figure 2.2.

(3) Obtain an expression for the velocity of the fluid, u, flowing through the annular space between two concentric cylinders whose radii and length are R_1, R_2 and L, respectively

(Figure 2.18). Assume that the pressure differential at the cylinder is Δp, and that the kinematic viscosity of the fluid is v. Consider also that the flow is laminar and the boundary layer is fully developed.

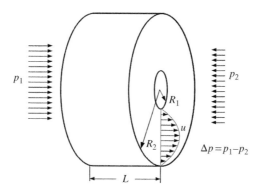

Figure 2.18   Flow between two concentric cylinders

(4)   Starting from the expression found for the fluid speed, u, in Exercise 3, obtain Eq. (2.62).

(5)   Consider the leakage between a stationary piston and its case, q_L, as illustrated in Figure 2.19. The pressure differential, Δp, is 170 bar and the kinematic viscosity and density of the fluid are 68 cSt and 870 kg/m^3, respectively.

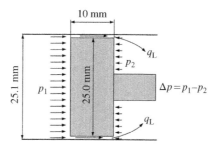

Figure 2.19   Leakage in a hydraulic cylinder

(a) Plot the velocity, u(m/s), and the Reynolds number, Re (based on the piston gap, $R_2 - R_1$), as functions of the radial distance, r, for r varying between R_1 and R_2.

(b) Obtain the maximum value of the speed, u, and the Reynolds number, Re, and calculate the corresponding radial distance, r_{max}.

(c) Calculate the leakage flow, q_L(l/min), through the piston gap.

(6)   Plot the volumetric flow (leakage), q_L, through the 0.05 mm clearance of the hydraulic cylinder shown in Figure 2.19, as a function of the kinematic viscosity, v, considering a pressure differential, Δp, of 170 bar. Compare your results with the volumetric efficiency curve shown in Figure 2.2. Assume that the flow is laminar and that the viscosity varies from 30 to 100 cSt. The fluid density is 870 kg/m^3.

(7) Consider a pipe with an inner diameter of 10 mm. Plot the variation of the Reynolds number with the internal flow for an ISO VG 32 ($v = 32$ cSt), an ISO VG 46 ($v = 46$ cSt) and an ISO VG 68 ($v = 68$ cSt) hydraulic fluid, respectively. Assuming that the Reynolds number limit for turbulence is 2300, what is the maximum flow that guarantees a laminar regime in each case? The fluid density can be taken as 870 kg/m^3.

References

[1] Merritt HE (1967) Hydraulic control systems. John Wiley & Sons, USA.
[2] Troyer D (2002) Machinery lubrication: understanding absolute and kinematic viscosity, http://www.machinerylubrication.com/Read/294/absolute-kinematic-viscosity. Accessed July 2013.
[3] Sauer-Danfoss (2010) Hydraulic fluids and lubricants, Technical Information, USA.
[4] Michael PW, Herzog SN, Marougy TE (2000) Fluid viscosity selection criteria for hydraulic pumps and motors. International Exposition for Power Transmission and Technical Conference, NCFP paper I00-9.12, Chicago, IL, USA.
[5] Ivantysyn J, Ivantysynova M (2003) Hydrostatic pumps and motors. Tech-Books International, New Delhi.
[6] Finnemore E, Franzini J (2001) Fluid mechanics with engineering applications, 10th Ed., McGraw Hill, USA.
[7] Balasubramanian K (2003) Smart bulk modulus sensor, MSc Thesis, University of Florida, USA.
[8] Bowns DE, Worton-Griffiths J (1972) The dynamic characteristics of a hydrostatic transmission system. Proceedings of the Institution of Mechanical Engineers, 186: 755–773.
[9] Stringer J (1976) Hydraulic system analysis: an introduction, John Wiley & Sons, USA.
[10] Akers A, Gassman M, Smith R (2006) Hydraulic power system analysis. CRC Press, USA.
[11] Hearn EJ (1997) Mechanics of materials 1, 3rd Ed., Elsevier, UK.
[12] Timoshenko S (1940) Strength of materials Part II: Advanced theory and problems, 2nd Ed., D Van Nostrand Company, USA.
[13] Parker-Hannifin (2012) EO-Tubes and Pipes for fittings and flanges, Industrial and Mobile Applications–Marine and Offshore Applications, USA.
[14] Manring N (2005) Hydraulic control systems. John Wiley & Sons, USA.
[15] Contitech (2012) Technical Information. http://www.contitech.de/pages/produkte/schlauchleitungen/industrieleitungen/downloads/fluid_tech-info_en.pdf. Accessed 15 April 2012.
[16] Streeter VL (1962) Fluid mechanics. McGraw-Hill, US.
[17] Kreyszig E (2006) Advanced engineering mathematics. John Wiley & Sons, USA.
[18] ESSO – Imperial oil (2010) Product data-sheet UNIVISTM N, USA.
[19] Fox RW, McDonald AT, Pritchard PJ (2004) Introduction to fluid mechanics, 6th Ed., John Wiley & Sons, USA.
[20] Doddannavar R, Barnard A (2005) Practical hydraulic systems. Operation and troubleshooting for engineers and technicians. Elsevier Science & Technology Books, UK.
[21] Esposito A (1980) Fluid power with applications, 4th Ed., Prentice Hall, USA.
[22] Pourmovahed A, Beachley NH, Fronczak FJ (1992) Modeling of a hydraulic energy regeneration system-part 1: analytical treatment. The Journal of Dynamic Systems, Measurement, and Control, 114: 155–159.
[23] Dorey RE (1988) Modelling of losses in pumps and motors. First Bath International Fluid Power Workshop, University of Bath, UK, September 1988, pp. 71–97.
[24] McCandlish D, Dorey RE (1984) The mathematical modeling of hydrostatic pumps and motors. Proceedings of the Institution of Mechanical Engineers, 198B(10): 165–174.

[25] Tessmann RK, Melief HM, Bishop RJ (2006) Basic hydraulic pump and circuit design. In: Totten, GE (ed) Handbook of lubrication and tribology Vol. I: application and maintenance, 2nd Ed., CRC Press, USA.

[26] Wilson WE (1946) Rotary-pump theory. ASME Transactions, 68: 371–384.

[27] Jung D, Kim H, et al. (2005) Experimental study on the performance estimation efficiency model of a hydraulic axial piston motors, Proceedings of the 6th JFPS International Symposium on Fluid Power, TSUKUBB, 2A3-1: 284–290.

[28] Tessmann RK (1979) A leakage path model for a hydraulic pump. The BFPR Journal, 12(1): 5–9.

[29] Wilson WE (1949) Performance criteria for positive-displacement pumps and fluid motors. ASME Transactions, 71(2): 115–120.

3

Hydrostatic Pumps and Motors

This chapter deals with the main components of hydrostatic transmissions: pumps and motors. We begin with a formal definition of hydrostatic and hydrodynamic pumps and motors. After discussing some basic concepts, a short description of the most common types of hydrostatic pumps and motors is provided. We close the chapter with a detailed discussion about pump and motor efficiency.

3.1 Hydrostatic and Hydrodynamic Pumps and Motors

In order to provide a proper definition of hydrostatic pumps and motors, we must first consider the general state of stress of a particle in a fluid flow, given by a 3×3 tensor as in Eq. (C.5), Appendix C:

$$[\tau] = \begin{bmatrix} \tau_{xx} & \tau_{xy} & \tau_{xz} \\ \tau_{yx} & \tau_{yy} & \tau_{yz} \\ \tau_{zx} & \tau_{zy} & \tau_{zz} \end{bmatrix} \tag{3.1}$$

As seen in Section C.2, if the fluid is static, the general stress tensor (3.1) reduces to (see Eqs. (C.9))

$$[\tau] = \begin{bmatrix} -p & 0 & 0 \\ 0 & -p & 0 \\ 0 & 0 & -p \end{bmatrix} \tag{3.2}$$

where p is called *hydrostatic pressure*, given that it is directly associated with the static equilibrium of the fluid. The negative sign indicates that p is always oriented towards the centre of the infinitesimal fluid element.

With Eqs. (3.1) and (3.2) in mind, we may now proceed to give a formal definition of hydrostatic and hydrodynamic pumps and motors. Consider the single-piston pump shown in Figure 3.1, similar to the one illustrated in Figure 1.18. Consider also that the fluid is incompressible and that there are no leakages in the pump. Although the practical applications for

Hydrostatic Transmissions and Actuators: Operation, Modelling and Applications, First Edition.
Gustavo Koury Costa and Nariman Sepehri.
© 2015 John Wiley & Sons, Ltd. Published 2015 by John Wiley & Sons, Ltd.
Companion Website: www.wiley.com/go/costa/hydrostatic

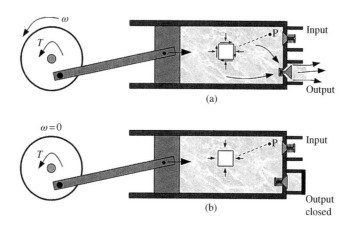

Figure 3.1   Single-piston hydrostatic pump: (a) output port open and (b) output port closed

this type of pump are very limited, the derived definitions can be applied to every other hydrostatic pump.

The operation of the piston pump shown in Figure 3.1 is quite straightforward. When the crank rotates counter-clockwise, the piston moves to the right (Figure 3.1(a)) and the volume within the pump is reduced. Fluid is then expelled through the output nozzle (input and output nozzles open and close with the help of two valves, as illustrated in the figure). Observe that during the pumping phase, a general state of tension, described by Eq. (3.1), develops in at least one arbitrary point, P, within the pressurized chamber.

Let us imagine that the output nozzle is blocked, as shown in Figure 3.1(b). All the mass of fluid within the pressurized chamber will stop moving, and a hydrostatic state of tension (Eq. (3.2)) will develop in every point P inside the pressurized volume. We can state the following:

> In a perfectly sealed hydrostatic pump, whenever the output nozzle is closed, every point inside the pressurized volume undergoes a hydrostatic state of tension.

Now, consider the centrifugal pump shown in Figure 3.2. When the fluid is pumped out, we can also state that a generic point, P, inside the pressurized chamber is under a general state of tension, just as in the case of the piston pump shown in Figure 3.1. However, even if the output nozzle is blocked, the rotor will still be able to rotate, and the fluid in the vicinity of the rotor blades will continue to undergo shear stresses. As a result, a hydrostatic state of tension will not exist in every point of the pressurized chamber.

> In a perfectly sealed hydrodynamic pump, there will always be shear stresses in at least one point of the fluid volume within its pressurized chamber, even if the output flow is blocked.

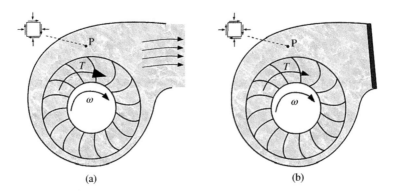

Figure 3.2   Typical hydrodynamic pump: (a) output port open and (b) output port closed

Hydrostatic pumps and motors are well suited for fluid power circuits because once the fluid at the high-pressure line 'stops', the whole line stays under the same pressure. The hydraulic fluid then becomes a medium through which force can be transmitted from the piston to the element that is blocking the flow. On the other hand, in hydrodynamic pumps, such as the centrifugal pump in Figure 3.2, when the output flow is blocked, part of the energy coming from the rotor dissipates in the form of fluidic friction due to the tangential stresses that develop within the fluid mass. There is no way to bring the whole bulk of fluid to a full stop, and therefore the fluid will never become a force transmission medium. Alternatively, one can say that in a perfectly sealed hydrostatic pump, blocking the output flow will immediately halt the pumping element; in hydrodynamic pumps, the pumping element will not stop even if the output nozzle is blocked.

A similar reasoning can be used for the definition of hydrostatic and hydrodynamic motors. The following statement is valid for hydrostatic motors:

> In a perfectly sealed hydrostatic motor, whenever the output shaft is stopped by the action of an external force, every point inside its pressurized volume will be characterized by a hydrostatic state of tension.

To prove this statement, let us imagine that the single-piston pump of Figure 3.1 is operating as a motor, as shown in Figure 3.3. Observe that in order to make it possible for the pump to operate as a motor, the input and output valves have been replaced by two externally controlled shut-off valves: V_1 and V_2. During the motoring cycle, on the first half of the crank revolution, valve V_1 remains open while V_2 is closed. The input nozzle is connected to a pressure source, p, and a force, F, appears on the piston. As long as the crank is free to rotate (Figure 3.3(a)), the motor shaft will turn with a speed, ω, and a torque, T. The state of tension of a generic point, P, inside the pressurized chamber in motoring mode will be given by the generic tensor (3.1). Now, imagine that we stop the motor shaft with

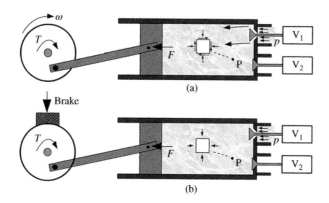

Figure 3.3 Single-piston hydrostatic motor: (a) unlocked shaft and (b) locked shaft

the aid of an external brake, as shown in Figure 3.3(b). The mass of fluid inside the pressurized chamber will stop flowing, and the state of tension at point P becomes hydrostatic (Eq. (3.2)). Observe that in spite of the motor shaft being halted, torque is still produced as long as the pressure at the motor input is maintained.

Figure 3.4 shows a typical hydrodynamic motor (or turbine). As the fluid flows from left to right (parallel arrows), it changes its direction after hitting the turbine blades. The change in the velocity vectors produces a torque, according to the Second Law of Newton, which is then transferred to the motor shaft. As long as the shaft is free to rotate, it moves with a torque T and a speed ω, extracting energy from the incoming flow (Figure 3.4(a)). Although we can stop the turbine by applying a brake to the motor shaft (Figure 3.4(b)), the flow is still free to pass through the turbine blades. Therefore, the state of tension at a generic point P in the pressurized side of the turbine never becomes hydrostatic and is always given by

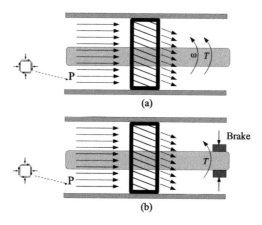

Figure 3.4 Hydrodynamic motor (turbine): (a) unlocked shaft and (b) locked shaft

Eq. (3.1). Note that torque is still produced at the motor shaft as long as fluid keeps flowing through the blades.

By observing the hydrodynamic motor shown in Figure 3.4, we may state the following:

> In a perfectly sealed hydrodynamic motor, there will always be shear stresses in at least one point of the fluid volume in its pressurized chamber, even if its output shaft is forcedly halted.

Remarks

1. Hydrostatic pumps and motors have also been defined as those for which the Pascal law applies [1].[1]
2. Hydrostatic pumps and motors are also known as *displacement machines*,[2] based on the fact that they displace a fixed amount of fluid in every rotation.
3. It is relatively common to find the terms 'hydraulic' and 'hydrostatic' being indistinguishably used in fluid power (see, for instance Refs. [2–4]). In a sense, it is not wrong to use 'hydraulic pumps', for example, when discussing hydrostatic pumps, since hydrostatics is actually a subset of hydraulics. However, in this book, whenever we refer to pumps, we use the term 'hydrostatic'. When discussing hydrostatic motors, given that hydrodynamic motors are naturally known as 'turbines', the term 'hydraulic motor' will be preferred.

3.2 Hydrostatic Machine Output

3.2.1 Average Input–Output Relations

As we saw in Chapter 1, pumps deliver hydraulic power (pressure and flow), while motors produce mechanical power (torque and angular speed). The relations between the mechanical power input and the hydraulic power output in a pump were given in Eqs. (1.16) and (1.20) for the ideal case where no volumetric or mechanical losses were present and are repeated here for convenience:

$$\begin{cases} q_p = D_p \omega_p \\ \Delta p_p = p_{po} - p_{pi} = \dfrac{T_p}{D_p} \end{cases} \tag{3.3}$$

---

[1] The Pascal law states that in a fluid at rest in a closed container, a pressure change in one part is transmitted without loss to every portion of the fluid and to the walls of the container.

[2] Or 'positive displacement'.

where p_{po} and p_{pi} are the pressures at the output and input ports of the pump, respectively; q_p and Δp_p are the output flow and pressure elevation at the pump; ω_p and T_p are the angular speed and the torque at the pump shaft; and D_p is the pump displacement.

Similarly, for the motor (Eqs. (1.17) and (1.22)), we have

$$\begin{cases} \omega_m = \dfrac{q_m}{D_m} \\[2mm] T_m = \left(p_{mi} - p_{mo}\right)D_m = \Delta p_m D_m \end{cases} \tag{3.4}$$

Observe that we have used the subscripts 'p' and 'm' to differentiate pump and motor. Therefore, in Eqs. (3.4), p_{mi} and p_{mo} are the pressures at the motor input and output ports; q_m and Δp_m are the input flow and pressure drop at the motor; ω_m and T_m are the angular speed and the torque at the motor shaft; and D_m is the motor displacement.[3]

We have seen that in variable-displacement pumps and motors, the displacements, D_p and D_m, can be externally changed. As an example, Figure 3.5 illustrates the operation of a variable-displacement pump. In the figure, the single-piston pump illustrated in Figure 3.1 has been modified to include a means of altering the piston stroke. A hydraulic cylinder, A, whose rod is pinned to the bar L connecting the crank to the piston, makes it possible to alter the crank length in a way that the piston stroke becomes a function of the crank radius. As the crank radius changes from R_{max} to R_{min}, the volume corresponding to the piston stroke changes from V_{max} to V_{min} and, as a consequence, the pump displacement also changes. In the hypothesis that the crank radius became zero, the pump stroke would also become zero, and no output flow would exist.

The variable-displacement pump shown in Figure 3.5 could be turned into a variable-displacement motor by connecting the input nozzle to a pressurized line and changing the

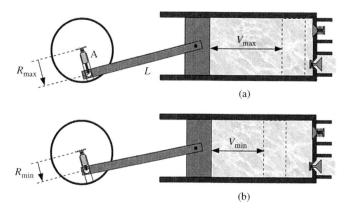

Figure 3.5   Variable-displacement pump at (a) maximum and (b) minimum displacement

---

[3] A formal proof of the torque relations in (3.3) and (3.4) will be given in Section 3.4.4.

input and output valves into shut-off valves, as shown in Figure 3.3. Observe that in this case, if the displacement became zero, the piston would not be able to move, blocking the hydraulic line connected to the motor input.

3.2.2   Instantaneous Pump Output

At a first glance, Eqs. (3.3) and (3.4) can mask the actual behaviour of hydrostatic pumps and motors. For example, assume that the displacement chamber[4] volume of the piston pump shown in Figure 3.1 is 1/4 l. This means that a volume of 1/4 l/rev is delivered by the pump, that is, the pump displacement, D_p, is 1/4 l/rev. If the shaft is connected to a prime mover rotating at a constant speed of 1000 rpm, the first relation in (3.3) states that the pump flow will also be constant and equal to 250 l/min. After a brief look at the pump design, however, we conclude that this cannot be true.

3.2.2.1   Output Flow Pattern

Figure 3.6 represents the pump shown in Figure 3.1 in more details. In the figure, we see that the velocity of the piston, v_p, is only positive during half a cycle of the crank revolution ($\pi \leq \theta \leq 2\pi$) where fluid is expelled from the displacement chamber into the circuit (Figure 3.6(b)). At the first half of the cycle ($0 \leq \theta \leq \pi$), fluid is admitted into the pump, as shown in Figure 3.6(a).

We can write x in Figure 3.6(b) as

$$x = R\cos(2\pi - \theta) + L\cos(\alpha) = R\cos(\theta) + L\cos(\alpha) \tag{3.5}$$

Expressing $L\cos(\alpha)$ in Eq. (3.5) as a function of the crank angle, θ, and the radius, R, gives

$$x = R\cos(\theta) + \sqrt{L^2 - R^2\sin^2(\theta)} \tag{3.6}$$

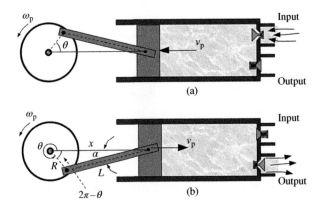

Figure 3.6   Single-piston pump: (a) suction and (b) discharge

---

[4] By displacement chamber, we mean every volumetric space in the pump or motor that is filled in and subsequently emptied out of the same portion of hydraulic fluid as the shaft rotates.

We now write the angle θ as the product between the prime mover speed, ω_p, and the time, t, and obtain

$$x = R\cos(\omega_p t) + \sqrt{L^2 - R^2\sin^2(\omega_p t)} \tag{3.7}$$

The instantaneous flow at the displacement chamber,[5] q_t, is given by

$$q_t = \frac{dV}{dt} \tag{3.8}$$

where V is the volume of the displacement chamber measured at time t.

Using the chain rule, Eq. (3.8) can be developed as follows:

$$q_t = \frac{dV}{d\theta}\frac{d\theta}{dt} = \frac{dV}{d\theta}\omega_p = \omega_p\left(A\frac{dx}{d\theta}\right) \tag{3.9}$$

where A is the area of the piston.

Substituting x as given by Eq. (3.6) into Eq. (3.9), after developing the derivative $dx/d\theta$ (we leave this derivation to the student as an exercise), we obtain for $\theta = \omega_p t$:

$$q_t = -\omega_p\left[RA\sin\left(\omega_p t\right) + \frac{R^2 A\sin(\omega_p t)\cos(\omega_p t)}{\sqrt{L^2 - R^2\sin^2(\omega_p t)}}\right] \tag{3.10}$$

Figure 3.7(a) plots the curve $q_t(t)$ for the particular case where $R = 5.0\,\text{cm}$, $L = 25.0\,\text{cm}$, $A = 50\,\text{cm}^2$ and $\omega_p = 1.0\,\text{rad/s}$ ($\cong 9.56\,\text{rpm}$). The negative portion of the curve corresponds to the suction phase, whereas the positive portion represents the pump discharge. Figure 3.7(b) shows the instantaneous output flow, q_{pt}, which in this case corresponds to the positive part of the curve $q_t(t)$, that is

$$q_{pt} = \begin{cases} q_t & \text{if } q_t \geq 0 \\ 0 & \text{if } q_t < 0 \end{cases} \tag{3.11}$$

The pulsating nature of the output flow, q_{pt}, can also be understood as follows: given that the volume of the displacement chamber increases during suction and decreases during discharge, the derivative dV/dt must change from negative to positive at some point during the crank rotation. Since the volume V is a continuous function of the time t, we can infer that q_t will smoothly alternate between a negative minimum to a positive maximum value. In other words, the displacement chamber will only output flow during a fraction of the shaft revolution. Given that the output flow of every hydrostatic pump is composed of the sum of individual flows coming from the displacement chambers, q_{pt} will always be composed of discrete pulsating flows and, as a result, will never be constant.[6]

---

[5] Depending on the angular position of the piston crank, q_t can be positive (discharge) or negative (suction). Obviously, only the discharge can be regarded as the pump output.

[6] Note that in Figure 3.7(a), the flow through the pump is discontinuous, as suction and discharge happen at different moments. This is one of the main differences between hydrostatic and hydrodynamic pumps. In the latter case, the flow that enters the pump equals the flow that leaves the pump at any moment in time (considering an ideal situation where losses do not exist and the fluid is incompressible).

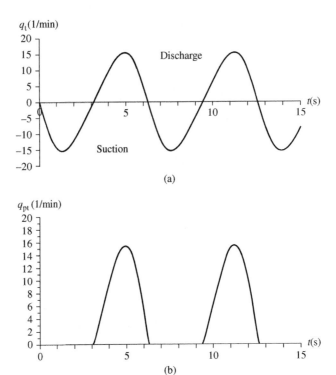

Figure 3.7   (a) Displacement chamber flow and (b) output flow

Although we cannot eliminate the oscillations in the output flow, q_{pt}, it is possible to make the flow smoother by adding more displacement chambers to the pump, as long as they do not work in phase with one another. For example, consider the combination of N single-piston pumps operating in parallel, as in Figure 3.8, for $N = 3$. Assume that each pump is dephased in relation to the other by a constant angle, ϕ, at the crank (ϕ is taken as 120° in Figure 3.8). The displacement flow at the ith pump, q_t^i, will be given by (see Eq. (3.10)):

$$q_t^i = -\omega_p \left[RA \sin\left(\theta_i\right) + \frac{R^2A \sin(\theta_i)\cos(\theta_i)}{\sqrt{L^2 - R^2\sin^2(\theta_i)}} \right] \qquad (3.12)$$

where

$$\theta_i = \omega_p t + \left(\frac{2\pi}{N}\right)(i - 1), \quad i = 1, 2, \ldots, N \qquad (3.13)$$

Observe that in the particular arrangement shown in Figure 3.8, two pistons will always be either suctioning or expelling fluid concurrently. Figure 3.9 shows the individual displacement flows, q_t^1, q_t^2 and q_t^3, and output flow, q_{pt}, for the pump illustrated in Figure 3.8. For convenience, we have adopted the same cylinder dimensions and prime mover rotation of the previous example.

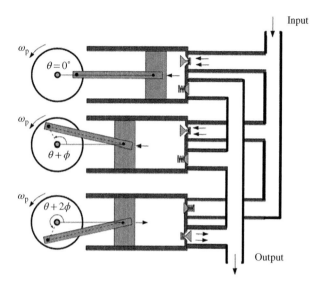

Figure 3.8   Three piston pumps operating in parallel

The oscillatory intensity in the pump output can be measured by the *flow pulsation coefficient*, σ_q, defined as [5]:

$$\sigma_q(\%) = \left(\frac{q_{\text{pt max}} - q_{\text{pt min}}}{q_{\text{p}}} \right) \times 100 = \left(\frac{q_{\text{pt max}} - q_{\text{pt min}}}{D_{\text{p}}\omega_{\text{p}}} \right) \times 100 \qquad (3.14)$$

where $q_{\text{pt min}}$, $q_{\text{pt max}}$ and q_{p} are the minimum, maximum and average[7] flows, as shown in Figure 3.9.

In this book, whenever we want to refer to the temporal behaviour of a specific output, we use subscript 't'. If this subscript is not present, it is understood that we are referring to the average value instead. Thus, we refer to the average pump flow as q_{p} and the instantaneous pump flow as q_{pt}. The same rule is valid for motors, that is, T_{m} is the average output torque and T_{mt} is the instantaneous output torque. Sometimes, we also change the way in which some variables are represented to avoid ambiguity. For instance, we have eliminated the subscript 'p' from the displacement chamber flow, q_{t}, in Eq. (3.8), even though it is a pump-related variable. We deliberately did this to avoid confusion between the displacement chamber flow and the output flow, for which the subscript 'p' had been reserved (Eq. (3.11)).

Depending on the number of parallel cylinders, different output flow patterns can be obtained. For instance, Figure 3.10 extends the arrangement shown in Figure 3.8 to five and six cylinders, respectively. Interestingly, the five-cylinder pump shows a smaller flow pulsation when compared to the six-cylinder pump. In fact, this has been generalized to other

---

[7] Note that the average flow q_{p}, in Eq. (3.14), is the same flow that appears in Eq. (3.3).

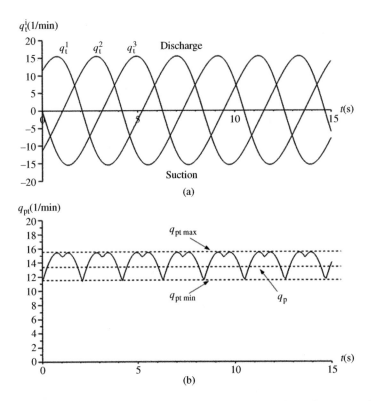

Figure 3.9 Displacement chamber flow (a) and output flow (b) for the three-piston pump illustrated in Figure 3.8

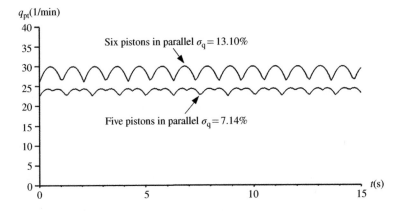

Figure 3.10 Output flow for five and six pistons in parallel

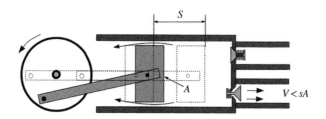

Figure 3.11 Influence of leakages on the output flow

types of piston pumps, causing engineers to choose not to build pumps with an even number of pistons.[8]

3.2.2.2 Volumetric Losses

In the derivation of Eq. (3.12), it was implicit that in every filling–emptying cycle of the displacement chambers, the volume of suctioned fluid and the volume of discharged fluid were the same and coincided with the inner displacement chamber volume, V. Strictly speaking, this could only be true if

1. there were no leakages in the pump;
2. the fluid were not compressible;
3. the fluid viscosity and density were insignificant.

In what follows, we present an explanation for each of these three statements.

Figure 3.11 illustrates how leakage interferes with the output flow. In the figure, we see one of the pump pistons discharging fluid into a pressurized line. The extreme positions of the pistons as the crank rotates half a revolution are shown as dashed lines. The volume of fluid, V, expected to be expelled through the output nozzle is sA (A is the piston area and s is the piston stroke). However, given that part of the flow leaks through the piston–case clearance, we actually have $V < sA$, as shown in the figure.

The influence of the fluid compressibility on the output flow is illustrated in Figure 3.12. During the first half of the crank revolution, the fluid inside the displacement chamber is at a relatively small pressure, p_i (Figure 3.12(a)). Let us, then, disregard the fluid elasticity at this stage and assume that the fluid volume, V_{in}, coming into the displacement chamber at suction phase is approximately equal to the displaced volume sA. During discharge, the pressure inside the displacement chamber must be raised to a much higher level (from p_i to p_o), and since the fluid is compressible, the piston must be displaced a (small) distance, Δs, and compress an equally small volume of fluid, ΔV ($\Delta s = \Delta V/A$), before any output is perceived. Note that ΔV relates to the pressure differential, Δp ($\Delta p = p_o - p_i$), through the compressibility equation (2.4), $\Delta V = -V\Delta p/\beta$. The discharged fluid volume, V_{out}, is

---

[8] We must remember that Figure 3.10 is a result of a theoretical analysis, which does not take factors such as leakage and fluid compressibility into consideration. In this aspect, the conclusion that the odd number of pistons is better than the even number of pistons has been questioned (see, e.g. Ref. [6]).

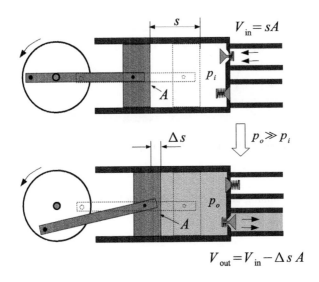

Figure 3.12   Influence of fluid compressibility on the output flow

therefore reduced by ΔsA, as shown in Figure 3.12(b). In the hypothetical situation of an incompressible fluid, $\beta \to \infty$ and $\Delta V \to 0$.

Lastly, Figure 3.13 illustrates how the oil properties can influence the flow coming into the displacement chamber when the piston travels at a high speed, v_p. In Figure 3.13(a), we picture the case where the fluid is incompressible and inviscid. As the piston is displaced to the left by $v_p\Delta t$, a low-pressure zone is formed inside the cylinder, creating a pressure differential between the suction line and the displacement chamber, $p_s - p_{ia}$. This pressure differential pushes fluid into the chamber and, given that the fluid is incompressible, no density gradient is produced along the flow path. The absence of viscous losses guarantees a high flow into the displacement chamber even at a small pressure differential.

The situation where the fluid is viscous and elastic is shown in Figure 3.13(b). Note the following: (a) because of the fluid elasticity, the density changes from ρ_s to ρ_i ($\rho_i < \rho_s$) so that the mass of fluid admitted into the displacement chamber is actually smaller (a smaller mass of input fluid results in a reduced discharged flow), and (b) due to the viscous losses, a higher pressure differential must be created between the displacement chamber and the suction line, that is, $p_{ib} < p_{ia}$. This can become dangerous, as the value of p_{ib} can become as low as the vapour pressure of the fluid, leading to cavitation. Cavitation not only reduces the mass of fluid during admission but also causes an excessive wear of the mechanical parts when the vaporized mass is rapidly compressed during discharge (implosion).

3.2.2.3   Volumetric Efficiency

Equation (3.3), which relates the average output flow, q_p, to the prime mover angular speed, ω_p, can be seen as a definition of the average output flow of a hypothetical perfectly sealed pump operating with an inviscid and incompressible fluid. We remember that 'displacement'

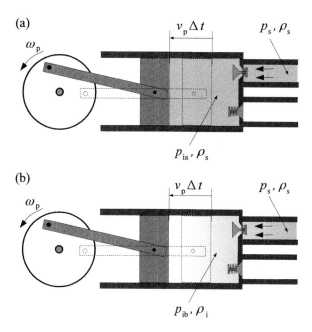

Figure 3.13 Operation with an (a) inviscid and incompressible fluid in contrast with a (b) viscous and compressible fluid

was specified on these grounds in Section 1.4.1 (no volumetric losses). However, given that flow losses will always occur during the pumping process, it is pertinent here to define the volumetric efficiency of the pump, η_p^v, as[9]

$$\eta_p^v = \frac{D_p \omega_p - q_{pL}}{D_p \omega_p} = 1 - \frac{q_{pL}}{D_p \omega_p} \tag{3.15}$$

where q_{pL} represents the total 'wasted' flow resulting from the sum of all the volumetric losses in the pump.

Although we have defined q_{pL} in Eq. (3.15) as the sum of all volumetric losses in the pump, it is common to consider only the losses related to leakage [5, 7]. On the other hand, the ISO 4391:1983 Standards also include fluid compression in their definition of volumetric efficiency:

$$\eta_p^v(\text{ISO}) = \frac{D_p \omega_p - q_{pLe} - q_{pLi} - (D_p \omega_p - q_{pLe} - q_{pLi})\dfrac{\Delta p}{\beta}}{D_p \omega_p} \tag{3.16}$$

where q_{pLe} and q_{pLi} are the external and internal leakage flows, respectively, and Δp is the pressure differential between the input and output ports.

---

[9] See Figure 2.2 for a previous discussion on volumetric efficiency.

The fluid compressibility term in Eq. (3.16) is usually very small because of the large value of the bulk modulus, β. For example, for a considerably high-pressure differential of 500 bar (500×10^5 N/m^2), the estimated compression losses for an equally high output flow of 100 l/min (0.00167 m^3/s) are

$$(D_p \omega_p - q_{pLe} - q_{pLi})\frac{\Delta p}{\beta} = \frac{0.00167 \times (500 \times 10^5)}{17 \times 10^8} = 0.00005\,\frac{m^3}{s} = 0.05\,\frac{l}{min}$$

Considering these low values for the compressibility losses, in this book we define q_{pL} in Eq. (3.15) as the sum of the external and internal leakage, disregarding the compressibility term in Eq. (3.16). Therefore, we use the following definition for the volumetric efficiency of the hydrostatic pump[10]:

$$\eta_p^v = \frac{D_p \omega_p - q_{pLe} - q_{pLi}}{D_p \omega_p} \tag{3.17}$$

3.2.3   Instantaneous Motor Output

The operation of a single-piston motor is relatively more complex when compared to the operation of a single-piston pump. For example, even with a precise control of the valves V_1 and V_2 in Figure 3.3, and assuming a constant input flow q_m, problems will inevitably occur at the piston head-ends where the piston speed changes direction. At those points, no torque can be produced at the motor shaft and, as a consequence, the crank will not be turned by the incoming flow. Thus, it will be necessary to help the crank continuing its rotation with a flywheel. An alternative solution would be to add more pistons to the motor, creating a multichambered unit, with the pistons dephased of a constant angle, ϕ. In this case, there would always be torque being output at the motor shaft. However, for reasons similar to those explained in Section 3.2.2, the delivered torque will still be oscillatory, as will be shown in the following section.

3.2.3.1   Output Torque Pattern

Consider the single-piston motor shown in Figure 3.3 and its more detailed view in Figure 3.14. Assuming that the motor input is at a pressure, p, the force acting on the piston will be pA. The friction force between the piston and the case, F_r, will be taken as constant in our study.[11] To make things even simpler, we also disregard the masses of the moving parts.

During the first half of the crank revolution (Figure 3.14(a)), hydraulic power from the pressure source is converted into useful power at the motor shaft. The conversion is not 100% effective though, because part of this energy dissipates into heat due to the work of the friction force, F_r. On the second half of the crank revolution (Figure 3.14(b)), the motor discharges at a lower pressure, p_0 that equals the atmospheric pressure in this particular example (zero bar gauge). This time, only the friction force, F_r, produces work-dissipating

---

[10] See Ref. [8] for a similar approach.
[11] As will be seen in Chapter 7, this is a fairly rough simplification. However, this assumption is sufficient for our purposes here.

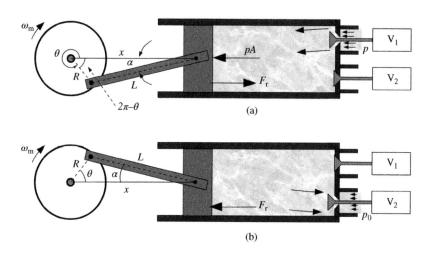

Figure 3.14 Single-piston motor: (a) motoring mode and (b) discharge mode

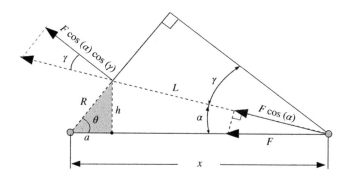

Figure 3.15 Geometry details of the crank-and-piston mechanism

energy from the motor. From Figure 3.14, we see that the magnitude of the resultant force, F, acting on the piston, is given by

$$
\begin{cases}
F = pA - F_r, & \pi < \theta < 2\pi \\
F = F_r, & 0 < \theta < \pi
\end{cases}
\tag{3.18}
$$

Figure 3.15 shows the geometric details of the crank-and-piston mechanism illustrated in Figure 3.14(b). Given that dynamic effects are being disregarded, the instantaneous output torque can be obtained as a function of the crank angle, θ, through a simple force balance. In addition, with the resultant force being defined by Eq. (3.18), the following analysis can be applied to the whole cycle of the crank revolution ($0 < \theta < 2\pi$).

From Figure 3.15,

$$
T_{m\theta} = FR \cos(\alpha) \cos(\gamma)
\tag{3.19}
$$

where $T_{m\theta}$ is the output torque at the motor shaft when the crank angle is θ.

The angles γ, α and θ can be related by the following expression:

$$\gamma = \left(\frac{\pi}{2} - \theta\right) - \alpha \tag{3.20}$$

With γ given by Eq. (3.20), we can rewrite $\cos(\gamma)$ in Eq. (3.19) as a function of the α and θ:

$$\cos(\gamma) = \cos\left[\left(\frac{\pi}{2} - \theta\right) - \alpha\right] = \cos\left(\frac{\pi}{2} - \theta\right)\cos(\alpha) + \sin\left(\frac{\pi}{2} - \theta\right)\sin(\alpha) \tag{3.21}$$

Equation (3.21) can be further simplified to

$$\cos(\gamma) = \sin(\theta)\cos(\alpha) + \cos(\theta)\sin(\alpha) \tag{3.22}$$

Observing Figure 3.15, we can write for $\sin(\alpha)$:

$$\sin(\alpha) = \frac{h}{L} = \frac{R\sin(\theta)}{L} \tag{3.23}$$

A well-known trigonometric relation gives

$$\cos(\alpha) = \sqrt{1 - \sin^2(\alpha)} = \sqrt{1 - \left(\frac{R}{L}\right)^2 \sin^2(\theta)} \tag{3.24}$$

The combination of Eqs. (3.19), (3.22) and (3.24) allows us to obtain the output torque for the single-piston motor as a function of the crank angle, θ, the radius, R, and the bar length, L:

$$T_{m\theta} = fFR \tag{3.25}$$

The factor, f, in Eq. (3.25) is given by

$$f = \sqrt{1 - \left(\frac{R}{L}\right)^2 \sin^2(\theta)} \left\{ \sin(\theta)\sqrt{1 - \left(\frac{R}{L}\right)^2 \sin^2(\theta)} + \cos(\theta)\left[\frac{R\sin(\theta)}{L}\right] \right\} \tag{3.26}$$

Similar to what was done in Section 3.2.2, we can extend the previous analysis to a motor with N parallel pistons working out of phase by a constant angle, ϕ. In such cases, the torque at the motor shaft will be the summation of the torque coming from the individual displacement chambers:

$$T_{m\theta} = R\sum_{i=1}^{N}(Ff_i) \tag{3.27}$$

where

$$f_i = \sqrt{1 - \left(\frac{R}{L}\right)^2 \sin^2(\theta_i)} \left\{ \sin(\theta_i)\sqrt{1 - \left(\frac{R}{L}\right)^2 \sin^2(\theta_i)} + \cos(\theta_i)\left[\frac{R\sin(\theta_i)}{L}\right] \right\} \tag{3.28}$$

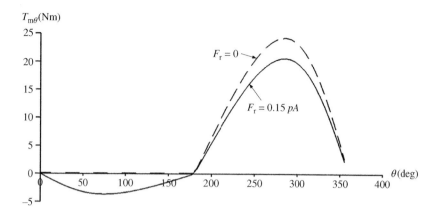

Figure 3.16   Torque output as a function of the crank angle for a complete revolution (single piston)

and θ_i is given by

$$\theta_i = \theta + \left(\frac{2\pi}{N}\right)(i-1), \quad i = 1, 2, \ldots, N \tag{3.29}$$

Figure 3.16 shows the motor torque for a single-piston pump ($N = 1$ in Eq. (3.27)). The dashed curve represents the motor output for the ideal case where $F_r = 0$, while the solid curve represents the case where a constant friction force $F_r = 0.15pA$ is applied. The negative portion of the curve (solid line) represents the phase when the piston is being decelerated by the friction force (motor discharge). It is part of a good motor design to reduce friction forces as much as possible so that the torque available at the motor shaft is increased.

Figure 3.17 shows the torque output for two multiple-piston motors. Similar to what was done in Figure 3.10, we have chosen a five-piston and a six-piston motor for this example ($N = 5$ and $N = 6$, in Eqs. (3.27)–(3.29)). The friction force was maintained as constant in both cases ($F_r = 0.15pA$). Note that because of the presence of other displacement chambers, there is always torque being produced at the motor shaft.

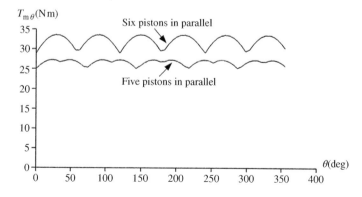

Figure 3.17   Torque output as a function of the crank angle for a complete revolution (multiple pistons)

3.2.3.2 Mechanical Efficiency

Among the terms on the right-hand side of Eq. (3.27), only F contains a reference to dissipative forces. In fact, if no friction existed, there would be no negative torque (resisting torque) at the motor shaft. Thus, as far as mechanical losses are concerned, the motor would be considered ideal. It is, therefore, pertinent here to identify the mechanical efficiency as the ratio between the average actual output torque, $T_m|_{\text{real}}$, and the average theoretical torque obtained from a motor with no internal friction, $T_m|_{\text{ideal}}$. Considering a time interval, Δt, we have from Eqs. (3.27) and (3.18):

$$\eta_m^m = \frac{T_m|_{\text{real}}}{T_m|_{\text{ideal}}} = \frac{\dfrac{1}{\Delta t}\int_0^{\Delta t}\left\{R\sum_{i=1}^{N}\left[(pA - F_r)\right]f_i\right\}dt}{\dfrac{1}{\Delta t}\int_0^{\Delta t}\left[R\sum_{i=1}^{N}(pAf_i)\right]dt} = 1 - \frac{\int_0^{\Delta t}\left\{\sum_{i=1}^{N}\left[(F_r f_i)\right]\right\}dt}{\int_0^{\Delta t}\left[\sum_{i=1}^{N}(pAf_i)\right]dt} \tag{3.30}$$

As expected, if no friction is present in the motor ($F_r = 0$), the mechanical efficiency becomes 100%.

The ISO 4391:1983 Standards uses the term *mechanical–hydraulic efficiency* to include the torque losses generated within the viscous flows occurring inside the motor and define mechanical efficiency as:

$$\eta_m^m(\text{ISO}) = \frac{T_m|_{\text{real}}}{\Delta p_m D_m} \tag{3.31}$$

In this book, we adopt Eq. (3.31) as the definition of mechanical efficiency for hydrostatic motors.[12]

3.2.4 Further Efficiency Considerations

Although we have used a pump example to define volumetric efficiency and a motor example to define mechanical efficiency, it is possible to use these concepts interchangeably between pump and motors. For instance, if we disregard compressibility losses, the volumetric efficiency of the motor can be defined as

$$\eta_m^v = \frac{D_m \omega_m}{D_m \omega_m + q_{mL}} = \frac{1}{1 + q_{mL}/D_m \omega_m} \tag{3.32}$$

where $q_{mL} = q_{mLe} + q_{mLi}$ is the sum of the external and internal leakage flows inside the motor, q_{mLe} and q_{mLi}, respectively. Therefore, the volumetric efficiency of the motor expresses the ratio between the (smaller) input flow required by a hypothetical motor without any volumetric losses and the actual (bigger) input flow needed to compensate for the motor leakages.

---

[12] Note that Eqs. (3.30) and (3.31) express the same concept.

Similarly, the mechanical efficiency of the pump can be expressed by

$$\eta_p^m = \frac{\Delta p_p D_p}{T_p|_{real}}$$ (3.33)

where $T_p|_{real}$ is the actual input torque at the pump shaft and $\Delta p_p D_p$ is a theoretical torque that would be required if there were no mechanical losses (note that $\Delta p_p D_p < T_p|_{real}$).

At this stage, the student might be wondering whether the efficiencies defined in this section can relate to the general efficiency definition given by Eq. (1.8). To answer this question, it is important to give a formal definition for the energetic efficiency of pumps and motors. This will be done in Section 3.4 when we perform an energy balance over pumps and motors at steady-state regime. Now, it is important to have an overview of the most usual types of hydrostatic pumps and motors before we move on, which is the subject of the following section.

3.3   Hydrostatic Pump and Motor Types

The crank-and-piston mechanism of the multicylinder pump shown in Figure 3.8 is not practical, as it would usually result in bulky and large units. However, the general idea of pistons operating in parallel remains valid and has been used as a basis for industrial pump and motor designs. We therefore begin this section with two piston-based models for pumps and motors that have been extensively employed in the fluid power industry – namely, the radial piston design and the axial piston design.

3.3.1   Radial Piston Pumps and Motors

Figure 3.18 shows a typical fixed-displacement radial piston pump where a rotary cam plays the role of the crank in Figure 3.1. In this particular configuration, springs keep the cylinder rods in contact with the cam, while the eccentric shaft rotates at an angular speed ω_p. Each displacement chamber connects to the input and output lines alternately through two check valves placed at the piston head-ends. The output flow can be modified by altering the rotor eccentricity, which results in a change of the pump displacement. The pump output flow will be the sum of the individual flows coming from each displacement chamber.

Radial piston motors are more complex to manufacture when compared to radial piston pumps. We present two different designs here. The first one is similar to the pump shown in Figure 3.18 and is composed of a rotating eccentric shaft, together with radially displaced cylinders, as shown in Figure 3.19. Differently from the radial pump illustrated in Figure 3.18, the eccentric is now a circular cam orbiting around the centre of rotation of the motor shaft. Note that the input and output lines must connect to the cylinders in a sequence, making it necessary to change the flow path continuously as the motor shaft rotates.[13] In the example shown in Figure 3.19, only one cylinder is connected to the high input pressure, p_i,

---

[13] A rotating distribution valve synchronized with the motor shaft can be used for this purpose.

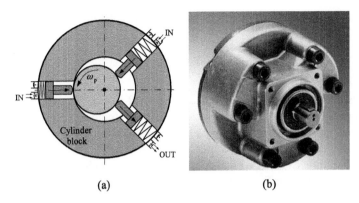

(a)   (b)

Figure 3.18   Schematic representation (a) and picture (b) of a radial piston pump (part b is a courtesy of Bosch-Rexroth Corp.)

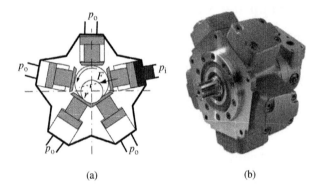

(a)   (b)

Figure 3.19   Schematic representation (a) and picture (b) of a radial piston motor with an eccentric shaft (part b is a courtesy of Bosch-Rexroth Corp.)

while the others are connected to the output line at a lower pressure, p_o ($p_o \ll p_i$). The cylinder connected to the input line presses against the rotating cam with a force F that originates a torque Fr on the motor shaft (see Figure 3.19(a)). As the motor rotates, other displacement chambers are sequentially connected to the pressure p_i, so that torque is constantly being produced at the motor shaft.[14]

Figure 3.20 shows another design for a radial piston motor where the pistons inside the cylinder block press against an outer stroke ring. The cylinders have rollers at the tip of their rods, and as they travel between two cam lobes (e.g. A and C), their pistons move in both directions symmetrically to the inflection point, B. Two regions – one between points A and B and another between points B and C – are then defined between the two cams, where the cylinders can be connected to the high- and low-pressure lines, respectively (the same happens at every other two consecutive cams in the motor). If a reversal of rotation is desired,

---

[14] It is possible to connect more than one cylinder to the high-pressure line depending on the pump design.

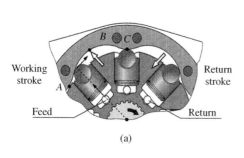

Figure 3.20 Schematic representation (a) and picture (b) of a radial piston motor with an outer stroke ring (courtesy of Bosch-Rexroth Corp.)

the connection to the high- and the low-pressure lines can be switched. Note that there is no need for a rotating valve this time, given that the fluid connections remain stationary.

3.3.2 Axial Piston Pumps and Motors

3.3.2.1 General Considerations

The operational principle of axial piston pumps and motors can be better visualized with the help of Figure 3.21 in which a series of hydraulic pistons slide on an inclined surface. The pistons are altogether placed inside a cylinder block (or cylinder barrel), which, in the figure, rotates counter-clockwise at an angular speed, ω. As the cylinder block rotates, it drags the pistons, causing them to slide on the inclined surface at a speed, v, while moving up and down at a (variable) speed, v_p. We assume that the piston rods never lift off of the inclined surface so that the displacement chamber of each cylinder expands as the piston goes down the plane on one half of the revolution and subsequently retracts on the other half. The two halves of

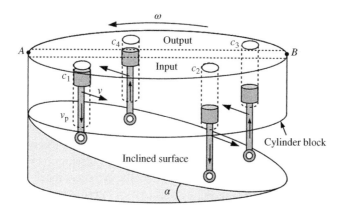

Figure 3.21 Axial piston arrangement of the cylinders

the path, marked by the points A and B, located at the extremities of the cylinder barrel, delineate the two regions through which fluid can either enter the displacement chambers (input) or be discharged to the hydraulic circuit (output).

In pumping mode, the cylinder block drives the cylinders down and up the plane in a counter-clockwise motion, as shown in Figure 3.21. As the cylinders move from A to B, they expand, causing fluid to be sucked into their chambers, and then subsequently discharge on the other half of the revolution when they travel from B to A. In the particular case of Figure 3.21, cylinders c_1 and c_2 are expanding, while cylinders c_3 and c_4 are retracting. If the rotation of the cylinder block is reversed, the flow is also inverted, that is, the input port becomes the output port and vice versa.

In motoring mode, the input region in Figure 3.21 is connected to a pressure source. The pressure creates a force on the rods of cylinders c_1 and c_2, which then move down the inclined surface and drive the cylinder block that is connected to the motor shaft. In the meantime, cylinders c_3 and c_4 absorb energy from the motor and move up the inclined surface, discharging hydraulic fluid at the output region. Note that if the flow through the motor is inverted – that is, the higher pressure is applied to the output region – the rotation of the motor is reversed.

3.3.2.2 Swashplate Type

Figure 3.22 shows a typical axial piston pump where a swashplate plays the role of the inclined surface in Figure 3.21. The pump shaft in Figure 3.22(a) is rigidly connected to the cylinder block that drags the pistons on the swashplate as it turns. We observe that the output flow depends on the swashplate inclination, which can often be adjusted (making the pump a variable-displacement pump). The way in which the displacement in an axial piston pump is controlled varies according to each manufacturer. Generally, there are two possibilities: manual or hydraulic. For example, in the scheme represented in Figure 3.22(a), the swashplate angle is controlled by means of a single-action hydraulic cylinder. Other configurations include using two hydraulic cylinders to produce a higher torque on the swashplate [9]. In a typical variable-displacement pump, the swashplate angle may be varied by ±15° or ±18° [10].

The advantages of using a hydraulic cylinder to change the swashplate angle are many. One advantage is to turn the control of the hydrostatic pump into a closed-loop control, making it possible for the pump to adjust its output flow according to some feedback from the circuit. Another benefit comes from the fact that a big torque is usually required to either turn the swashplate or keep it in a steady position. In addition, vibrations in the circuit are usually felt by the operator who is in charge of controlling the swashplate lever. This situation aggravates with bigger sized pumps, in such a way that, in general, mechanical swashplate operation is not found in pumps whose displacement is larger than 45–50 cm^3/rev [10].

The port plate shown in Figure 3.23(a) is responsible for the division between the input and output regions. Two cuts in the plate connect the input and output ports with the cylinders. The cuts are interrupted at the two extremities where the cylinders reach the minimum and maximum strokes (points A and B in Figure 3.21). Such interruption is necessary in order to prevent a cylinder from communicating with both high- and low-pressure regions at the same time, causing an undesired backflow from the output to the input port.

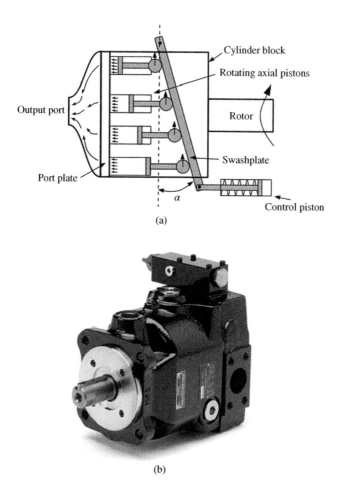

(a)

(b)

Figure 3.22   Schematic representation and picture (courtesy of Parker-Hannifin Corp.) of a swashplate pump

We note that because of the interruption in the fluid flow through the port plate cuts, the fluid gets momentarily trapped around region A (Figure 3.23(a)), and because of the inclination of the swashplate, a compression happens in the small space of time when the piston is passing from the output region (HA) to the input region (LA), as seen in Figure 3.23(b). As a result, a small backflow due to the decompression of the trapped fluid occurs as the piston moves from the high-pressure area, HA, into the low-pressure area, LA. This phenomenon is associated with the so-called *commutation loss* and was observed by Newton in 1947 [2] as he, according to Ref. [11], carried out one of the first hydrostatic transmission analysis published on a scientific journal.

It is important to observe that the pistons in swashplate pumps undergo a considerable radial load because of the swashplate inclination. Figure 3.24 is an excerpt from Figure 3.22 where we see the forces acting on the piston and rod of an individual cylinder as it moves up and around the swashplate with a linear speed, *v*. The vertical reaction by the swashplate

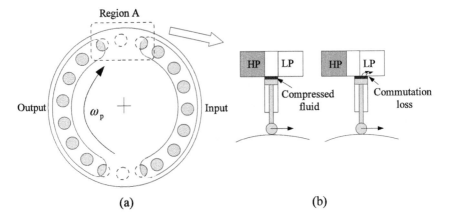

Figure 3.23   Port plate (a) and commutation loss (b)

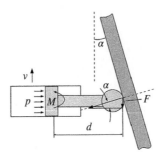

Figure 3.24   Forces acting on the swashplate pump piston

onto the rod, F, produces a torque Fd that must be counterbalanced by a momentum, M, at the piston. The piston is responsible for both the pump sealing and the maintenance of the momentum balance. This dual role of the piston (sealing and bearing) considerably increases the demands on the pump design [12]. Note that the reaction from the swashplate on the piston rod increases with the pressure, p, in the displacement chamber.

We observe that the swashplate pump can be reversed into a motor by connecting one of the ports to a high-pressure source. As a result, the mechanical construction, the physical characteristics and the nomenclature that we have used for pumps can naturally be applied to motors as well.

The angular speed of a swashplate motor can be controlled by the variation of the swashplate angle, α, in much the same way as we control the pump output flow. The smaller the swashplate angle, the faster the angular speed of the rotor becomes, as long as α is not zero. For $\alpha = 0$, there is no torque produced on the cylinder block and the motor stops. Figure 3.25 shows the schematic representation of a swashplate motor.

To help us visualize the dependence of the angular speed, ω_m, on the swashplate angle, α, consider an individual cylinder in the cylinder block. From Figure 3.25, we see that the

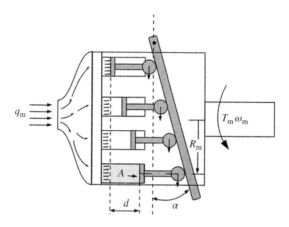

Figure 3.25 Schematic representation of the swashplate motor

volume V_i, displaced by an individual cylinder during one complete revolution, is given by

$$V_i = Ad = 2AR_m \tan(\alpha) \tag{3.34}$$

where A is the piston area, d is the piston stroke and R_m is the distance between the shaft axis and the centre of the cylinder.

In a motor with N cylinders, the total volumetric displacement, D_m, will be given by

$$D_m = \sum_{i=1}^{N} V_i = 2NAR_m \tan(\alpha) \tag{3.35}$$

Generally, motor manufacturers state the minimum and the maximum swashplate angles, α_{min} and α_{max}, in their catalogues. The maximum and minimum displacements, D_m^{max} and D_m^{min}, can then be written as

$$\begin{cases} D_m^{min} = 2NAR_m \tan\left(\alpha_{min}\right) \\ D_m^{max} = 2NAR_m \tan(\alpha_{max}) \end{cases} \tag{3.36}$$

By substituting the term $2NAR_m$, obtained from the second equation in (3.36), into Eq. (3.35), we have

$$D_m = \left[\frac{D_m^{max}}{\tan\left(\alpha_{max}\right)} \right] \tan(\alpha) \tag{3.37}$$

Therefore, if we disregard the volumetric losses, we can then write the average input flow, q_m, as

$$q_m = \omega_m \left[\frac{D_m^{max}}{\tan\left(\alpha_{max}\right)} \right] \tan(\alpha) \tag{3.38}$$

Remember that, since D_m^{max} in Eq. (3.38) is given in units of volume per revolution, the motor speed, ω_m, must be given in revolutions per unit time. Considering a constant input flow, q_m, it is clear from Eq. (3.38) that as the swashplate angle, α, decreases, the shaft speed, ω_m, increases. In addition, if we make $\alpha = 0$, the input flow, q_m, must also become zero, causing the input line to be blocked.[15]

3.3.2.3 Bent-Axis Type

A variation of the swashplate design can be seen in Figure 3.26 in which the operation of bent-axis pump/motor is illustrated.[16] The basic difference between the swashplate and the bent-axis design is that in bent-axis pumps and motors, it is the cylinder block, B, that is inclined relatively to the shaft, S. The swivel angle, β, determines the pump/motor displacement, and the shaft, S, which rotates along with the block B, can either drive the pistons (pump) or be driven by the pistons (motor).

The bent-axis design has the advantage of not overloading the cylinder pistons, as shown in Figure 3.27 in the details of the forces acting on a single cylinder. Observe that, differently from the swashplate design shown in Figure 3.24, no momentum is generated at the piston. This can be explained by the fact that due to the spherical contact between the piston rod and the shaft, the reaction force, F, necessarily acts along the centreline of the cylinder. Given that the piston does not need to counterbalance any torque generated by the pressure, seal rings can be used between the piston and the case. As a result, volumetric and mechanical losses are generally smaller when compared to swashplate units and the swivel angle can achieve values higher than $45°$. However, despite these advantages, bent-axis pumps and

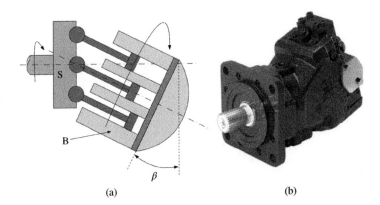

(a)   (b)

Figure 3.26 Schematic representation (a) and picture (b) of a typical bent-axis motor (part b is a courtesy of Danfoss Power Solutions)

---

[15] This is a totally undesirable situation and should be avoided.

[16] Strictly speaking, given the fact the pistons are not oriented along the pump/motor shaft, bent-axis pumps and motors should not be regarded as 'axial piston' machines. However, since their operational principle is exactly the same as axial piston pumps and motors, we have included them in this section as well. Bent-axis pumps and motors have frequently been considered as a subset of axial piston pumps and motors (see, e.g. Refs. [5, 13]).

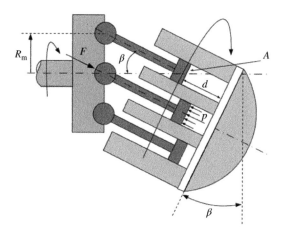

Figure 3.27   Forces acting on the bent-axis motor piston

motors are usually bulkier than similar sized swashplate units. They are also relatively costly
and present a more complex design requirement [12].

Let us establish a relationship between the flow and the swivel angle, similar to what was
done for swashplate motors in Eq. (3.38). Observing Figure 3.27, we see that the volume V_i,
displaced by an individual cylinder during one complete rotation, is given by

$$V_i = Ad = 2AR_m\sin(\beta) \tag{3.39}$$

In a motor with N cylinders, the total volumetric displacement, D_m, will be

$$D_m = \sum_{i=1}^{N} V_i = 2NAR_m\sin(\beta) \tag{3.40}$$

Note that the only difference between Eqs. (3.40) and (3.35) is that for bent-axis motors, the
sine of the tilt angle is used; whereas in swashplate units, the tangent of the swashplate angle
appears instead. Thus, we may skip the intermediate steps and write the following equation
for the motor flow, q_m, in the absence of volumetric losses[17]:

$$q_m = \omega_m \left[\frac{D_m^{max}}{\sin\left(\beta_{max}\right)} \right] \sin(\beta) \tag{3.41}$$

where the motor speed must be given in revolutions per unit of time (see comments after
Eq. (3.38)).

Before we move on to the next type of axial piston pumps and motors, it is interesting
to observe that swashplate pumps and motors react more rapidly to a change in the swashplate angle when compared with bent-axis units. To see that, consider the pump versions

---

[17] In spite of the fact that Eq. (3.41) has been developed for motors, it can be equally applied for bent-axis pumps.

of Eqs. (3.38) and (3.41) at their first derivatives. For a constant angular speed, ω_p, for swashplate pumps we have

$$
\begin{cases}
q_p = \left[\dfrac{\omega_p D_p^{max}}{\tan\left(\alpha_{max}\right)} \right] \tan(\alpha) \\[4ex]
\dfrac{dq_p}{d\alpha} = \left[\dfrac{\omega_p D_p^{max}}{\tan\left(\alpha_{max}\right)} \right] [1 + \tan^2(\alpha)]
\end{cases}
\tag{3.42}
$$

Similarly, for bent-axis pumps:

$$
\begin{cases}
q_p = \left[\dfrac{\omega_p D_p^{max}}{\sin\left(\beta_{max}\right)} \right] \sin(\beta) \\[4ex]
\dfrac{dq_p}{d\beta} = \left[\dfrac{\omega_p D_p^{max}}{\sin\left(\beta_{max}\right)} \right] \cos(\beta)
\end{cases}
\tag{3.43}
$$

If we assume that $\alpha_{max} = \beta_{max}$ in Eqs. (3.42) and (3.43), the following relation can be obtained, considering the same maximum displacements, D_p^{max}, for the swashplate and the bent-axis units:

$$
\frac{dq_p}{d\alpha} = \frac{dq_p}{d\beta} \left[\frac{1 + \tan^2(\alpha)}{\cos(\beta)} \right]
\tag{3.44}
$$

The term between brackets in Eq. (3.44) is greater than 1 for $\alpha > 0$ and $\beta > 0$. The immediate implication is that the output flow of a swashplate pump varies with the inclination of the angle, α, at a higher rate compared to the rate change that occurs in an equivalent bent-axis pump. A similar reasoning can be used for swashplate motors in the sense that the variation of the torque on the motor shaft is more sensitive to the variation of the swashplate angle when compared to responses in bent-axis units [14].

3.3.2.4 Floating-Cup Type

Another type of axial piston pumps and motors that has been recently developed by Innas B.V.[18] [15, 16] is schematically shown in Figure 3.28(a). In floating-cup units, the piston rods are solidly connected to a rotor and the cylinder cases (cups) tilt. The piston edges are round to accommodate the tilting of the cups, and the symmetrical arrangement of the cups relative to the rotor makes the whole set hydrostatically balanced [15].

The operation of the floating-cup pump shown in Figure 3.28 can be understood as follows. The shaft is solidly connected to the rotor and the pistons so that the whole set rotates together. The pistons drag the cylinder barrels around, sliding within the cups and causing the displacement chambers to expand and retract during a complete revolution. The input

---

[18] http://www.innas.com.

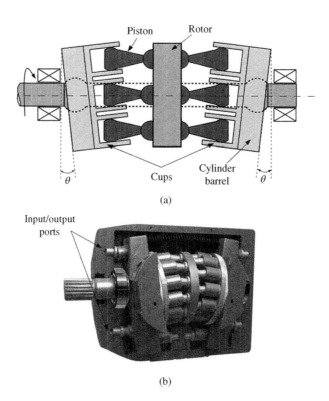

Figure 3.28 Schematic representation (a) and picture (b) of a floating-cup pump (part b is a courtesy of Innas B.V.)

and output lines (not shown in the figure) can then be connected to the two halves of the rotor through the ports indicated in Figure 3.28(b).

One important feature of the floating-cup design is that the cups are not solidly connected to the barrel to accommodate the elliptic deviation that occurs when the barrel is tilted.[19] In addition, because of the mechanical construction, tilting is limited to small angles, which is compensated for by a larger number of displacement chambers. For instance, it has been reported that floating-cup machines operate with 24 (12×2) cylinders, almost three times more than a typical bent-axis unit with nine cylinders [12].

The advantages and disadvantages of floating-cup pumps and motors in relation to swash-plate and bent-axis units have been much debated. On the one hand, the larger number of displacement chambers clearly produces a less oscillatory output. On the other hand, volumetric losses due to the increase in the number of sliding surfaces may become an issue (although some recent reports have shown otherwise [15]). A deeper analysis on the theme is beyond the scope of this book and the interested student can consult Refs. [12, 15, 16] for more information.

---

[19] This can be explained as follows: as the barrel tilts of an angle θ, the projection of the cylinders on the barrel surface will necessarily describe an ellipsis.

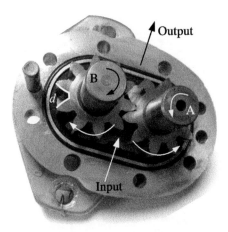

Figure 3.29   External gear pump

3.3.3   Gear Pumps and Motors

Gear pumps and motors are very common in hydraulic circuits. Normally, two designs are used: the external and the internal–external gear design. A third category is the gerotor,[20] which can also be perceived as a subset of the internal–external gear design [17]. Since we are focusing on operational aspects, it is sufficient to cover two representative models in this section. Thus, we divide this topic into external gear units and gerotors.

3.3.3.1   External Gear Pumps and Motors

The most popular type of hydrostatic pump is the external gear pump (Figure 3.29). The displacement chambers of an external gear pump are the spaces between the gear teeth and the pump case (d, in the figure). As the mover gear, A, rotates counter-clockwise, the fluid gets trapped inside the space between the gear teeth and the pump case and is pushed around, as indicated by the white arrows in the figure.

Apart from being fixed-displacement units, external gear pumps have limited flow because of the relatively small volumetric chambers. However, their mechanical construction is simple, and they have a relatively low-cost option when compared to other designs.

External gear pumps can operate as motors as well. For example, suppose that the output port of the pump in Figure 3.29 is connected to a high-pressure source. The action of the pressure on the gear teeth will generate a resultant torque, T, as seen in Figure 3.30, which shows the pressure p acting on teeth 1, 2 and 3 at the input side of the motor. If we consider the teeth areas, S, that are exposed to the input pressure, p, as identical, the forces on teeth 1, 2 and 3 will be the same, as will be the corresponding torques T_1, T_2 and T_3, acting on gear A. Considering an average radius, R, a torque balance on gear A, gives

$$T = -T_1 - T_2 + T_3 = -pSR - pSR + pSR = -pSR \qquad (3.45)$$

---

[20] The name 'gerotor' is derived from 'Generated Rotor'.

Figure 3.30 Torque generation on an external gear motor

The direction of rotation of gear A in Figure 3.30 will necessarily be clockwise, given that the resulting torque is negative (note that we are considering a clockwise torque as negative in Eq. (3.45)). As a result, gear B will rotate counter-clockwise. Note that any of the two gears in Figure 3.30 can be chosen to drive the output shaft.

3.3.3.2 Gerotor Pumps and Motors

In gerotor pumps, the inner gear (external teeth gear) drives the outer gear (internal teeth gear), which is allowed to rotate freely inside the pump case, as shown in Figure 3.31. The inner gear has one less tooth than the outer gear, and as both gears rotate clockwise, each displacement chamber (cavity formed between gears, as shown in the figure) expands during half a rotation (input side) and retracts during the other half (output side).

Figure 3.32 shows the motor operation of the gerotor, where the input side is now exposed to the higher pressure, p. The input pressure acts on the internal gear teeth and produces

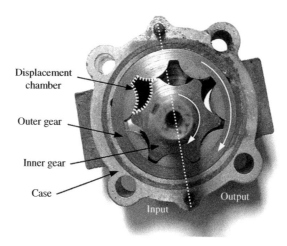

Figure 3.31 Gerotor pump

Figure 3.32   Gerotor motor

a torque, T, on the rotor. As a result, the motor shaft moves clockwise, as indicated in the figure. Note that, because of the internal geometry, the torque generated at every displacement chamber will be clockwise oriented (that was not the case with the external gear motor shown in Figure 3.30, where the torque T_3 was contrarily oriented in relation to the torques T_1 and T_2). Gerotor motors are usually indicated for low-speed and high-torque applications [18].

3.3.4   Vane Pumps and Motors

The vane pump, shown in Figure 3.33, has two input regions and two output regions. As the pump shaft rotates clockwise, the vanes in the rotor slots are pressed against the pump case, mainly by action of the centrifugal force. In some models, the pressure at the pump output is also used to force the vanes against the case, while in some configurations springs

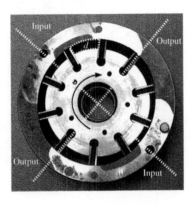

Figure 3.33   Vane pump

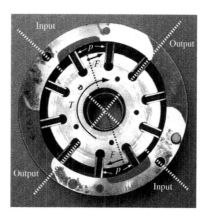

Figure 3.34   Vane motor

can also be used to keep the vanes in contact with the case. In the figure, the displacement chamber, d, is expanding and is therefore receiving fluid from the input port. Note that, as soon as d enters the following quadrant, it begins to retract and discharges fluid through the output port so that there is a change from input to output in every quarter of a revolution.

By pressurizing the input regions of the vane pump shown in Figure 3.33, we obtain a motor. Figure 3.34 illustrates what happens when the input pressure, p, acts on the limiting vanes of two symmetrically located displacement chambers. Because of the elliptic format of the case, each limiting vane of a displacement chamber has a different area exposed to the pressure, p. A resultant force, F, is then produced, which creates a binary, T, on the motor shaft. Note that, because of the motor symmetry, the shaft does not get radially loaded in this particular design.

3.3.5   Digital Displacement Pumps and Motors

A relatively recent technology, denominated *digital displacement* [4, 19], has claimed to significantly improve the efficiencies of hydrostatic pumps and motors.[21] In order to understand the digital displacement concept, first consider the general output of a hydrostatic pump, as shown in Figure 3.9. We saw that the output flow in Figure 3.9(b) is the sum of the individual output flows coming from all the displacement chambers (positive parts of the curves shown in Figure 3.9(a)). We also saw that in variable-displacement machines, we can alter the input/output flows by reducing or increasing the displacement of the individual chambers altogether. For example, in an axial piston pump, we can modify the output flow by changing the swashplate angle α (Figure 3.22(a)). Note that, no matter how small the swashplate angle becomes, every active displacement chamber will contribute to the total output flow. In other words, all the active cylinders will be submitted to the high pressure

---

[21] 'Digital displacement hydraulic power technology' is a registered trademark of Artemis Intelligent Power Ltd (http://www.artemisip.com), currently owned by Mitsubishi Power Systems Europe (MPSE). The term 'digital displacement' does not indicate a new type of pump and motor *per se* but rather refers to a new way in which pump and motor displacements are controlled.

at the output/input line. The same is valid for every type of hydrostatic pump or motor we have studied so far.

For an example of how digital displacement technology helps improving the efficiency of pumps and motors, we recall from the previous chapter that pressure differentials were responsible for the volumetric losses in hydrostatic machines (see Eq. (2.95)). Therefore, the larger the number of pistons exposed to a high-pressure differential, the higher the volumetric losses become. For instance, consider an axial piston pump for which the swashplate angle has been set to a very small value (small output flow). If the internal and external leakages become significant due to the pressure differential between the output and the input lines, it can be easily concluded from Eq. (3.17) that the volumetric efficiency can be greatly reduced in this case.

The idea behind the digital displacement technology is simple: modify the number of active cylinders so that the number of displacement chambers that are exposed to the high pressure of the output/input line can be changed as needed. This concept is better explained with the help of Figure 3.35, which shows a piston pump composed of five in-line cylinders[22] driven by a common crank shaft that receives power from a prime mover (not shown in the figure). The cylinders are connected to the input and output lines through high-speed solenoid valves,[23] i_1 through i_5 and o_1 through o_5, respectively. Note that the pump itself has a fixed displacement, and the output flow is controlled by opening and closing the valves in a way that not all the cylinders get exposed to the high pressure of the output line all the time. The total leakage through the cylinder walls can then be reduced, especially at low flow regimes when fewer cylinders are connected to the high-pressure line. In addition, given that friction forces on the pump/motor bearings and other sliding surfaces are themselves

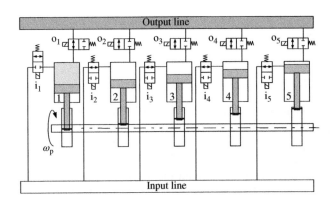

Figure 3.35   Schematic representation of a digital displacement pump

---

[22] Interestingly, digital displacement pumps and motors have originally been conceived as radial piston machines (http://www.artemisip.com/our-technology, accessed in September 2013).

[23] High-speed solenoid valves constitute a technological challenge to digital displacement machines. On the one hand, they must be sufficiently large not to cause an excessive energy loss due to flow throttling. On the other hand, they must be light enough to have a fast response time and produce low electric heating. In addition, because of the very fast shift between high- and low-pressure states of the fluid, compressibility losses play an important role in their design as well. More on the theme of high-speed valves can be found, for example, in Ref. [20] and the references therein.

proportional to the pressure acting on the pistons, mechanical efficiency is also improved by the idling of some displacement chambers. Valve control is therefore the heart of the system, since it is responsible for the actuation of each individual piston.

Consider, for example, cylinder 2 of the pump shown in Figure 3.35. Depending on the status of the valves i_2 and o_2, we can have one of the four situations represented in Table 3.1, which also shows the pressure inside the cylinder and the particular status of the piston movement: up (ascending) or down (descending).

From Table 3.1, we see that cylinder 2 can either supply or extract flow from the output line at a high pressure. Similarly, suction and discharge from the input line can occur with a low pressure inside the cylinder chamber, which can be isolated from the output line by the corresponding valve.

The means by which the output flow can be controlled depends on the opening sequence of the valves. As an example, for the five-cylinder pump shown in Figure 3.35, the sequence of operation illustrated in Figure 3.36, will substantially reduce the output flow because only one cylinder at a time will be connected to the output line. In the figure, the charge–discharge cycle for each of the five cylinders is represented by a sinusoidal curve. By opening and closing the correct valves, it is possible to connect only one cylinder to the high-pressure line while venting the others. In the particular scheme shown in Figure 3.36, cylinders 1, 4, 2, 5 and 3 become sequentially active, while the others are disconnected from the output line.

Table 3.2 shows the status of the valves for the output shown in Figure 3.36.

Another way of representing the sequence shown in Table 3.2 is by assigning the number 1 to the active cylinder and 0 to the others [21]. With this simpler representation, the binary

Table 3.1 Possible operations for the second cylinder

Valve i_2	Valve o_2	Piston direction	Pressure	Action
Open	Closed	Down	Low	Suction from input line
Open	Closed	Up	Low	Discharge to input line
Closed	Open	Up	High	Discharge to output line
Closed	Open	Down	High	Suction from output line

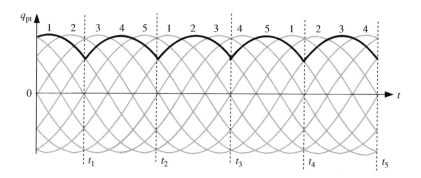

Figure 3.36 Example of a digitally controlled output flow

Table 3.2 Digital displacement sequence for the pump represented in Figure 3.36

Time (t)	i_1	i_2	i_3	i_4	i_5	o_1	o_2	o_3	o_4	o_5
$0 < t < t_1$	**Closed**	Open	Open	Open	Open	**Open**	Closed	Closed	Closed	Closed
$t_1 < t < t_2$	Open	Open	Open	**Closed**	Open	Closed	Closed	Closed	**Open**	Closed
$t_2 < t < t_3$	Open	**Closed**	Open	Open	Open	Closed	**Open**	Closed	Closed	Closed
$t_3 < t < t_4$	Open	Open	Open	Open	**Closed**	Closed	Closed	Closed	Closed	**Open**
$t_4 < t < t_5$	Open	Open	**Closed**	Open	Open	Closed	Closed	**Open**	Closed	Closed

sequence for the example above becomes

$$10000 - 00010 - 01000 - 00001 - 00100 - 10000 \cdots$$

Different output flows can be obtained for different opening sequences of the solenoid valves. For example, Figure 3.37 shows two identical and dephased sequences. The difference between the two sequences is the initial time, t_i, where the solenoid valve associated with cylinder 1 (o_1 in Figure 3.35) is first connected to the high-pressure line. In this particular example, the output flow can be controlled by choosing a convenient starting time for the sequential operation of the valves (for $t_{i2} > t_{i1}$, we obtain $q_{p2} < q_{p1}$). Note that, once the first valve is activated, the sequence is as follows:

$$10000 - 00100 - 01000 - 00010 - 00100 - 00001 \cdots$$

From Figure 3.37, we can conclude that the oscillations on the output flow depend on the number of cylinders and on the opening sequence of the valves. We can see that control possibilities are huge.[24] However, programming the opening sequence of the valves is no trivial task. For further information, the interested student can consult Ref. [23] and the references therein.

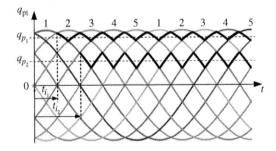

Figure 3.37 Digital output flow control

---

[24] Although the examples given in Figures 3.36 and 3.37 show only one piston connected to the high-pressure line at a time, it does not have to be the case. In fact, we can have more than one piston activated at a sequence step and also connect the output valves to different branches of the circuit so that part of the pump can operate as a motor, absorbing energy from the hydraulic circuit [22].

The same observations made for pumps can be extended to motors. Within motors, the displacement chambers are 'digitally connected' to the high-pressure line, causing a series of momentary torque pulses at the motor shaft. Again, the non-used displacement chambers can be completely idled to reduce volumetric and mechanical losses.

3.4 Energy Losses at Steady-State Operation

We have indirectly introduced the subject of energy losses in hydrostatic pumps and motors in Sections 3.2.2 and 3.2.3 when we defined the mechanical and volumetric efficiency of pumps and motors (see Eqs. (3.17), (3.31)–(3.33)). Given that both mechanical and volumetric losses ultimately translate into power dissipation, it is not difficult to see that there must be a connection between mechanical and volumetric efficiencies and the definition given by Eq. (1.8) in Chapter 1.

In this section, we perform a simple thermodynamic analysis on a typical hydrostatic machine (pump or motor) to obtain some detailed expressions for the pump and motors efficiencies. Note that we are now concerned with efficiency as defined by Eq. (1.8), that is, the ratio between the output and input powers at a pump or a motor. We begin by slightly changing the original notation in Eq. (1.8) into a more general form and writing the efficiency,[25] η, as

$$\eta = \frac{\dot{E}_o}{\dot{E}_i} \tag{3.46}$$

where $\dot{E}_o$ and $\dot{E}_i$ represent the output and input energy rates at the pump or the motor.

> The efficiency, as defined by Eq. (3.46), must be obtained when the pump/motor is at steady-state regime, that is, when $\dot{E}_o$ and $\dot{E}_i$ no longer vary in time.[26]

3.4.1 Energy Balances

In order to create a representative model of a generic pump and motor, we begin by observing that in both cases, two regions can be readily identified: an input region where the displacement chambers expand, and an output region where the displacement chambers retract. Figure 3.38 shows a simplified planar representation of a pump and a motor, where two cylinders, A and B, are driven together on two opposite inclined planes. In Figure 3.38(a), where a motor is represented, it is the pressure acting on the piston of the cylinder A that is responsible for the movement of the block, C, to the left, while cylinder B retracts and discharges into the output line. In Figure 3.38(b), an external force (not shown in the figure) moves the block to the right and the device works as a pump, sucking fluid from the input

---

[25] Usually referred to as *overall efficiency* to differentiate it from the volumetric and mechanical efficiencies defined earlier.

[26] The ISO 4391:1983 Standards define the steady-state operation conditions as being the 'conditions in which relevant parameters do not change substantially after a period for stabilization'. Therefore, we can think of efficiency as being dependent on the average values of flow and pressure when the pump/motor is operating at steady-state regime.

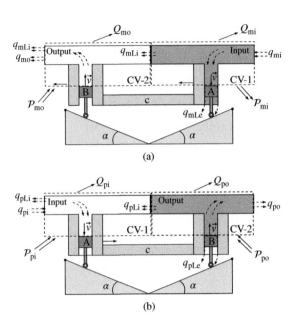

Figure 3.38 Energy balance in a generic hydrostatic motor (a) and pump (b)

region (cylinder A) and pumping into the output region (cylinder B). The following symbols can be identified in the figure:

1. q_{pLi}, q_{mLi}: pump and motor internal leakages;
2. q_{pLe}, q_{mLe}: pump and motor external leakages;
3. q_{pi}, q_{mi}: pump and motor input flows;
4. q_{po}, q_{mo}: pump and motor output flows;
5. Q_{pi}, Q_{po}: heat losses at the pump through the input and output regions;
6. Q_{mi}, Q_{mo}: heat losses at the motor through the input and output regions;
7. P_{pi}, P_{po}: power input to the pump through the input and output regions;
8. P_{mi}: power output from the motor through the input region;
9. P_{mo}: power input to the motor through the output region.

Two different control volumes have been chosen in Figure 3.38, one corresponding to the input regions (CV-1) and the other corresponding to the output regions (CV-2). The energy flows through these control volumes differ in each case. For instance, at the input region of the motor (Figure 3.38(a)), the cylinder on the right absorbs power from the incoming flow and outputs power to the motor shaft[27] (P_{mi}). On the other hand, the cylinder on the

---

[27] Strictly speaking, there is no 'shaft' in Figure 3.38. However, we may think of this figure as a simpler representation of an axial piston machine for which a rotating shaft can be applied.

left (motor output) absorbs power from the shaft (P_{mo}), as it resists the movement of the cylinder block.

Concerning the pump (Figure 3.38(b)), both input and output control volumes, CV-1 and CV-2, receive power from the shaft. In other words, the input power is split between the processes of suctioning (P_{pi}) and expelling the hydraulic fluid (P_{po}). Note that the energy that comes with the input flow, q_{pi}, is also added to the input power, P_{pi}.

In what follows, we consider the fluid as incompressible. We also assume that pump and motor external leakages only happens between the high-pressure chambers and the cases. And specifically regarding the pump, we also note that the internal leakages, q_{pLi}, flowing from the higher pressure chamber into the lower pressure chamber eventually escape through the input port, as indicated in Figure 3.38(a).[28]

An energy balance over the motor and the pump at steady-state regime gives

$$\text{Motor :} \begin{cases} q_{mi}e_{mi} - q_{mLi}e_{mi} - q_{mLe}e_{mi} - P_{mi} - Q_{mi} = 0 \\ q_{mLi}e_{mo} + P_{mo} - Q_{mo} - q_{mo}e_{mo} - q_{mLi}e_{mo} = 0 \end{cases} \tag{3.47}$$

$$\text{Pump :} \begin{cases} q_{pi}e_{pi} + q_{pLi}e_{pi} + P_{pi} - Q_{pi} - q_{pLi}e_{pi} = 0 \\ P_{po} - q_{pLi}e_{po} - Q_{po} - q_{pLe}e_{po} - q_{po}e_{po} = 0 \end{cases} \tag{3.48}$$

In Eqs. (3.47) and (3.48), the subscripts mi, mo, pi and po refer to the motor and pump input and output ports, and e is the total energy per unit volume, given by

$$e = \frac{\rho v^2}{2} + p + \rho u + \rho g h \tag{3.49}$$

where ρ is the fluid density, v is the average cross-sectional velocity of the fluid, p is the absolute pressure, u is the internal energy per unit mass, g is the acceleration of gravity and h is the relative height of the considered location.

Equations (3.47) and (3.48) can be combined as

$$T_m\omega_m = (q_{mi}e_{mi} - q_{mo}e_{mo}) - (q_{mLe} + q_{mLi})e_{mi} - Q_m \tag{3.50}$$

$$T_p\omega_p = (q_{po}e_{po} - q_{pi}e_{pi}) + (q_{pLe} + q_{pLi})e_{po} + Q_p \tag{3.51}$$

where $T_m\omega_m$ is the output power of the motor, $T_p\omega_p$ is the input power at the pump, $Q_m = Q_{mi} + Q_{mo}$ and $Q_p = Q_{pi} + Q_{po}$ are the motor and pump mechanical losses, which ultimately translate into heat losses.

---

[28] This comes from the observation that if there were no internal leakages at the pump, the incoming flow would completely fill the low-pressure chamber. Given that the fluid is incompressible, any extra flow would have to either leak through the piston and the case or leave the control volume through the input port. Since we are assuming that there is no leakage between the lower pressure chamber and the case, we are therefore left with the second option.

3.4.2   Overall Efficiencies

Three terms can be clearly identified on the right-hand side of Eq. (3.51): the hydraulic power produced by the pump, $q_{po}e_{po} - q_{pi}e_{pi}$; the energy related to volumetric losses, $(q_{pLe} + q_{pLi})e_{po}$; and the heat (mechanical) losses Q_p. Thus, considering the input power, $T_p\omega_p$, we can write for the pump efficiency (see Eqs. (3.46) and (3.51)):

$$\eta_p = \frac{q_{po}e_{po} - q_{pi}e_{pi}}{T_p\omega_p} = \frac{q_{po}e_{po} - q_{pi}e_{pi}}{(q_{po}e_{po} - q_{pi}e_{pi}) + (q_{pLe} + q_{pLi})e_{po} + Q_p} \qquad (3.52)$$

The efficiency of a hypothetical pump with no internal or external leakage and no mechanical losses (heat losses) can be obtained from Eq. (3.52):

$$\eta_p = \frac{q_{po}e_{po} - q_{pi}e_{pi}}{q_{po}e_{po} - q_{pi}e_{pi}} = 100\% \qquad (3.53)$$

Similarly, we identify three terms on the right-hand side of Eq. (3.50): the input hydraulic power, $q_{mi}e_{mi} - q_{mo}e_{mo}$; the energy related to volumetric losses, $(q_{mLe} + q_{mLi})e_{mi}$; and the heat (mechanical) losses Q_m. Therefore, considering the output, $T_m\omega_m$, the motor efficiency will be given by (see Eqs. (3.46) and (3.50)):

$$\eta_m = \frac{T_m\omega_m}{q_{mi}e_{mi} - q_{mo}e_{mo}} = \frac{(q_{mi}e_{mi} - q_{mo}e_{mo}) - (q_{mLe} + q_{mLi})e_{mi} - Q_m}{q_{mi}e_{mi} - q_{mo}e_{mo}} \qquad (3.54)$$

Therefore, the efficiency of a motor without any losses where all the available energy coming from the fluid is turned into mechanical power at the shaft is

$$\eta_m = \frac{q_{mi}e_{mi} - q_{mo}e_{mo}}{q_{mi}e_{mi} - q_{mo}e_{mo}} = 100\% \qquad (3.55)$$

3.4.3   Simplified Efficiency Equations

Some simplifications can be made concerning energy flows in the efficiency equations. For example, the difference between the gravitational and kinetic energies at the input and output ports can be disregarded given that, in general, no significant variations exist in the mean fluid speed and the geometric location at the input and output regions. If we also assume that the fluid temperature is constant, the internal energies at the input and output will be the same. These assumptions allow us to write Eq. (3.49) as $e = p$. Thus, the equations for the pump and motor efficiencies become:

$$\eta_p = \frac{q_{po}P_{po} - q_{pi}P_{pi}}{T_p\omega_p} = \frac{q_{po}P_{po} - q_{pi}P_{pi}}{(q_{po}P_{po} - q_{pi}P_{pi}) + (q_{pLe} + q_{pLi})P_{po} + Q_p} \qquad (3.56)$$

$$\eta_m = \frac{T_m\omega_m}{q_{mi}P_{mi} - q_{mo}P_{mo}} = \frac{(q_{mi}P_{mi} - q_{mo}P_{mo}) - (q_{mLe} + q_{mLi})P_{mi} - Q_m}{q_{mi}P_{mi} - q_{mo}P_{mo}} \qquad (3.57)$$

Equations (3.56) and (3.57) match the ISO 4391:1983 definitions of efficiency, which are given by

$$\eta_p(\text{ISO}) = \frac{q_{po}P_{po} - q_{pi}P_{pi}}{T_p\omega_p} \tag{3.58}$$

$$\eta_m(\text{ISO}) = \frac{T_m\omega_m}{q_{mi}P_{mi} - q_{mo}P_{mo}} \tag{3.59}$$

As mentioned at the beginning of this section, there must be a relationship between volumetric, mechanical and overall efficiencies, given that they are all related to energetic losses in the pump/motor. We deal with this matter in the following section.

3.4.4 Efficiency Relations

Consider a perfect hydrostatic pump/motor with no internal or external leakage and no mechanical losses. Under these conditions, pump and motor efficiencies become 100% and we can write Eqs. (3.58) and (3.59) as

$$1 = \frac{q_{po}P_{po} - q_{pi}P_{pi}}{T_p\omega_p} \tag{3.60}$$

$$1 = \frac{T_m\omega_m}{q_{mi}P_{mi} - q_{mo}P_{mo}} \tag{3.61}$$

Because of the fact that no leakages are being considered, we can write that $q_{po} = q_{pi} = D_p\omega_p$ and $q_{mo} = q_{mi} = D_m\omega_m$. Therefore, Eqs. (3.60) and (3.61) become

$$T_p\omega_p = D_p\omega_p(P_{po} - P_{pi}) = D_p\omega_p\Delta P_p \tag{3.62}$$

$$T_m\omega_m = D_m\omega_m(P_{mi} - P_{mo}) = D_m\omega_m\Delta P_m \tag{3.63}$$

From Eqs. (3.62) and (3.63), it follows that the torques at the pump and motor shafts for a 100% efficient machine are given by

$$\begin{cases} T_p = D_p\Delta P_p \\ T_m = D_m\Delta P_m \end{cases} \tag{3.64}$$

Note that we have now formally proven the torque relations in Eqs. (3.3) and (3.4).

An interesting result comes from the multiplication of the pump and motor volumetric and mechanical efficiencies (Eqs. (3.17), (3.31)–(3.33)):

$$\begin{cases} \eta_p^v\eta_p^m = \dfrac{(D_p\omega_p - q_{pLe} - q_{pLi})\,\Delta P_p}{\omega_p T_p|_{\text{real}}} \\[4mm] \eta_m^v\eta_m^m = \dfrac{\omega_m T_m|_{\text{real}}}{(D_m\omega_m + q_{mLe} + q_{mLi})\Delta P_m} \end{cases} \tag{3.65}$$

In Eqs. (3.65), $\omega_p T_p|_{\text{real}}$ and $\omega_m T_m|_{\text{real}}$ can be identified with the right-hand sides of Eqs. (3.50) and (3.51). Therefore, we have

$$
\begin{cases}
\eta_p^v \eta_p^m = \dfrac{(D_p \omega_p - q_{pLe} - q_{pLi}) \Delta p_p}{(q_{po} p_{po} - q_{pi} p_{pi}) + (q_{pLe} + q_{pLi}) p_{po} + Q_p} \\[4mm]
\eta_m^v \eta_m^m = \dfrac{(q_{mi} p_{mi} - q_{mo} p_{mo}) - (q_{mLe} + q_{mLi}) p_{mi} - Q_m}{(D_m \omega_m + q_{mLe} + q_{mLi}) \Delta p_m}
\end{cases}
\tag{3.66}
$$

where the simplifying assumption $e = p$ has been used.

In Eqs. (3.66), we can identify the terms $D_p \omega_p - q_{pLe} - q_{pLi}$ and $D_m \omega_m + q_{mLe} + q_{mLi}$ with the pump output flow, q_{po}, and the motor input flow, q_{mi}, respectively. If we also make $\Delta p_p = p_{po} - p_{pi}$ and $\Delta p_m = p_{mi} - p_{mo}$, we obtain for $\eta_p^v \eta_p^m$ and $\eta_m^v \eta_m^m$:

$$
\begin{cases}
\eta_p^v \eta_p^m = \dfrac{q_{po} p_{po} - q_{po} p_{pi}}{(q_{po} p_{po} - q_{pi} p_{pi}) + (q_{pLe} + q_{pLi}) p_{po} + Q_p} \\[4mm]
\eta_m^v \eta_m^m = \dfrac{(q_{mi} p_{mi} - q_{mo} p_{mo}) - (q_{mLe} + q_{mLi}) p_{mi} - Q_m}{q_{mi} p_{mi} - q_{mi} p_{mo}}
\end{cases}
\tag{3.67}
$$

Now, let us compare Eqs. (3.67) with Eqs. (3.56) and (3.57), which give the pump and motor overall efficiencies. Note that the denominators of the expressions for η_p and $\eta_p^v \eta_p^m$ are the same. The same applies to the numerators of the expressions for η_m and $\eta_m^v \eta_m^m$. We also observe that the following inequalities hold:

$$
\begin{cases}
q_{po} p_{po} - q_{pi} p_{pi} < q_{po} p_{po} - q_{po} p_{pi}, & \text{because } q_{pi} > q_{po} \\[2mm]
q_{mi} p_{mi} - q_{mo} p_{mo} > q_{mi} p_{mi} - q_{mi} p_{mo}, & \text{because } q_{mi} > q_{mo}
\end{cases}
\tag{3.68}
$$

The inequalities (3.68), together with Eqs. (3.67), (3.56) and (3.57), lead us to conclude that

$$
\begin{cases}
\eta_p^v \eta_p^m > \eta_p \\[2mm]
\eta_m^v \eta_m^m < \eta_m
\end{cases}
\tag{3.69}
$$

Expressions (3.69) are very important because they relate the overall efficiencies to the volumetric and mechanical efficiencies of a hydrostatic pump or motor. Strictly speaking, we see that by multiplying the mechanical and volumetric efficiencies of a pump we are overestimating its overall efficiency, while in a motor the overall efficiency is underestimated. However, we know that whenever a pump is tested for performance, it is usual to connect the low-pressure (suction) port to the tank so that the products $q_{pi} p_{pi}$ and $q_{po} p_{pi}$ become very low when compared to $q_{po} p_{po}$. The same happens to the motor when in connection to the discharge port. With these considerations, we can write the following approximate equations:

$$
\begin{cases}
q_{po} p_{po} - q_{pi} p_{pi} \cong q_{po} p_{po} - q_{po} p_{pi} \cong q_{po} p_{po} \\[2mm]
q_{mi} p_{mi} - q_{mo} p_{mo} \cong q_{mi} p_{mi} - q_{mi} p_{mo} \cong q_{mi} p_{mi}
\end{cases}
\tag{3.70}
$$

From Eqs. (3.70), (3.67), (3.56) and (3.57), we approximate η_p and η_m as:

$$\begin{cases} \eta_p = \eta_p^v \eta_p^m \\ \eta_m = \eta_m^v \eta_m^m \end{cases} \tag{3.71}$$

Equations (3.71) have been widely accepted as a means of interrelating mechanical, volumetric and overall efficiencies, and constitute a reasonable and very useful approximation.

3.5 Modelling Pump and Motor Efficiencies

Generally, pump and motor manufacturers provide volumetric and overall efficiency charts in their catalogues. The efficiency curves are usually obtained for a set of fixed values of some important parameters, such as the pressure differential between the pump/motor input and output ports, for example.

Efficiency charts are very important in the design of any hydraulic circuit since they equip the designer with the tools to choose the best pump/motor for each situation. However, it is frequently more useful to have a mathematical relationship, for example between efficiency and the angular speed or pressure than to have the same information in a graphical form. Given that efficiency equations are not in general provided by the manufacturers, we must approximate the graphical data with a suitable interpolation equation. Two paths can be followed: we can perform a convenient data fitting (using a generic polynomial function), or we can try to adjust a pre-existing mathematical model to the experimental data. Although the second option will also require some data fitting in the end, it gives us a better physical insight into the problem and also verifies the potential validity of the adopted mathematical model. Despite the fact that there are a considerable number of different mathematical models available today (see Section 2.2.5), in this book we choose a simple approach and use Eqs. (2.95) and (2.100) to develop general expressions for pump and motor efficiencies. We begin by having a look at the way efficiency data are presented by pump and motor manufacturers.

3.5.1 Performance Curves

Pump and motor manufacturers usually publish experimental efficiency data for their products in their catalogues. Information regarding volumetric and overall efficiency curves is most common, although it is possible to find information about mechanical efficiency as well. In any case, we can always obtain the mechanical efficiencies from volumetric and overall efficiency data using Eqs. (3.71).

Figure 3.39 shows three typical sets of volumetric efficiency curves provided by two different pump/motor manufacturers [24–26]. In the first set (Figure 3.39(a)), the volumetric efficiencies of a Parker-Hannifin F11-5 motor [24] are plotted against the angular speed for two pressure differentials, Δp. In the second and third sets of efficiency curves (Figures 3.39(b) and (c)), the volumetric and overall efficiencies of a Sauer-Danfoss Series

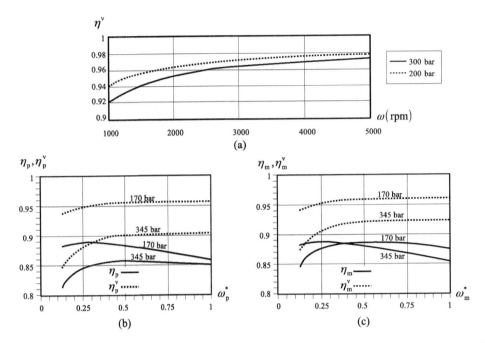

Figure 3.39 Typical volumetric efficiency curves: (a) bent-axis pump/motor (reproduced courtesy of Parker-Hannifin Corp.); (b, c) axial piston pump and motor (reproduced courtesy of Danfoss Power Solutions)

40 pump (a) and motor (b) [25, 26] are given as a function of the ratio between the actual shaft speed and a referential rated speed[29] value, ω^*.

Sometimes, manufacturers include the actual input/output flows in their catalogues instead of giving volumetric efficiency values. For instance, Figure 3.40 shows a Bosch-Rexroth N series external gear motor [27] chart in which the motor input flow is given as a function of the angular speed of the motor shaft for different motor sizes (displacements). If needed, efficiencies can be obtained for every point in the curves by dividing the values of the ideal flow (angular speed times displacement) by the corresponding flow value at the ordinate axis.

Another way of presenting the overall efficiencies is shown in Figure 3.41, where a set of level curves for the same pump represented in Figure 3.39(b) is shown. This way of showing data is useful when we want to select a region of maximum efficiency where we want our pump or motor to operate. For example, despite the fact that Figure 3.41 refers to the same pump represented in Figure 3.39(b), it is in Figure 3.41 that we can best visualize the pressure and speed ranges for which the overall efficiency is greater than 88%.

Most of the time, we need additional information to be able to effectively use the performance curves. For example, we need to know the value of the rated speeds in Figures 3.39(b) and (c) and in Figure 3.41 to be able to visualize how the efficiencies change with the shaft speed. Moreover, if a volumetric efficiency curve is given, it is necessary to know

---

[29] The ISO 4391:1983 Standards define rated conditions as the 'steady state conditions for which a component or system is recommended as a result of specified testing'. The rated characteristics are, in general, given in the manufacturer catalogues.

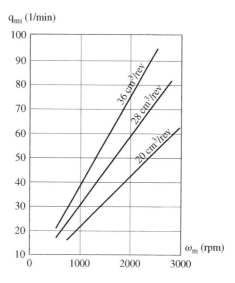

Figure 3.40 Input flow as a function of the angular speed for an external gear motor (reproduced courtesy of Bosch-Rexroth Corp.)

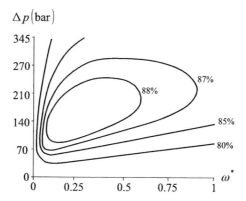

Figure 3.41 Overall efficiency level curves for the Sauer-Danfoss series 40 axial piston pump (reproduced courtesy of Danfoss Power Solutions)

the pump/motor displacements to obtain the actual output/input flows for a given pressure differential and speed.[30]

Table 3.3 contains additional information for the Sauer-Danfoss Series 40 pumps and motors whose performance curves have been represented in Figures 3.39(b) and (c), and 3.41. These pump and motor models will be used as a basis for the examples and case studies of this book.[31]

In Table 3.3, ω_R is the rated speed, p_{cont} is the maximum continuous pressure and p_{max} is the maximum instantaneous pressure.

---

[30] See Eqs. (3.15) and (3.32) for the relations between the volumetric efficiencies and the actual output/input flows.

[31] There is no particular reason for choosing these pump and motor models other than convenience. Thus, the same design procedures used in the examples to come can be applied to any pump and motor from any manufacturer, as long as performance curves are available.

Table 3.3 Danfoss series 40 pumps and motors

	Pumps							
Model	M25PV	M35PV	M44PV	M46PV	M25PT	M35PT	M44PT	M46PT
Configuration	Single variable pump				Tandem variable pump			
D_p (cm^3/rev)	24.6	35.0	43.5	46.0	24.6×2	35.0×2	43.5×2	46.0×2
ω_R (rpm)	4000	3600	3300	4000	4000	3600	3300	4000
	Motors							
Model	M25MF	M35MF	M44MF	–	–	M35MV	M44MV	M46MV
Configuration	Fixed displacement					Variable displacement		
D_m (cm^3/rev)	25.0	35.0	44.0	–	–	35.0	44.0	46.0
ω_R (rpm)	4000	3600	3300	–	–	3600	3300	4000
J_0 (kg/m^{-1})	0.0018	0.0033	0.0032	–	–	0.0033	0.0032	0.0050
	Common features							
P_{cont}	210 bar							
P_{max}	354 bar							

Reproduced courtesy of Danfoss Power Solutions

   In the following sections, we show how to express the data given in performance curves through mathematical equations. The student should be aware that this is not an easy task, and there is no single approach to the problem. Nevertheless, as already mentioned in the introduction of this section, we have chosen a relatively simple approach that is sufficient for the purposes of this book and for many practical problems involving hydrostatic pumps and motors.[32] The student should be able to apply the same procedures to other pump/motor models found in the market.

3.5.2   Volumetric Efficiency Modelling

3.5.2.1   Basic Equations

If we assume that the pressure differentials between the high- and low-pressure ports and between the high-pressure port and the case are approximately the same, we can write the total leakage flow at the pump/motor, q_L, as a function of the pressure differential, Δp (see Eq. (2.95)):

$$q_L = K_s \left(\frac{\Delta p}{\mu} \right) \tag{3.72}$$

In Eq. (3.72), we have represented the leakages by the general variable q_L, instead of using different notations for the pump and the motor (q_{mL} and q_{pL}). The same has been adopted for K_s and Δp. The reason is to simplify the notation as much as possible to avoid unnecessary repetitions. It will also be clear from the context whether we are referring to a pump or a motor.

---

[32] Modelling pump and motor efficiencies has been the subject of much research over the years. In this aspect, it is not by any means a universally established theme (see, e.g. Refs. [1, 28, 29] and references therein). Therefore, the student who wants to pursue further knowledge on the matter is encouraged to consult the references quoted within the text. Our intention is to present only one line of thought among the many different approaches. As will be seen along the way, there will always be room for improvement as no specific theory can be regarded as perfect. In any case, as observed in Figures 3.39–3.41, the worst choice to make is to assume pump and motor efficiencies as constants.

By substituting q_L given by Eq. (3.72) into Eqs. (3.15) and (3.32), after some mathematical manipulations, we obtain

$$\eta_p^v = 1 - \varphi \left(\frac{\Delta p^*}{D^* \omega^*} \right) \tag{3.73}$$

$$\eta_m^v = \frac{1}{1 + \varphi \left(\dfrac{\Delta p^*}{D^* \omega^*} \right)} \tag{3.74}$$

In Eqs. (3.73) and (3.74), D^*, ω^* and Δp^* are relative values of the displacement, speed and pressure differential[33]:

$$\Delta p^* = \frac{\Delta p}{\Delta p_r}, \quad \omega^* = \frac{\omega}{\omega_r} \quad \text{and} \quad D^* = \frac{D}{D_r} \tag{3.75}$$

where Δp_r, ω_r and D_r are the reference values of the pressure differential, angular speed and displacement. Typical choices are the maximum design pressure (Δp_{max}), the rated speed (ω_R) and the maximum displacement (D_{max}). The coefficient, φ, on the other hand, is given by

$$\varphi = \frac{K_s \Delta p_r}{\mu D_r \omega_r} \tag{3.76}$$

Remarks

1. The coefficient K_s depends on the pressure differential, Δp, in a very complex manner (see remarks after Eq. (2.100)). In fact, the pressure differential affects the inner forces between the moving parts, which causes strains that end up affecting the clearance dimensions. In the end, it is very difficult to precisely calculate the values of the involved deformations, and we usually need to recur to elaborate numerical methods to approach the problem.
2. The viscosity of the fluid (μ) depends on the temperature at the pump and motor leaks. We know that temperature elevations are caused by the inner friction at the fluid itself and between the moving parts, which relates directly to the angular speed of the shaft, ω. As a result, the fluid viscosity ultimately depends on the angular speed. Note, however, that the relation between μ and ω is not easy to determine because of the complex processes involved in heat generation inside the tiny pump and motor gaps.
3. Typically, manufacturers test variable-displacement pumps and motors at their maximum displacement. As a result, we can make $D^* = 1$ in Eqs. (3.73) and (3.74) when making comparisons with catalogue data.

---

[33] Note that the product must be positive for Eqs. (3.73) and (3.74) to make sense. It is possible, however, that negative displacements associated with a flow inversion arise in numerical simulations; in such cases, the absolute value of $D^* \omega^*$ should be used instead.

4. Variable-displacement units change their mechanical configuration with the displacement, D^*. The mechanical configuration, on the other hand, directly influences the force balances between the moving parts, and as a result the coefficient K_s in Eq. (3.76) should also be affected. Therefore, strictly speaking, the coefficient φ in Eq. (3.76) should be a function of Δp^*, ω^* and D^* [28]. Interestingly, the influence of displacement on volumetric losses is usually minimal [29] and may be disregarded.

Given that the coefficient φ in Eqs. (3.73) and (3.74) was not constant, McCandlish and Dorey [29] suggested a data-fitting procedure to match the theory to the experimental results. The basic idea was to assume that φ could change along and between constant–pressure curves (see Figures 3.39(a) and (b)). We explain this method in the following section.

3.5.2.2 Data Fitting

Figure 3.42 shows a typical volumetric efficiency chart for a hydrostatic pump/motor, where the efficiency has been plotted against the relative angular speed, ω^*, for two relative pressure differentials between the input and output ports, Δp_A^* and Δp_B^* ($\Delta p_B^* > \Delta p_A^*$). As mentioned earlier, these curves are usually obtained for $D^* = 1$. Therefore, with the coordinates of each point 'i' on the efficiency curves, the correspondent coefficient φ_i can be easily calculated with the help of Eqs. (3.73) and (3.74).

The procedure used here can be described as follows. First, we consider the variation of φ with the relative speed ω^* along a constant–pressure curve. A linear relation could be used:

$$\varphi_j = a_0 + a_1 \omega^* \tag{3.77}$$

where φ_j is taken along a constant–pressure curve whose relative pressure differential is Δp_j^* (if the notation used in Figure 3.42 is followed, we have that $j = A, B$).

In order to determine the constants a_0 and a_1 in Eq. (3.77), we substitute the experimental values of Δp^*, ω^*, D^* and η^v, obtained at two distinct points (or nodes) of a

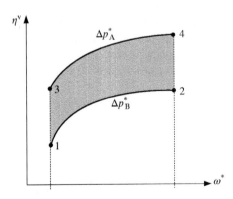

Figure 3.42   Typical volumetric efficiency curves for a hydrostatic pump/motor

constant–pressure curve, into Eqs. (3.73) and (3.74). Then we calculate the values of φ_j at the selected points, which, together with Eq. (3.77), form a simple system of linear equations on a_0 and a_1. For example, let ω_1^*, ω_2^*, φ_1 and φ_2 be the values of the relative angular speed ω^* and coefficient φ_j at nodes 1 and 2, in Figure 3.42. From Eq. (3.77), we have

$$\begin{cases} \varphi_1 = a_0 + a_1\omega_1^* \\ \varphi_2 = a_0 + a_1\omega_2^* \end{cases} \tag{3.78}$$

Solving Eqs. (3.78) for a_0 and a_1, we obtain

$$\begin{cases} a_1 = \dfrac{\varphi_1 - \varphi_2}{\omega_1^* - \omega_2^*} \\ a_0 = \varphi_1 - \left(\dfrac{\varphi_1 - \varphi_2}{\omega_1^* - \omega_2^*} \right) \omega_1^* \end{cases} \tag{3.79}$$

The next step is to account for the dependence of φ on the pressure differential. This can be done by assuming a linear variation of the coefficient φ between two pressure levels (e.g. Δp_A^* and Δp_B^* in Figure 3.42). Therefore, suppose that we have already obtained φ for each constant–pressure curve in Figure 3.42, as follows:

$$\varphi_A = a_0^A + a_1^A\omega^* \quad \text{and} \quad \varphi_B = a_0^B + a_1^B\omega^* \tag{3.80}$$

Assuming that φ grows linearly from φ_A to φ_B, we can write that (see Figure 3.43):

$$\varphi = \varphi_A + (\Delta p^* - \Delta p_A^*)\left(\frac{\varphi_B - \varphi_A}{\Delta p_B^* - \Delta p_A^*} \right) \tag{3.81}$$

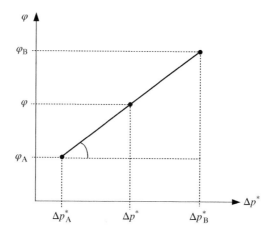

Figure 3.43 Linear interpolation of φ based on the relative pressure differential

It is possible to show that the substitution of φ_A and φ_B, obtained from Eqs. (3.80) into Eq. (3.81), gives

$$\varphi = b_0 + b_1 \omega^* + b_2 \Delta p^* + b_3 \omega^* \Delta p^* \tag{3.82}$$

where

$$\begin{cases} b_0 = \dfrac{\Delta p_B^* a_0^A - \Delta p_A^* a_0^B}{\Delta p_B^* - \Delta p_A^*} \\[3mm] b_1 = \dfrac{\Delta p_B^* a_1^A - \Delta p_A^* a_1^B}{\Delta p_B^* - \Delta p_A^*} \\[3mm] b_2 = \dfrac{a_0^B - a_0^A}{\Delta p_B^* - \Delta p_A^*} \\[3mm] b_3 = \dfrac{a_1^B - a_1^A}{\Delta p_B^* - \Delta p_A^*} \end{cases} \tag{3.83}$$

The four constants, b_0 through b_3, can then be determined by applying Eqs. (3.79) to each constant–pressure curve and substituting the resulting constants a_0^A, a_0^B, a_1^A and a_1^B into Eqs. (3.83). Alternatively, we can substitute the expression for φ, given by Eq. (3.82), into Eqs. (3.73) and (3.74) and build the following matrix equation whose elements can be obtained directly from the volumetric efficiency curves:

$$\begin{pmatrix} 1 & \omega_1^* & \Delta p_1^* & \omega_1^* \Delta p_1^* \\ 1 & \omega_2^* & \Delta p_2^* & \omega_2^* \Delta p_2^* \\ 1 & \omega_3^* & \Delta p_3^* & \omega_3^* \Delta p_3^* \\ 1 & \omega_4^* & \Delta p_4^* & \omega_4^* \Delta p_4^* \end{pmatrix} \begin{pmatrix} b_0 \\ b_1 \\ b_2 \\ b_3 \end{pmatrix} = \begin{pmatrix} c_1 \\ c_2 \\ c_3 \\ c_4 \end{pmatrix} \tag{3.84}$$

where each constant c_i ($i = 1, \dots, 4$), on the right-hand side, is given by

$$\begin{cases} c_i = \left(\dfrac{D^* \omega_i^*}{\Delta p_i^*} \right) (1 - \eta_i^v), & \text{for pumps} \\[4mm] c_i = \left(\dfrac{D^* \omega_i^*}{\Delta p_i^*} \right) \left(\dfrac{1}{\eta_i^v} - 1 \right), & \text{for motors} \end{cases} \tag{3.85}$$

Remarks

1. Equation (3.82) could have been obtained by writing the coefficient φ as the product of two independent linear functions, $f(\Delta p^*)$ and $g(\omega^*)$, as

$$\varphi = f(\Delta p^*) g(\omega^*) = (d_0 + d_1 \Delta p^*)(e_0 + e_1 \omega^*) \tag{3.86}$$

2. The idea of splitting the unknown coefficients into separate functions of the independent variables was proposed in Ref. [29]. Different interpolation functions can be used, depending on the pump and motor designs. For instance, for gear pumps, it was reported that the following relation (n is a constant) was appropriate [28, 29]:

$$\varphi = d_0(\Delta p^*)^n[e_0 + e_1\omega^* + e_2(\omega^*)^2] \tag{3.87}$$

Strictly speaking, the volumetric efficiency models given by Eqs. (3.73), (3.74), (3.82) and (3.83) should be valid only within the region where the coefficient φ was interpolated. In our case, this would correspond to the grey area in Figure 3.42. However, in some situations, the behaviour of the dependent variable is very predictable, to the extent that we might feel confident to use our models even outside the interpolation region. This is the case with the general volumetric efficiency curves shown in Figure 3.42 where we see that the two curves show a very similar progression, indicating that volumetric efficiency both decreases with pressure and increases with speed. Such behaviour can also be observed in the actual volumetric efficiency curves shown in Figures 3.39(b) and (c).

Using efficiency models outside the interpolation range must be done with care, and we should always prefer to obtain as much experimental data as possible [29]. However, depending on the actual trend of the experimental curves and the experience of the designer, it is in fact possible to obtain valuable information from the theoretical model outside the catalogued pressure and speed limits.

3.5.2.3 Case Study

Consider the volumetric efficiency curves shown in Figures 3.39(b) and (c) and reproduced again in Figure 3.44 along with the coordinates of the nodes[34] 1, 2, 3 and 4.

For the sake of comparison, let us first assume that φ is constant in Eqs. (3.73) and (3.74). Therefore, a single value of ω^*, η^v and Δp^* is needed in each case. Making $D^* = 1$ in Eqs. (3.73) and (3.74), we obtain

$$\begin{cases} \varphi = \left(\dfrac{\omega^*}{\Delta p^*}\right)(1 - \eta^v), & \text{for pumps} \\[4mm] \varphi = \left(\dfrac{\omega^*}{\Delta p^*}\right)\left(\dfrac{1}{\eta^v} - 1\right), & \text{for motors} \end{cases} \tag{3.88}$$

For instance, if we substitute the coordinates of points 1 in Figures 3.44(a) and (b) into Eqs. (3.88), we have

$$\begin{cases} \varphi = (1 - 0.847)\left(\dfrac{0.125}{345/345}\right) = 0.01913\,, & \text{for pumps} \\[4mm] \varphi = \left(\dfrac{1}{0.870} - 1\right)\left(\dfrac{0.125}{345/345}\right) = 0.01868\,, & \text{for motors} \end{cases} \tag{3.89}$$

where we have written Δp^* explicitly as $\Delta p^* = \Delta p/345$.

---

[34] The node coordinates have been obtained by visual inspection.

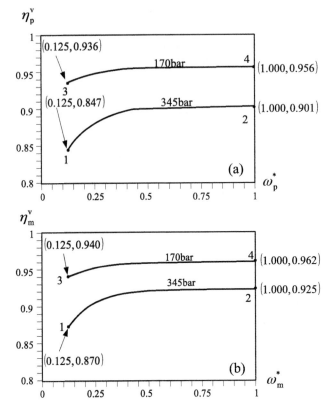

Figure 3.44 Data selection on the volumetric efficiency curves for the Sauer-Danfoss series 40 (a) pump and (b) motor

The resulting efficiency models for the pump and the motor become

$$\begin{cases} \eta_p^v = 1 - \dfrac{0.01913\Delta p^*}{D^*\omega^*} = 1 - \dfrac{0.01913\Delta p}{D^*\omega^*345} \\ \eta_m^v = \dfrac{1}{1 + \dfrac{0.01868\Delta p^*}{D^*\omega^*}} = \dfrac{1}{1 + \dfrac{0.01868\Delta p}{D^*\omega^*345}} \end{cases} \tag{3.90}$$

Observe that we have chosen to write the final model as a function of the pressure differential, Δp, instead of the relative pressure differential, Δp^*, in Eqs. (3.90).

Figure 3.45 shows the efficiency curves obtained from Eqs. (3.90) for the pump and the motor operating at maximum displacement ($D^* = 1$). We also show the original curves for comparison. Observe that, although the theoretical models were able to predict the overall

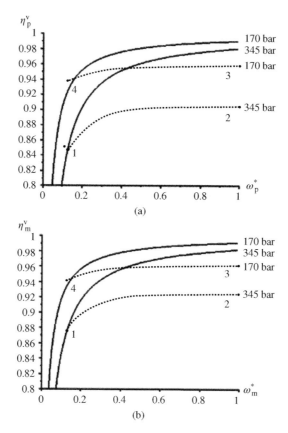

Figure 3.45 Comparison between the constant coefficient model (solid curves) and the experimental data (dashed curves) for (a) pumps and (b) motors

behaviour of the efficiency curves, the results only matched the experimental data at point 1. This suggests that the variation of φ with ω^* and Δp^* cannot be disregarded and a higher degree approximation should be used instead.

Let us now use Eq. (3.82) to obtain the coefficient φ and write a model for the volumetric efficiencies. The coefficients b_0 through b_3 can be obtained through solving the matrix equation (3.84) or by direct substitution of the nodal values of ω^*, η^v and Δp^* into Eqs. (3.79) and (3.83). Let us choose the later method in this example. Here are the steps:

1. Obtain the values of the coefficient φ at each selected node on the volumetric efficiency curves for the pump and the motor. By isolating φ in Eqs. (3.73) and (3.74) and entering the nodal values shown in Figure 3.44, note that Δp^* in Eqs. (3.73) and (3.74) can be

either 0.493 (170/345) or 1 (345/345) depending on the constant–pressure curve considered. For the pump, we have

$$
\begin{cases}
\varphi_1 = \left(\dfrac{0.125}{1}\right)(1 - 0.847) = 0.019 \\[2mm]
\varphi_2 = \left(\dfrac{1}{1}\right)(1 - 0.901) = 0.099 \\[2mm]
\varphi_3 = \left(\dfrac{0.125}{0.493}\right)(1 - 0.936) = 0.016 \\[2mm]
\varphi_4 = \left(\dfrac{1}{0.493}\right)(1 - 0.956) = 0.089
\end{cases}
\tag{3.91}
$$

and for the motor:

$$
\begin{cases}
\varphi_1 = \left(\dfrac{0.125}{1}\right)\left(\dfrac{1}{0.870} - 1\right) = 0.019 \\[2mm]
\varphi_2 = \left(\dfrac{1}{1}\right)\left(\dfrac{1}{0.925} - 1\right) = 0.081 \\[2mm]
\varphi_3 = \left(\dfrac{0.125}{0.493}\right)\left(\dfrac{1}{0.940} - 1\right) = 0.016 \\[2mm]
\varphi_4 = \left(\dfrac{1}{0.493}\right)\left(\dfrac{1}{0.962} - 1\right) = 0.080
\end{cases}
\tag{3.92}
$$

2. With the values of φ_1 through φ_4 given by (3.91) and (3.92) and using Eqs. (3.79), we calculate the constants a_0^A, a_1^A, a_0^B and a_1^B, needed for Eqs. (3.83). In order to avoid unnecessary repetitions (the whole process can be easily done with the help of a spreadsheet program), we only show how to calculate the constants a_0^A and a_1^A, relative to the curve at 345 bar ($\Delta p^* = 1$) for the pump. Thus, from Eqs. (3.79) and (3.91), we have

$$
\begin{cases}
a_1^A = a_1^{345} = \dfrac{0.019 - 0.099}{0.125 - 1} = 0.091 \\[2mm]
a_0^A = a_0^{345} = 0.019 - 0.091 \times 0.125 = 0.008
\end{cases}
\tag{3.93}
$$

If the same procedure is repeated for the curve at 170 bar, we obtain $a_0^B = a_0^{170} = 0.006$ and $a_1^B = a_1^{170} = 0.083$. The same can be done for the motors, where the constants will be $a_0^A = a_0^{345} = 0.010$, $a_1^A = a_1^{345} = 0.071$, $a_0^B = a_0^{170} = 0.007$ and $a_1^B = a_1^{170} = 0.073$.

3. Finally, we substitute all the constants calculated in the last item into Eqs. (3.83) to obtain b_0 through b_3. In the end, the following interpolation equations can be written for the pump and the motor, respectively (Eq. (3.82)):

$$
\begin{cases}
\varphi = 0.0039 + 0.0759\omega^* + 0.0038\Delta p^* + 0.0154\omega^*\Delta p^*, & \text{for pumps} \\[2mm]
\varphi = 0.0044 + 0.0749\omega^* + 0.0053\Delta p^* - 0.0035\omega^*\Delta p^*, & \text{for motors}
\end{cases}
\tag{3.94}
$$

Equations (3.94) together with Eqs. (3.73) and (3.74) constitute the modelling equations for the volumetric efficiencies.

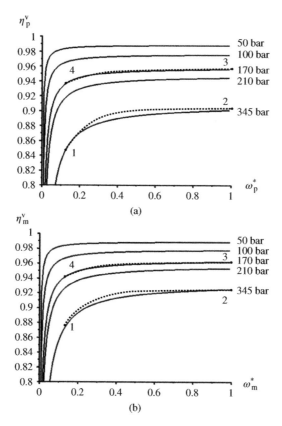

Figure 3.46 Extrapolated efficiency curves and comparison between the theoretical model (solid curves) and the experimental data (dashed curves) for (a) pumps and (b) motors

Figure 3.46 shows the theoretical efficiency curves for the pump and the motor, obtained from Eqs. (3.73), (3.74) and (3.94), for some selected pressure differentials at maximum displacement ($D^* = 1$). We have also included the original curves for $\Delta p = 170\,\text{bar}$ and $\Delta p = 345\,\text{bar}$ shown in Figure 3.44 for comparison. In contrast to the constant φ model, the theoretical curves now show a strong agreement with the experimental data provided by the manufacturer. The curves also show the extrapolated values for ω^* and Δp^* outside the original interpolation region limited by points 1, 2, 3 and 4.

Figure 3.47(a) shows the extrapolated values of the pump volumetric efficiencies at low relative speeds for pressure differentials of 170 and 345 bar. We see that in both cases, the volumetric efficiencies become zero below a minimal angular speed of the pump shaft. This can be explained by the fact that there is a point at which all the flow produced by the pump gets lost through the leakages. In such case, no flow is output by the pump, resulting in 'dead zones', as indicated in the figure.[35] Given that leakages increase with pressure, the dead

---

[35] It is important to note that efficiency is a function of the pump flow and not of the speed alone. In this aspect, the same dead zones observed in Figure 3.47(a) would also be observed for small pump displacements in a constant–speed pump.

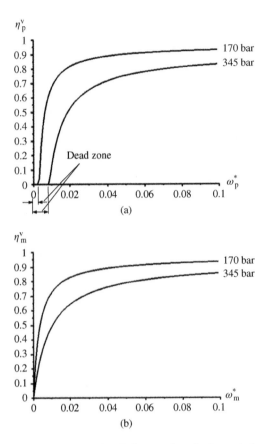

Figure 3.47 Extrapolation of the volumetric efficiency values for low relative speeds of the (a) pump and the (b) motor

zone is obviously larger for higher pressure differentials. Similarly, Figure 3.47(b) shows the extrapolated values of the motor volumetric efficiencies. Note that in contrast to those of pumps, the motor efficiency curves pass through the origin at every pressure differential. We can conclude from this fact that if the motor shaft stops ($\omega_m^* = 0$), there will still be leakage losses ($q_{mL} > 0$). In fact, from Eq. (3.32), we have

$$\lim_{\omega_m^* \to 0}(\eta_m^v) = \lim_{\omega_m^* \to 0}\left(\frac{D_m\omega_m^*\omega_r}{D_m\omega_m^*\omega_r + q_{mL}}\right) = \frac{0}{0 + q_{mL}} = 0 \qquad (3.95)$$

3.5.3 Overall Efficiency Modelling

3.5.3.1 Basic Equations

We have already mentioned at the beginning of Section 3.5 that pump and motor manufacturers usually show volumetric and overall efficiencies in their catalogues. On the other hand, when modelling pump and motor overall efficiencies, we should begin by writing a suitable equation for the mechanical efficiencies as functions of the pressure differentials and the shaft speeds. Consider, therefore, the mechanical efficiencies of pump and motor, given by Eqs. (3.31) and (3.33), respectively. Note that we can write the actual torques at the pump and motor, $T_p|_{real}$ and $T_m|_{real}$, in terms of the torque losses, T_L, as follows:

$$\begin{cases} T_p\Big|_{real} = \Delta pD + T_L \\ T_m|_{real} = \Delta pD - T_L \end{cases} \tag{3.96}$$

Substituting T_L given by Eq. (2.100) into Eqs. (3.96), we obtain

$$\begin{cases} T_p\Big|_{real} = \Delta pD + K_\omega(\mu\omega) + T_f \\ T_m|_{real} = \Delta pD - K_\omega(\mu\omega) - T_f \end{cases} \tag{3.97}$$

Now, from Eqs. (3.31), (3.33) and (3.97), we write

$$\eta_p^m = \frac{\Delta pD}{\Delta pD + K_\omega(\mu\omega) + T_f} = \frac{1}{1 + \dfrac{K_\omega(\mu\omega)}{\Delta pD} + \dfrac{T_f}{\Delta pD}} \tag{3.98}$$

$$\eta_m^m = \frac{\Delta pD - K_\omega(\mu\omega) - T_f}{\Delta pD} = 1 - \frac{K_\omega(\mu\omega)}{\Delta pD} - \frac{T_f}{\Delta pD} \tag{3.99}$$

Equations (3.98) and (3.99) can be written as functions of the relative speed, ω^*, displacement, D^* and pressure differential, Δp^*, as

$$\eta_p^m = \frac{1}{1 + \phi_\omega\left(\dfrac{\omega^*}{\Delta p^*D^*}\right) + \dfrac{\phi_f}{\Delta p^*D^*}} \tag{3.100}$$

$$\eta_m^m = 1 - \phi_\omega\left(\frac{\omega^*}{\Delta p^*D^*}\right) - \frac{\phi_f}{\Delta p^*D^*} \tag{3.101}$$

where the coefficients ϕ_ω and ϕ_f are given by

$$\phi_\omega = \frac{K_\omega\mu\omega_r}{D_r\Delta p_r} \quad \text{and} \quad \phi_f = \frac{T_f}{\Delta p_rD_r} \tag{3.102}$$

The overall efficiencies of the pump and the motor can then be obtained from Eqs. (3.100), (3.101) and (3.71):

$$\eta_p = \frac{\eta_p^v}{1 + \phi_\omega\left(\dfrac{\omega^*}{\Delta p^*D^*}\right) + \dfrac{\phi_f}{\Delta p^*D^*}} \tag{3.103}$$

$$\eta_m = \eta_m^v\left[1 - \phi_\omega\left(\frac{\omega^*}{\Delta p^*D^*}\right) - \frac{\phi_f}{\Delta p^*D^*}\right] \tag{3.104}$$

Remarks

1. It has been reported that torque losses reduce with displacement [29]. This makes sense, for example, in the case of axial piston pumps and motors, where we see that the friction between the pistons and the cylinder block increases with the swashplate angle, α (see Figure 3.24). As a result, the coefficient ϕ_ω should vary with D^*. Moreover, such variation is not insignificant. For instance, it has been shown that at small displacements, torque losses can be reduced as much as 50% in relation to the torque losses at full displacement [29].

2. Pump and motor manufacturers do not typically provide performance curves at different displacements in their catalogues. However, we know from the last remark that mechanical losses are usually expected to reach a maximum at maximum displacement. Thus, since we are modelling the overall efficiency based on the maximum displacement, we can expect our final equations to give the smallest overall efficiency possible within the displacement range. In other words, modelling pump and motor overall efficiencies based on $D^* = 1$ will underestimate the actual overall efficiency at $D^* < 1$, which is, in the long run, a safe approach.

3. The coefficient ϕ_f depends on the external load acting on the pump/motor shaft, and since the external load affects the pressure differential between the input and output ports of the pump/motor (see Eqs. (3.3) and (3.4)), it is reasonable to assume that ϕ_f depends on Δp^* as well. In this book, we consider that ϕ_f is constant at a constant–pressure curve.

Similar to the coefficient φ_j, in Eq. (3.77), ϕ_ω in Eqs. (3.103) and (3.104) can be written as a function of the relative speed, ω^*, for each pressure level Δp_j^*. On the other hand, we have also assumed that ϕ_f is constant along a constant–pressure curve. Therefore, we may use the following approximations for ϕ_ω and ϕ_f at $\Delta p^* = \Delta p_j^*$:

$$\begin{cases} \phi_{\omega j} = a_0 + a_1\omega^* + a_2(\omega^*)^2 \\ \phi_{fj} = b_0 \end{cases} \qquad (3.105)$$

where $\phi_{\omega j}$ and ϕ_{fj} are the values of the coefficients ϕ_ω and ϕ_f at the pressure level $\Delta p^* = \Delta p_j^*$ and a_0 through b_0 are constants. Note that because of the relatively higher complexity of the overall efficiency model, we have chosen a quadratic approximation for ϕ_ω instead of the linear interpolation we had used for φ_j, in Eq. (3.77).

Substituting $\phi_{\omega j}$ and ϕ_{fj} given by Eqs. (3.105) into the overall efficiency definitions (3.103) and (3.104), we obtain

$$\eta_p = \frac{\eta_p^v}{1 + \dfrac{[a_0 + a_1\omega^* + a_1(\omega^*)^2]\omega^*}{\Delta p^* D^*} + \dfrac{b_0}{\Delta p^* D^*}} \qquad (3.106)$$

$$\eta_m = \eta_m^v \left\{ 1 - \frac{[a_0 + a_1\omega^* + a_1(\omega^*)^2]\,\omega^*}{\Delta p^* D^*} - \frac{b_0}{\Delta p^* D^*} \right\} \qquad (3.107)$$

Considering that the values of ω^*, Δp^*, η^v and η are known at node 'i' (i.e. ω_i^*, Δp_i^*, η_i^v and η_i) and selected on a constant–pressure curve, $\Delta p^* = \Delta p_j^*$, it is convenient to rewrite Eqs. (3.106) and (3.107) as

$$a_0 \omega_i^* + a_1 (\omega_i^*)^2 + a_2 (\omega_i^*)^3 + b_0 = \Delta p_i^* c_i \tag{3.108}$$

where

$$\begin{cases} c_i = D^* \left(\dfrac{\eta_i^v}{\eta_i} - 1 \right), & \text{for pumps} \\[4mm] c_i = D^* \left(1 - \dfrac{\eta_i}{\eta_i^v} \right), & \text{for motors} \end{cases} \tag{3.109}$$

Four nodes are needed to determine the constants a_0, a_1, a_2 and b_0 in Eq. (3.108). Let us label these nodes 1, 2, 3 and 4 and assume that we have already modelled the volumetric efficiencies so that η_i^v is known at each point. We can then write the following matrix equation, based on Eqs. (3.108) and (3.109):

$$\begin{bmatrix} \omega_1^* & (\omega_1^*)^2 & (\omega_1^*)^3 & 1 \\ \omega_2^* & (\omega_2^*)^2 & (\omega_2^*)^3 & 1 \\ \omega_3^* & (\omega_3^*)^2 & (\omega_3^*)^3 & 1 \\ \omega_4^* & (\omega_4^*)^2 & (\omega_4^*)^3 & 1 \end{bmatrix} \begin{pmatrix} a_0 \\ a_1 \\ a_2 \\ b_0 \end{pmatrix} = \Delta p_j^* \begin{pmatrix} c_1 \\ c_2 \\ c_3 \\ c_4 \end{pmatrix} \tag{3.110}$$

Equation (3.110) can be solved for the constants a_0 through b_0, which can then be substituted into Eqs. (3.105) to obtain the coefficients ϕ_ω and ϕ_f, relative to the pressure differential Δp_j^*. The next step is to perform another linear interpolation on Δp^* between the pressure levels Δp_A^* and Δp_B^* as follows (see Eq. (3.81) for a similar situation):

$$\begin{cases} \phi_\omega = \phi_{\omega A} + (\Delta p^* - \Delta p_A^*) \left(\dfrac{\phi_{\omega B} - \phi_{\omega A}}{\Delta p_B^* - \Delta p_A^*} \right) \\[4mm] \phi_f = \phi_{fA} + (\Delta p^* - \Delta p_A^*) \left(\dfrac{\phi_{fB} - \phi_{fA}}{\Delta p_B^* - \Delta p_A^*} \right) \end{cases} \tag{3.111}$$

From Eqs. (3.110) and (3.105), we obtain the coefficients $\phi_{\omega A}$, $\phi_{\omega B}$, ϕ_{fA} and ϕ_{fB} for two individual constant–pressure curves at Δp_A^* and Δp_B^*, and use Eqs. (3.111) to get the final expressions[36] for ϕ_ω and ϕ_f.

3.5.3.2 Case Study

Consider the pumps and motors represented by the performance curves in Figures 3.39 (b) and (c). Figure 3.48 shows four nodes randomly chosen on the overall efficiency curves

---

[36] Note that the procedure for obtaining the coefficients ϕ_ω and ϕ_f is lengthy and highly prone to numerical errors. A computer program could be extremely helpful in this case.

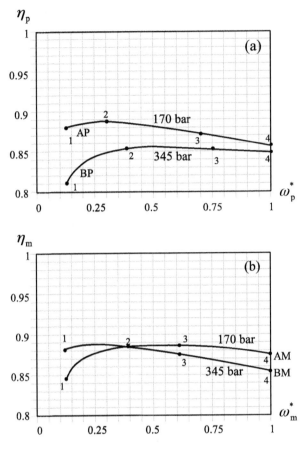

Figure 3.48 Data selection on the overall efficiency curves for the pump and motor represented in Figures 3.39(b) and (c), respectively

at 170 and 345 bar. The curves have been labelled AP and BP (pumps) and AM and BM (motors) so that they can be more easily referenced later.

The nodal coordinates for Figures 3.48(a) and (b) are listed in Table 3.4.

In what follows, we use Eqs. (3.110) and (3.105) to obtain the coefficients $\phi_{\omega j}$ and ϕ_{fj} for each curve shown in Figure 3.48 (AP, BP, AM and BM). Later, we perform a final linear interpolation on pressure using Eqs. (3.111). In order to avoid repetitive calculations, we limit ourselves to showing just part of the process, leaving the verification of the other equations to the student as an exercise.

Table 3.4 Nodal values of the overall efficiencies (Figure 3.48)

Node	AP ($\Delta p^* = 0.493$)		BP ($\Delta p^* = 1$)		AM ($\Delta p^* = 0.493$)		BM ($\Delta p^* = 1$)	
	ω^*	η	ω^*	η	ω^*	η	ω^*	η
1	0.125	0.881	0.125	0.812	0.125	0.845	0.125	0.881
2	0.300	0.888	0.380	0.855	0.390	0.885	0.390	0.885
3	0.700	0.872	0.750	0.854	0.620	0.886	0.620	0.875
4	1.000	0.859	1.000	0.850	1.000	0.875	1.000	0.854

Substituting the numerical values of ω_i^* and Δp_j^* for the curve AP into Eq. (3.110), we obtain[37]

$$
\begin{pmatrix}
0.125 & 0.016 & 0.002 & 1.000 \\
0.300 & 0.090 & 0.027 & 1.000 \\
0.700 & 0.490 & 0.343 & 1.000 \\
1.000 & 1.000 & 1.000 & 1.000
\end{pmatrix}
\begin{pmatrix}
a_0 \\ a_1 \\ a_2 \\ b_0
\end{pmatrix}
=
\begin{pmatrix}
0.031 \\ 0.034 \\ 0.047 \\ 0.056
\end{pmatrix}
\tag{3.112}
$$

The solution of Eq. (3.112) gives $a_0 = -0.0014$, $a_1 = 0.0571$, $a_2 = -0.0302$ and $b_0 = 0.0302$, which results in the following expressions for $\phi_{\omega AP}$ and ϕ_{fAP} (see Eq. (3.105)):

$$
\begin{cases}
\phi_{\omega AP} = -0.0014 + 0.0571\omega^* - 0.0302(\omega^*)^2 \\
\phi_{fAP} = 0.0302
\end{cases}
\tag{3.113}
$$

The same process can be repeated for the curve BP, in which case we obtain

$$
\begin{cases}
\phi_{\omega BP} = -0.0996 + 0.2060\omega^* - 0.0990(\omega^*)^2 \\
\phi_{fBP} = 0.0527
\end{cases}
\tag{3.114}
$$

Using Eqs. (3.111), we finally have for the pump:

$$
\begin{cases}
\phi_\omega = \phi_{\omega AP} + (\Delta p^* - 0.4928)\left(\dfrac{\phi_{\omega BP} - \phi_{\omega AP}}{0.5072}\right) \\[4mm]
\phi_f = \phi_{fAP} + (\Delta p^* - 0.4928)\left(\dfrac{\phi_{fBP} - \phi_{fAP}}{0.5072}\right)
\end{cases}
\tag{3.115}
$$

where $\phi_{\omega AP}$, ϕ_{fAP}, $\phi_{\omega BP}$ and ϕ_{fBP} are given by Eqs. (3.113) and (3.114), respectively.

The student can verify that the following equations are valid for the motor (curves AM and BM):

$$
\begin{cases}
\phi_{\omega AM} = -0.1377 + 0.2141\omega^* - 0.0957(\omega^*)^2 \\
\phi_{fAM} = 0.0639 \\
\phi_{\omega BM} = 0.2959 - 0.3328\omega^* + 0.1579(\omega^*)^2 \\
\phi_{fBM} = -0.0444
\end{cases}
\tag{3.116}
$$

Again, the final equations for ϕ_ω and ϕ_f are obtained from Eqs. (3.111):

$$
\begin{cases}
\phi_\omega = \phi_{\omega AM} + (\Delta p^* - 0.4928)\left(\dfrac{\phi_{\omega BM} - \phi_{\omega AM}}{0.5072}\right) \\[4mm]
\phi_f = \phi_{fAM} + (\Delta p^* - 0.4928)\left(\dfrac{\phi_{fBM} - \phi_{fAM}}{0.5072}\right)
\end{cases}
\tag{3.117}
$$

---

[37] The nodal values of the volumetric efficiencies needed for the coefficients, c_i, on the right-hand side of Eq. (3.110), were calculated using Eqs. (3.73) and (3.94). It can be shown that the following values are obtained for the points 1, 2, 3 and 4 of the curve AP, respectively, $\eta_1^v = 0.9361$, $\eta_2^v = 0.9494$, $\eta_3^v = 0.9548$ and $\eta_4^v = 0.9560$.

Table 3.5 Overall efficiency models for the pumps and motors represented in Figure 3.39

Equation	Pump	Motor
$\phi_{\omega A}, \phi_{fA}$	$\phi_{\omega A} = -0.0014 + 0.0571\omega^* - 0.0302(\omega^*)^2$ $\phi_{fA} = 0.0302$	$\phi_{\omega A} = -0.1377 + 0.2141\omega^* - 0.0957(\omega^*)^2$ $\phi_{fA} = 0.0639$
$\phi_{\omega B}, \phi_{fB}$	$\phi_{\omega B} = -0.0996 + 0.2060\omega^* - 0.0990(\omega^*)^2$ $\phi_{fB} = 0.0527$	$\phi_{\omega B} = 0.2959 - 0.3328\omega^* + 0.1579(\omega^*)^2$ $\phi_{fB} = -0.0444$
ϕ_ω, ϕ_f	$$\phi_\omega = \phi_{\omega A} + (\Delta p^* - 0.4928)\left(\frac{\phi_{\omega B} - \phi_{\omega A}}{0.5072}\right)$$ $$\phi_f = \phi_{fA} + (\Delta p^* - 0.4928)\left(\frac{\phi_{fB} - \phi_{fA}}{0.5072}\right)$$	
η	$$\eta = \frac{\eta^v}{1 + \dfrac{\phi_f}{D^*\Delta p^*} + \phi_\omega\left(\dfrac{\omega^*}{D^*\Delta p^*}\right)}$$	$$\eta = \eta^v\left[1 - \frac{\phi_f}{D^*\Delta p^*} - \phi_\omega\left(\frac{\omega^*}{D^*\Delta p^*}\right)\right]$$

Table 3.5 summarizes the modelling equations for the overall efficiencies of the pump and the motor represented in Figures 3.49(a) and (b).

Using the modelling equations listed in Table 3.5, we can plot the overall efficiency curves for the pump and the motor at 170 and 345 bar, as seen in Figure 3.49. We have also included the original efficiency curves (dashed lines) for comparison where we see a reasonably good agreement between the theoretical model and the catalogue data at the interpolated curves.

Figure 3.50 shows the efficiency level curves for the pump. Although the agreement between theory and experiment cannot be considered absolutely perfect, we still see that it is very acceptable. Most of all, the pattern of the efficiency curves obtained from the equations in Table 3.5 is correct, which indicates that the theoretical model is able to reproduce the results in a coherent manner.

3.5.4 Mechanical Efficiency

Although it is not usual to find mechanical efficiency charts in catalogues, we can easily extract that information from the interpolation equations obtained for the volumetric and overall efficiencies. We simply write the mechanical efficiencies of the pump and motor as

$$\begin{cases} \eta_p^m = \dfrac{\eta_p}{\eta_p^v} \\ \eta_m^m = \dfrac{\eta_m}{\eta_m^v} \end{cases} \tag{3.118}$$

Figures 3.51(a) and (b), which have been constructed with the help of Eqs. (3.118) and the interpolation functions for the volumetric and overall efficiencies developed in Sections 3.5.2 and 3.5.3, show the behaviour of the mechanical efficiency of the pump and the motor with the relative speeds for 170 and 345 bar, respectively. The dashed line in Figure 3.51(b) indicates an error in the efficiency (efficiency values cannot be greater

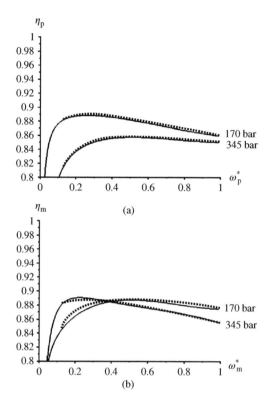

Figure 3.49 Comparison between the theoretical overall efficiency (solid curves) and the experimental data (dotted curves) for (a) pumps and (b) motors

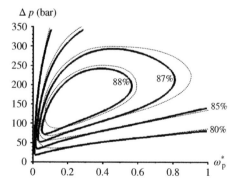

Figure 3.50 Comparison between the theoretical overall efficiency (solid curves) and the experimental data (dotted curves) for pumps

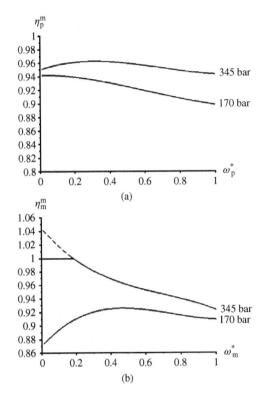

Figure 3.51 Mechanical efficiencies for the (a) pumps and the (b) motors represented by Figures 3.39(a) and (b)

than 1.0). These erroneous points are due to numerical imprecisions, which can occur from the choice of incorrect interpolation points (remember that we chose the interpolation points on the performance curves by visual inspection). It is important, therefore, to include some lines of code in the efficiency calculation routine to ignore efficiency values, which are greater than unity (we assume that we are using a computer program to plot the interpolated efficiency charts).

Exercises

(1) When defining hydrostatic and hydrodynamic pumps and motors, in Section 3.1, we implicitly assumed that the fluid was incompressible. Would those definitions equally apply to a compressible fluid?

(2) Obtain an expression for the average output flow, q_p, of the single-piston pump shown in Figure 3.6 in the particular case where $R = L$. Consider the angular speed of the crank, ω_p, as constant. Calculate the value of the average flow considering the following dimensional parameters: $R = 5.0\,\text{cm}$, $L = 5.0\,\text{cm}$, $A = 50.0\,\text{cm}^2$ and $\omega_p = 1.0\,\text{rad/s}$.

(3) A variable-displacement axial piston pump with a maximum displacement of 24.6 cm³/rev and a limiting swashplate angle of 15° is connected to a prime mover rotating at 1800 rpm. Plot the average output flow as a function of the swashplate angle as it varies from 0° to 15°. Repeat the process for a bent-axis pump with a swivel angle of 15° and compare the results.

(4) Consider the single-piston sliding along an inclined surface shown in Figure 3.52, and show that the height, h, at the point of contact between the piston rod and the inclined surface can be expressed as

$$h = R\tan(\alpha)[1 + \cos(\theta)]$$

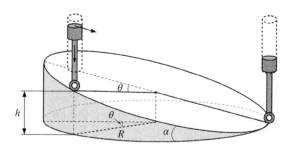

Figure 3.52   Single piston on an inclined surface

(5) Consider the input flow for a 20 cm³/rev gear motor at 210 bar, as shown in Figure 3.40, and draw the corresponding (approximate) volumetric efficiency curve.

(6) Show that the volumetric efficiency of the single-piston pump shown in Figure 3.53 is given by

$$\eta_p^v = 1 - \frac{C^3 p}{6L\mu RD\omega_p}$$

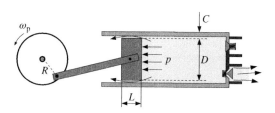

Figure 3.53   Single-piston pump

Tip: Consider the flow between the piston and the case as similar to a flow between two parallel plates with a depth πD and set apart by a distance C.

(7) Plot the curve $\eta_p^v \times \omega_p$ for the single-piston pump described in Exercise 6 using the following data:

- Fluid viscosity: 32 cSt
- Fluid density: 870 kg/m^3
- Pump dimensions: $R = 40$ mm, $D = 100$ mm, $L = 20$ mm, $C = 0.05$ mm
- Displacement chamber pressure: $p = 80$ bar

References

[1] Ivantysyn J, Ivantysynova M (2003) Hydrostatic pumps and motors. Tech-books International, New Delhi.
[2] Newton Jr., GC (1947) Hydraulic variable-speed transmissions as servomotors. Journal of The Franklin Institute, 243(6): 439–469.
[3] Parker-Hannifin (2005). Hydraulic pump division *XL* series, P1/PD models. Technical handbook – LTE-00049-1, USA.
[4] Payne GS, Kiprakis AE, Ehsan M, Rampen WHS, Chick JP, Wallace AR (2007). Efficiency and dynamic performance of Digital Displacement™ hydraulic transmission in tidal current energy converters. Proceedings of the IMechE, 221-A: 207–218.
[5] Rabie MG (2009) Fluid power engineering. McGraw-Hill, USA.
[6] Manring ND (2000) The discharge flow ripple of an axial-piston swash-plate type hydrostatic pump. Journal of Dynamic Systems, Measurement, and Control, 122: 263–268.
[7] Esposito A (1980) Fluid power with applications, 4th Ed., Prentice Hall, USA.
[8] Akers A, Gassman M, Smith R (2006) Hydraulic power system analysis. CRC Press, USA.
[9] Kugi A, Schlacher K, Aitzetmüller H, Hirmann G (2000). Modeling and simulation of a hydrostatic transmission with variable-displacement pump. Mathematics and Computers in Simulation, 53: 409–414.
[10] Klocke C (2011) Control terminology for hydrostatic transmissions. Proceedings of the 52nd National Conference on Fluid Power, Las Vegas, USA, Paper 23.3.
[11] Bowns DE, Worton-Griffiths J (1972) The dynamic characteristics of a hydrostatic transmission system. Proceedings of the Institution of Mechanical Engineers, 186: 755–773.
[12] Ivantysynova M (2008) Innovations in pump design – What are future directions? Proceedings of the 7th JFPS International Symposium on Fluid Power, Toyama, Japan, Paper OS1-3.
[13] Bosch-Rexroth (2004) Training mobile hydraulic-axial piston units: basic principles and function, RE 90 600/01.04, Germany.
[14] Linde (2012) Functional description of H-02 series pumps and motors: swashplate vs. bent-axis. Linde Hydraulics Corporation, USA.
[15] Achten PAJ (2005) Volumetric losses of a multi piston floating cup pump. Proceedings of the 50th National Conference on Fluid Power, NCFP-paper I05-10.2.
[16] Vael GEM, Achten PAJ, Brink TVD (2009). Efficiency of a variable displacement open circuit floating cup pump. Proceedings of the 11th Scandinavian International Conference on Fluid Power, SICFP'09, Linköping, Sweden.
[17] Fabiani M, Mancò S, Nervegna N, Rundo M et al. (1999) Modelling and simulation of gerotor gearing in lubricating oil pumps, SAE Technical Paper 1999-01-0626.
[18] Parker-Hannifin (2007). Gerotor motors: heavy duty high torque/low speed hydraulic motors. Technical Brochure, USA.
[19] Salter S (2005) Digital hydraulics for renewable energy. Proceedings: World Renewable Energy Conference, Aberdeen, UK.
[20] Katz A (2008). Design of a high speed hydraulic on/off valve, Masters Thesis, Worcester Polytechnic Institute, USA.
[21] Taylor J (2011) Digital displacement wind-turbine transmission. Presentation given at the 14th European Conference on Latest Technologies on Renewable Energy, Heriot-Watt University, Edinburgh, UK.
[22] Linjama M, Huhtala K (2009) Digital pump–motor with independent outlets. The 11th Scandinavian International Conference on Fluid Power, SICFP'09, Linköping, Sweden.
[23] Ehsan Md, Rampen W, Salter S (2000) Modeling of digital-displacement pump-motors and their application as hydraulic drives for nonuniform loads. ASME Journal of Dynamic Systems, Measurement, and Control, 122: 210–215.

[24] Parker-Hannifin (2013) Hydraulic motor/pump series F11/F12 fixed-displacement, Catalogue HY30-8249/UK.

[25] Sauer-Danfoss (2010) Series 40 axial piston motors technical information, S20L0636 Rev FD – October, USA.

[26] Sauer-Danfoss (2010) Series 40 axial piston pumps technical information, S20L0635 Rev EJ – October, USA.

[27] Bosch-Rexroth (2007) External gear motors F & N series. Catalogue RA 14 025/04.07, Bosch Rexroth Corporation, USA.

[28] Dorey RE (1988) Modelling of losses in pumps and motors. First Bath International Fluid Power Workshop, University of Bath, UK, September 1988, pp. 71–97.

[29] McCandlish D, Dorey RE (1984) The mathematical modeling of hydrostatic pumps and motors. Proceedings of the Institution of Mechanical Engineers, 198B(10): 165–174.

4

Basic Hydrostatic Transmission Design

The previous chapter was dedicated to the study of hydrostatic pumps and motors. In this chapter, we consider pump and motor combined in a hydrostatic transmission and present the basic steps for designing a hydrostatic transmission for a steady-state operation (we deal with transients in the following chapter). The focus is on the hydraulics of the transmission only. Thus, aspects related to the mechanical components, for example, will not be studied here.

The outline of the chapter is as follows: after briefly discussing some important points about hydrostatic transmission operation, we develop the efficiency expressions that will be used through the rest of the chapter. Two case studies are then presented to illustrate the process of choosing a suitable pump and motor for a typical hydrostatic transmission. In the first case study, we use a variable-displacement pump and a fixed-displacement motor, while in the second case study, a fixed-displacement pump is coupled with a variable-displacement motor. These two examples cover most of the typical issues involving steady-state transmission design. We then proceed to discuss volumetric and mechanical losses as information needed for the selection of the charge pump and the oil cooler.

4.1 General Considerations

4.1.1 Output Speed Control

Generally speaking, there are two distinct ways of controlling the motor speed in a hydrostatic transmission: change the pump or change the motor displacement. Both pump and motor displacements can be adjusted together, but this third case can be seen as a combination of the first two alternatives.

The way in which pressure is created in a hydrostatic transmission depends on the type of speed control that we choose for the motor. Figure 4.1 illustrates both situations. In the first case (variable-displacement pump and fixed-displacement motor), the resisting torque at the

Hydrostatic Transmissions and Actuators: Operation, Modelling and Applications, First Edition.
Gustavo Koury Costa and Nariman Sepehri.
© 2015 John Wiley & Sons, Ltd. Published 2015 by John Wiley & Sons, Ltd.
Companion Website: www.wiley.com/go/costa/hydrostatic

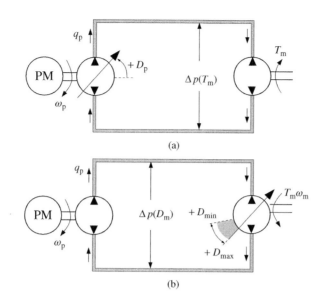

Figure 4.1 Motor speed control using the (a) pump displacement and the (b) motor displacement

motor, T_m, is responsible for creating the pressure differential, Δp, in the circuit. This is the reason that we used the functional notation, $\Delta p(T_m)$, meaning that Δp is a function of T_m in Figure 4.1(a). For instance, in some transmissions (see Case Study 1), the torque at the motor shaft is constant, which implies that Δp being constant as well. In such cases, altering the motor speed requires a variation of the power input at the pump (see Figure 1.38). As a rule of thumb, if a constant torque at the motor is required throughout the motor speed range, a fixed-displacement motor combined with a variable-displacement pump should be our first choice.

In fixed-displacement pumps and variable-displacement motor transmissions (Figure 4.1(b)), the pressure differential, Δp, is created by the speed control itself, which is now carried out at the motor. Therefore, Δp is a function of the motor displacement, D_m, as indicated in the figure. In this situation, it is possible to increase the angular speed of the motor shaft without altering the pump flow. However, torque is reduced as the angular speed grows higher because of the conservation of energy. As mentioned in Chapter 1, this configuration is not very practical, because it is impossible to set the transmission ratio to zero and therefore stop the motor shaft. Moreover, the pressure at the motor input can easily achieve prohibitive levels by reducing the motor displacement. On the other hand, this arrangement naturally allows for a wider range of transmission ratios when compared to fixed-displacement motor circuits.

4.1.2 Transmission Losses

Losses in hydrostatic transmissions are divided into volumetric and mechanical losses. Volumetric losses directly affect the motor speed, whereas mechanical losses influence the motor torque. Both of them affect the output power, which is a product of the speed and the torque, and therefore contribute to lowering down the overall transmission efficiency. The situation can be better explained with the illustration shown in Figure 4.2, which represents a simple power transmission between two hydraulic cylinders. Power is input to the left cylinder as the product between the applied force and the cylinder speed, $F_p v_p$, and is transmitted by the fluid to the right cylinder. In a sense, we can think of the circuit shown in Figure 4.2 as a hydrostatic transmission with the left cylinder playing the role of a pump and the right cylinder playing the role of a motor. Two situations are illustrated: an ideal one in which no losses exist and the input power is transferred completely to the cylinder on the right, and a real one in which both leakage and friction are present.

If the two cylinders in Figure 4.2 were identical, a non-leaking circuit (no volumetric losses) would result in $v_p = v_m$. However, if leakage is present, the flow into cylinder B will be smaller than the flow created by cylinder A, and, given that the two cylinders have the same cross-sectional area, there will be a reduction of the speed, v_m. Similarly, in the absence of friction (no mechanical losses), the force applied on the piston rod of cylinder A would be transferred completely to cylinder B. However, if there is friction, the force produced by the second cylinder will be smaller than the force input to the first cylinder, that is $F_m < F_p$. Since the output power is given by $F_m v_m$, it is obvious that both mechanical and volumetric losses affect the output power and therefore the overall transmission efficiency.

After this brief discussion about energy losses, a suitable definition of hydrostatic transmission efficiency will be discussed in detail in the following section.

4.2 Hydrostatic Transmission Efficiency

In the previous chapter, we performed an energy balance on a typical hydrostatic pump and a motor to obtain their overall efficiencies, given by Eqs. (3.52) and (3.54), which were later simplified to Eqs. (3.56) and (3.57) by considering only the pressure in Eq. (3.49). In this section, we apply the same energy balance concepts to obtain an expression for the overall efficiency of a typical hydrostatic transmission. Without loss of generality, we assume from the start that we can disregard the gravitational, internal and kinetic energy terms in Eq. (3.49) and write the hydraulic power at different points of the circuit as the product

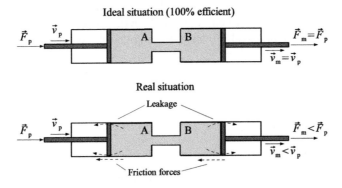

Figure 4.2   Mechanical and volumetric losses in a hydrostatic transmission

between the pressure and the volumetric flow. We can then proceed to perform an energy balance on the transmission circuit, as will be seen in the following section.

4.2.1 Energy Balance

Consider the closed-circuit transmission shown in Figure 4.3. In the figure, the heat losses originated from viscous forces and friction at the pump and the motor are represented by Q_p and Q_m, respectively. Number 1 is the point in the pump–motor circuit where the highest pressure occurs. Following clockwise, we reach point 4, where the pressure is at the lowest level. We remember that in the more complete representation shown in Figure 1.43, the line between the pump output and the motor input had additional connections to relief and check valves and that leakages could happen through those valves. Such leakages are represented by the term q_{cL} in Figure 4.3. The figure does not show any references to internal or external leakages in the pump or motor. They exist, and will soon be mentioned, but are not shown to ensure clarity. Following the convention adopted in the previous chapter, the flows into and out of the pump are represented by the subscripts 'pi' and 'po', respectively. Similarly, for the motors, the subscripts 'mi' and 'mo' are used.

The secondary circuit (charge circuit) is responsible for replenishing the external leakages. The charge pump input and output flows are represented by q_{ri} and q_{ro}, respectively. Note that, because of the internal and external leakages within the charge pump, the flow q_{ro} should be smaller than the flow q_{ri}. However, given that the pressure differential at the charge pump is relatively low, we may assume that $q_{ro} = q_{ri} = q_r$ in our calculations.

Throughout this chapter, the fluid will be considered incompressible.

In Figure 4.3, we observe that both main and charge pumps are driven by the same prime mover. Therefore, we can write the input power, $T_p\omega_p$, as

$$T_p\omega_p = P_p + P_r \tag{4.1}$$

where P_p and P_r represent the power input at the main pump and the charge pump, respectively.

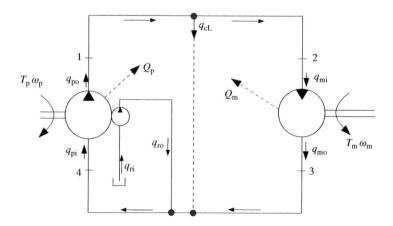

Figure 4.3 Flows in a typical hydrostatic transmission

In order to obtain an expression for the transmission efficiency, we must first perform an energy balance at the motor and the pump. Observing Figure 4.3, we can write for the motor:

$$p_{mo}q_{mo} = p_{mi}q_{mi} - p_{mi}q_{mLe} - p_{mi}q_{mLi} - T_m\omega_m - Q_m \tag{4.2}$$

Equation (4.2) states that a part of the incoming energy, $p_{mi}q_{mi}$, is turned into useful shaft power ($T_m\omega_m$), a part gets lost into the motor case ($p_{mi}q_{mLe}$) and through the internal leakages from the high-pressure to the low-pressure ports ($p_{mi}q_{mLi}$), and the rest dissipates in the form of heat (Q_m), leaving a remnant $p_{mo}q_{mo}$ at the motor output. The output power at the motor shaft, $T_m\omega_m$, can be readily obtained by a rearrangement of the terms in Eq. (4.2):

$$T_m\omega_m = p_{mi}q_{mi} - p_{mo}q_{mo} - (p_{mi}q_{mLe} + p_{mi}q_{mLi} + Q_m) \tag{4.3}$$

As for the energy balance at the main pump, we observe that the pump absorbs the input power P_p, which is equal to $T_p\omega_p - P_r$ (see Eq. (4.1)). A part of that energy is lost into heat because of the mechanical losses, Q_p. The energy at the pump output, $q_{po}p_{po}$, must also be subtracted from the losses through internal and external leakages, $q_{pLi}p_{po}$ and $q_{pLe}p_{po}$. On the other hand, the fluid that flows into the pump carries with it an amount of energy, $q_{pi}p_{pi}$. In the end, the energy balance at the pump will be expressed by the following equation:

$$q_{po}p_{po} = T_p\omega_p - P_r - q_{pLe}p_{po} - q_{pLi}p_{po} - Q_p + q_{pi}p_{pi} \tag{4.4}$$

The input power coming from the prime mover, $T_p\omega_p$, can then be obtained from Eq. (4.4):

$$T_p\omega_p = q_{po}p_{po} - q_{pi}p_{pi} + (q_{pLe}p_{po} + q_{pLi}p_{po} + Q_p + P_r) \tag{4.5}$$

In order to write an expression for the efficiency of the hydrostatic transmission, we first recall the general definition of efficiency, given by Eq. (1.8), and, rewritten later, in Eq. (3.46). We can then state that the efficiency of the hydrostatic transmission represented in Figure 4.3, η_{HT}, is given by the power output divided by the power input:

$$\eta_{HT} = \frac{T_m\omega_m}{T_p\omega_p} \tag{4.6}$$

If we substitute $T_m\omega_m$ and $T_p\omega_p$, given by Eqs. (4.3) and (4.5), into Eq. (4.6), we have

$$\eta_{HT} = \frac{p_{mi}q_{mi} - p_{mo}q_{mo} - (p_{mi}q_{mLe} + p_{mi}q_{mLi} + Q_m)}{q_{po}p_{po} - q_{pi}p_{pi} + (q_{pLe}p_{po} + q_{pLi}p_{po} + Q_p + P_r)} \tag{4.7}$$

We can also express η_{HT} from Eq. (4.7) in a slightly different manner:

$$\eta_{HT} = \left(\frac{p_{mi}q_{mi} - p_{mo}q_{mo}}{q_{po}p_{po} - q_{pi}p_{pi}}\right) \left\{ \frac{1 - \left[\dfrac{(q_{mLe} + q_{mLi})p_{mi} + Q_m}{p_{mi}q_{mi} - p_{mo}q_{mo}}\right]}{1 + \left[\dfrac{(q_{pLe} + q_{pLi})p_{po} + Q_p + P_r}{q_{po}p_{po} - q_{pi}p_{pi}}\right]} \right\} \tag{4.8}$$

If we disregard the power needed to drive the charge pump, P_r, the term between braces on the right-hand side of Eq. (4.8) can be compared with the product of the individual pump and motor efficiencies, given by Eqs. (3.56) and (3.57):

$$\frac{1 - \left[\dfrac{(q_{mLe} + q_{mLi})\,p_{mi} - Q_m}{q_{mi}p_{mi} - q_{mo}p_{mo}}\right]}{1 + \left[\dfrac{(q_{pLe} + q_{pLi})\,p_{po} + Q_p}{q_{po}p_{po} - q_{pi}p_{pi}}\right]} = \frac{\dfrac{(q_{mi}p_{mi} - q_{mo}p_{mo}) + (q_{mLe} + q_{mLi})p_{mi} - Q_m}{q_{mi}p_{mi} - q_{mo}p_{mo}}}{\dfrac{(q_{po}p_{po} - q_{pi}p_{pi}) + (q_{pLe} + q_{pLi})p_{po} + Q_p}{q_{po}p_{po} - q_{pi}p_{pi}}} = \eta_p \eta_m$$

$$(4.9)$$

It will be shown in the sequence that the term within parenthesis in Eq. (4.8) must be smaller than or equal to 1. Thus, if we compare the efficiency of the hydrostatic transmission, η_{HT}, given by Eq. (4.8), with the product of the pump and motor efficiencies, given by Eq. (4.9), we conclude that η_{HT} is always smaller than the product $\eta_p \eta_m$. Therefore, $\eta_p \eta_m$ can be seen as an upper limit for the efficiency of the closed-circuit transmission shown in Figure 4.3. This is an important conclusion because it allows us to estimate the best-case scenario of the efficiency for a given transmission without the need to perform a detailed analysis in the circuit.

4.2.2 Conduit Efficiency

The hydrostatic transmission efficiency as defined by Eq. (4.8) can be written as the product of three terms[1]:

$$\eta_{HT} = \eta_C \eta_p \eta_m \tag{4.10}$$

where η_p and η_m are the overall efficiencies of the pump and the motor, respectively, and η_C is given by

$$\eta_C = \frac{p_{mi}q_{mi} - p_{mo}q_{mo}}{q_{po}p_{po} - q_{pi}p_{pi}} \tag{4.11}$$

Observe that η_C, given by Eq. (4.11), is the ratio between the energy drop at the motor and the energy gain at the pump. Because of the circuit losses that occur along the hydraulic lines between the pump and the motor, it is clear that $\eta_C < 1$ in Eq. (4.11). Therefore, η_C indicates the *conduit efficiency* of the transmission because the energy drops between the pump and motor input and output ports depend only on conduit losses.[2] If no losses occurred, the energy coming from the pump would reach the motor unchanged and vice versa (i.e. $p_{po}q_{po} = p_{mi}q_{mi}$ and $p_{mo}q_{mo} = p_{pi}q_{pi}$), and we would have

$$\eta_C = \frac{p_{mi}q_{mi} - p_{pi}q_{pi}}{p_{mi}q_{mi} - p_{pi}q_{pi}} = 100\% \tag{4.12}$$

---

[1] See Ref. [1] for a similar definition of hydrostatic transmission efficiency.
[2] We must add here that losses will also take place in line connections such as elbows and junctions or any other element situated between the pump and the motor.

Equation (4.11) can be simplified if we make the reasonable assumption that no leakage happens in the hydraulic line connecting pump output to the motor input,[3] in which case, $q_{po} = q_{mi} = q_{12}$. In addition, we note that the pump input flow, q_{pi}, and the motor output flow, q_{mo}, can be written in terms of the pump/motor volumetric efficiencies and q_{12}, as follows: $q_{pi} = q_{12}/\eta_p^v$ and $q_{mo} = q_{12}\eta_m^v$. With these considerations, Eq. (4.11) becomes

$$\eta_C = \frac{p_{mi}q_{mi} - p_{mo}q_{mo}}{p_{po}q_{po} - p_{pi}q_{pi}} = \frac{p_{mi}q_{12} - \eta_m^v p_{mo}q_{12}}{p_{po}q_{12} - \frac{p_{pi}q_{12}}{\eta_p^v}} = \frac{p_{mi} - \eta_m^v p_{mo}}{p_{po} - \frac{p_{pi}}{\eta_p^v}} \qquad (4.13)$$

Looking carefully at Eq. (4.13), we observe that η_C increases if either η_m^v or η_p^v becomes smaller. We can therefore infer that η_C reaches a minimal value for the hypothetical situation where $\eta_p^v = \eta_m^v = 100\%$, in which case we have:

$$\eta_C = \frac{p_{mi} - p_{mo}}{p_{po} - p_{pi}} = \frac{\Delta p_m}{\Delta p_p} \qquad (4.14)$$

The ratio $\Delta p_m/\Delta p_p$ in Eq. (4.14) is a direct consequence of the mechanical–hydraulic losses at the hydraulic lines. For instance, from Eq. (2.69) and considering the localized pressure losses at tees, junctions and other connectors in lines 1–2 and 3–4, Δp_{L12} and Δp_{L34}, we can write the following relations for the pressures p_{mi} and p_{pi}:

$$\begin{cases} p_{mi} = p_{po} - f\left(\dfrac{L_{12}}{d_{12}}\right)\left(\dfrac{\rho \bar{u}_{12}^2}{2}\right) - \Delta p_{L12} \\[4mm] p_{pi} = p_{mo} - f\left(\dfrac{L_{34}}{d_{34}}\right)\left(\dfrac{\rho \bar{u}_{34}^2}{2}\right) - \Delta p_{L34} \end{cases} \qquad (4.15)$$

where ρ is the fluid density, f is the friction factor given by Eqs. (2.70) and (2.71) for laminar and turbulent flows, respectively: d_{12}, L_{12}, d_{34} and L_{34} are the inner diameters and the total lengths of lines 1–2 and 3–4, respectively, which may also include the equivalent lengths of localized elements and non-straight portions of the conduits.[4] Finally, $\bar{u}_{12}$ and $\bar{u}_{34}$ are the average speeds of the fluid within lines 1–2 and 3–4, respectively.

By adding both sides of Eqs. (4.15), we obtain the following general expression:

$$\Delta p_m = \Delta p_p - \left[\left(\frac{\rho}{2}\right)\sum\left(\frac{fL\bar{u}^2}{d}\right) + \sum \Delta p_L\right] \qquad (4.16)$$

where the summations encompass every stretch of conduit, bend and connector distributed along the lines.

---

[3] Remember that our analysis is focused at the steady-state regime and that we are not considering pressure overshoots, so no significant flow is expected through the relief valves between lines 1–2 and 3–4. We can then assume that $q_{cL} = 0$ in Figure 4.3.

[4] One usual way of accounting for non-straight portions of tubes and other elements for which the theory as developed in Section 2.2.2 cannot be applied is to compare the pressure drop at this particular element to the pressure drop at an equivalent straight tube with the same inner diameter, d. In other words, we simply assign an equivalent length, L_e, to parts of the hydraulic line and use Eq. (2.69) to obtain the corresponding pressure drop.

Substituting Δp_m, obtained from Eq. (4.16), into Eq. (4.14), we have

$$\eta_C = 1 - \frac{1}{\Delta p_p}\left[\left(\frac{\rho}{2}\right)\sum\left(\frac{fL\bar{u}^2}{d}\right) + \sum\Delta p_L\right] \qquad (4.17)$$

In the particular case where the inner conduit diameters are constant along the circuit, Eq. (4.17) simplifies to[5]

$$\eta_C = 1 - \frac{1}{\Delta p_p}\left[\left(\frac{f\rho\bar{u}^2}{2}\right)\sum\left(\frac{L}{d}\right) + \sum\Delta p_L\right] \qquad (4.18)$$

Equation (4.18) can be particularized for laminar flows, where the friction factor, f, is given by Eq. (2.70). If we also write the average speed, $\bar{u}$, as a function of the conduit flow, q_c, and the cross-sectional area, $\pi d^2/4$, we obtain

$$\eta_C = 1 - \left(\frac{128\nu\rho q_c}{\Delta p_p \pi d^3}\right)\sum\left(\frac{L}{d}\right) - \frac{\sum\Delta p_L}{\Delta p_p} \qquad (4.19)$$

In Eq. (4.19), we must remember that the summation $\sum(L/d)$ also includes minor pressure losses along the circuit.[6]

Remarks

1. Remember that Eq. (4.19) is only valid for laminar flow regimes. In the turbulent regime, pressure losses can be obtained by using f, given by Eq. (2.71) into Eqs. (4.17) or (4.18).
2. Because of the fact that the pressure differential at the pump, Δp_p, is usually much bigger than the pressure losses along the hydraulic lines, it is common to consider $\eta_C = 1$ in Eq. (4.10) and write the transmission efficiency η_{HT} as $\eta_{HT} = \eta_p\eta_m$.
3. A quick look at Eq. (4.19) shows that the conduit efficiency is considerably affected by the inner diameter of the hydraulic line, d. In fact, using small values for d may not only reduce the conduit efficiency but also increase the chance of the flow becoming turbulent (see Eq. (2.51)).

4.2.3   Minor Pressure Losses

Figure 4.4 shows two typical ways of connecting pump and motor in a hydrostatic transmission. In the first arrangement, the pump output is connected to the motor input through

---

[5] We have left the diameter, d, inside the summation in Eq. (4.18) for convenience, as will be soon made clear.
[6] It is usual to refer to pressure losses as major or minor losses. Major losses are those that occur along the conduits and minor losses are those that occur in localized parts such as bends and fittings [2]. Another type of minor loss occurs when there is an expansion or contraction at the cross-sectional area of the hydraulic line. These types of minor losses are very small in well-designed circuits where the correct fittings are used and assembly is carefully made and therefore will not be considered here.

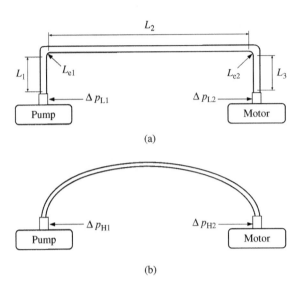

Figure 4.4 Typical ways to connect a pump and a motor in a transmission: (a) metal tubes and (b) flexible hoses (only one branch of the circuit is shown)

a metal tube, with two 90° bends at the upper corners, while in the second arrangement, a flexible hose is used to connect the pump output to the motor input.

In Figure 4.4, we see three typical sources of minor pressure losses in the circuit: straight tube fittings (Δp_{L1} and Δp_{L2}), hose fittings (Δp_{H1} and Δp_{H2}) and bends (L_{e1} and L_{e2}). Within the hose-fitting category, we can also identify the quick-disconnect couplings, which are very common in hydraulic circuits. Let us briefly analyse each one of the three sources.

4.2.3.1 Straight Tube Fittings

Today the technology for tube fittings has evolved to the point of minimizing pressure losses. Moreover, tubes and fittings are available in standard dimensions and are designed to provide a smooth match between the linked parts. As a result, in most cases, there is no interruption of continuity in a straight tube junction, and we can disregard the associated theoretical pressure losses if no more specific information has been provided by the manufacturer. For instance, Figure 4.5 illustrates a typical tube fitting [3], where we observe that there is no discontinuity between the tube and the body of the connection.

4.2.3.2 Permanent Hose Fittings

Hoses can be connected to the circuit either permanently or temporarily through a quick-disconnect coupling. Figure 4.6 shows a permanent coupling where the hose must be forced all the way against the fitting. Here, the ferrule is crimped against the hose so that it can be properly held in place. In this type of connection, there will always be a small difference between the inner diameter of the fitting and the inner diameter of the hose.

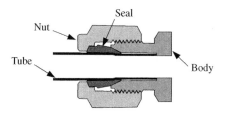

Figure 4.5   Typical tube fitting

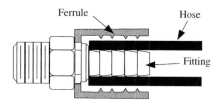

Figure 4.6   Typical hose fitting

However, in well-assembled connections, this difference in diameters should not introduce significant pressure losses and we can assume a continuity of the hydraulic line, just as we did in straight tube fittings.

4.2.3.3   Quick-Disconnect Hose Fittings

A very common type of hose fitting is the quick-disconnect coupling, which allows for a rapid disconnection of the hose from the hydraulic line when necessary. Figure 4.7 shows one typical design in which we see that both the plug and the socket contain an internal check valve to prevent the pressurized fluid leaking when the hose is not connected. This type of hose fitting is useful to quickly isolate the hydraulic parts for maintenance or in situations where different circuit configurations are desired (e.g. in training hydraulic units).

Because of the presence of check valves, it is easy to see that the situation with a quick-disconnect coupling shown in Figure 4.7 is likely to yield greater pressure losses. This is the reason why manufacturers usually provide pressure drop curves for their quick-disconnect fittings, as illustrated in Figure 4.8 (see, e.g. Ref. [4]).

4.2.3.4   Tube and Hose Bends

Pressure losses at bends are mainly a result of the change in fluid direction [5]. However, the relation between bend losses and the curvature r/d (Figure 4.9) is not well defined, as different investigators have obtained divergent experimental results [6]. Figure 4.9, for example, shows one particular experimental curve [7, 2], which can be used as a guide for pressure drop calculations in conduit bends in general. The curve gives the dimensionless equivalent length ratio, L_e/d, as a function of the relative curvature, r/d for a 90° bend.

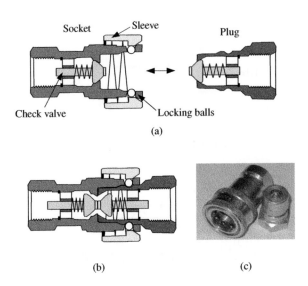

Figure 4.7 Typical quick-disconnect coupling (a) disconnected socket and plug, (b) connected coupling and (c) picture

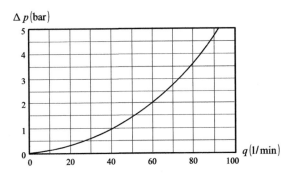

Figure 4.8 Typical pressure drop chart for a quick coupling with an ½ in. inner diameter (reproduced courtesy of RYCO Hydraulics Pty. Ltd.)

The same curve can be used approximately for 45° bends if we multiply the values of L_e/d by 0.4 [7].

In what follows, we exemplify the calculation of conduit efficiency for a typical hydrostatic transmission considering the two particular configurations shown in Figure 4.4.

4.2.4   Practical Application

Suppose that we need to connect an M25PV variable-displacement pump to an M25MF fixed-displacement motor (Table 3.3). The first thing that we need to know is the type of connection required at the pump and motor input and output ports. This piece of information

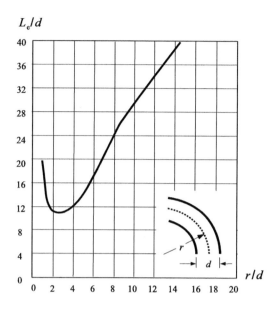

Figure 4.9   Equivalent length ratio for conduits at 90° bends

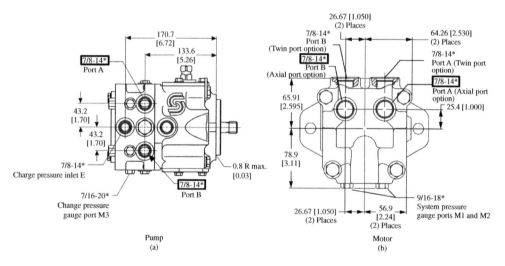

Figure 4.10   Views of the M25PV pump (a) and M25MF motor (b) showing the input and output ports A and B (courtesy of Danfoss Power Solutions)

can be obtained from the pump and motor manufacturer. For instance, Figures 4.10(a) and (b), obtained from the M25PV and M25MF catalogues [8, 9], identify the ports A and B (input and output ports of the pump and motor) as requiring a 7/8-14 SAE J514 straight thread O-ring port connection. In this specification, the outer thread diameter of the male connection is 7/8 in. (14 is the thread pitch, i.e. the number of threads per inch).

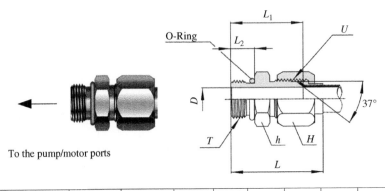

To the pump/motor ports

Tube O.D.		U (UNF)	T (UNF)	D	h	H	L	L_1	L_2
Inch	Metric								
1/8	3	5/16"-24	5/16"-24	1.6	11.1	9.5	34.7	26.9	7.5
3/16	4	3/8"-24	3/8"-24	3.2	12.7	11.0	37.3	27.9	7.5
1/4	6	7/16"-20	7/16"-20	4.4	14.2	14.2	40.3	31.2	9.1
5/16	8	1/2"-20	1/2"-20	6.0	15.8	16.0	41.6	31.2	9.1
3/8	10	9/16"-18	9/16"-18	7.5	17.4	17.4	44.9	33.0	9.9
1/2	12	314"-16	3/4"-16	9.9	22.2	22.2	50.3	37.6	11.1
5/8	16	7/8"-14	7/8"-14	12.3	25.4	25.4	59.8	43.2	12.7
3/4	18,20	11/16"-12	11/16"-12	15.5	31.7	32.0	65.5	50.0	15.1
7/8	22	13/16"-12	13/16"-12	18.0	34.9	35.0	68.1	50.5	15.1
1	25	15/16"-12	15/16"-12	21.5	38.1	38.0	70.8	51.8	15.1
1-1/4	32	15/8"-12	15/8"-12	37.5	47.6	50.8	73.6	55.1	15.1
1-1/2	38	17/8"-12	17/8"-12	33.0	53.9	57.0	86.9	60.2	15.1
2	50	21/2"-12	21/2"-12	45.0	69.8	73.0	98.1	70.6	15.1

Figure 4.11   Details of a 7/8-14 SAE J514 O-ring tube-case male fitting   (courtesy of HY-LOK Corporation)

Figure 4.11 shows the details of a 7/8-14 SAE J514 straight thread O-ring port connection obtained from a tube fittings manufacturer [10]. The outer diameter for the steel tube (OD) in this connection is 5/8 in. (~16 mm), as seen in the table adjacent to Figure 4.11. The left thread (T) connects to the pump/motor ports, as indicated.

Although the outer diameter of the tube is standardized (see Table 4.1), the inner diameter varies according to the operational pressure. The fitting shown in Figure 4.11, for example, has an inner diameter $D = 12.3$ mm, which is approximately equivalent to a 1.8 mm thick tube.[7] The tube is flared at its end to keep the connection steady.

Table 4.1 lists some recommended wall thicknesses for seamless stainless steel tubes [3]. Considering the maximum pressure of 345 bar (5004 psi) indicated in Table 3.3, we see that a 0.083 in. (2.1 mm) thick tube perfectly satisfies the maximum pressure requirement.

---

[7] Note that there may be a mismatch between the inner diameter of the fitting (12.3 mm) and the inner diameter of the tube, which depends on the chosen tube thickness. Such mismatch is usually small and can be disregarded.

Table 4.1 Typical wall thicknesses (in.) for a seamless stainless steel tube as a function of the outer diameter OD (in.) and (maximum) inner pressure (psi)

Tube OD	Wall thickness														
	0.010	0.012	0.014	0.016	0.020	0.028	0.035	0.049	0.065	0.083	0.095	0.109	0.120	0.134	0.156
1/16	5,600	6,900	8,200	9,500	12,100	16,800									
1/8						8,600	10,900								
3/16						5,500	7,000	10,300							
1/4						4,000	5,100	7,500	10,300						
5/16							4,100	5,900	8,100						
3/8							3,300	4,800	6,600						
1/2							2,600	3,700	5,100	6,700					
5/8								3,000	4,000	5,200	6,100				
3/4								2,400	3,300	4,300	5,000	5,800			
7/8								2,100	2,800	3,600	4,200	4,900			
1									2,400	3,200	3,700	4,200	4,700		
1¼										2,500	2,900	3,300	3,700	4,100	4,900
1½											2,400	2,700	3,000	3,400	4,000
2												2,000	2,200	2,500	2,900

Reproduced courtesy of Parker-Hannifin Corp.

The inner diameter of the chosen tube will therefore be 11.7 mm (0.6 mm smaller than D in Figure 4.11).

In what follows, we assume that the inner diameter of the tube (approximately 12 mm) remains constant along the hydraulic lines, including the fittings. Before we proceed, however, we still need to make a choice about the oil viscosity.

We have seen that oil viscosity changes with temperature. This is the reason why pump and motor manufacturers do not specify one single value for the recommended viscosity of the hydraulic fluid. For instance, considering the pump and motor we are using in this example, it is the manufacturer's recommendation that the viscosity of the fluid should stay within the range of 12–60 cSt [8, 9]. This means that the pressure losses will vary according to our fluid choice and its capacity to keep the viscosity constant within a certain temperature range (i.e. the *viscosity index*, see Section 2.1.1). One safe choice would be an ISO VG 32 hydraulic fluid (Table 2.2) whose viscosity is 32 cSt as this value is an approximation of the average value between the minimum and maximum viscosities allowed. We are going to use this fluid for all the examples in this chapter and assume that the viscosity remains constant in time.[8]

Figure 4.12 shows an example of a hydrostatic transmission connected by metal tubes (bend details are given in the centre of the figure). We assume that we do not have more specific information about the pressure drops at the fittings and disregard the losses at the connections.

According to Eq. (2.52), the maximum flow in the circuit, q_{cmax}, considering the laminar regime limit is given by

$$q_{cmax} = 3.47d = 3.47 \times 12 = 41.6\,\mathrm{l/min} = 41,600\,\mathrm{cm^3/min} \tag{4.20}$$

---

[8] This is a reasonable assumption in well-designed circuits where the fluid temperature is properly controlled during operation.

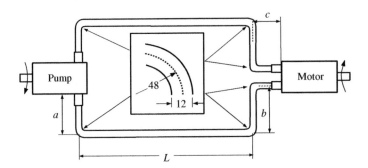

Figure 4.12   Example of metal tube lines in a hydrostatic transmission

For the maximum pump displacement of $24.6\,\text{cm}^3/\text{rev}$ (Table 3.3), the angular speed of
the pump shaft corresponding to the flow q_{cmax} is[9]

$$\omega_{\text{pmax}} = \frac{41,600}{24.6} = 1691\ \text{rpm} \tag{4.21}$$

From Eq. (4.21), we conclude that as long as $0 \le \omega_p \le 1691$ rpm, the flow regime in the
transmission lines is laminar.

Now, according to Figure 4.9, the equivalent length ratio, L/d, corresponding to each of
the six bends in Figure 4.12 is 12 (note that $r/d = 48/12 = 4$ in Figure 4.12). The conduit
efficiency, η_C, can then be obtained from Eq. (4.19) as

$$\eta_C = 1 - \left(\frac{128\nu\rho q_c}{\Delta p_p \pi d^3}\right)\left(\frac{2a + 2b + 2c + 2L}{d} + 6 \times 12\right) \tag{4.22}$$

Figure 4.13 shows the curves $\eta_C(q_c)$ for some selected values of the pressure differential, Δp_p, considering $d = 12$ mm, $a = 0.1$ m, $b = 0.15$ m, $c = 0.05$ m and $L = 1.0$ m in
Figure 4.12. We have also used the values of $870\,\text{kg/m}^3$ and $32 \times 10^{-6}\,\text{m}^2/\text{s}$ for the density, ρ, and the viscosity, ν, respectively.[10] As expected, the conduit efficiency is reduced
at smaller pressures. Nevertheless, we observe that the computed values are still relatively
high (>95%).

Let us consider yet another case where pump and motor are connected by long hoses as
shown in Figure 4.14. Assume that the length, L, of each branch of the circuit (high-pressure
and low-pressure lines) is 20 m. If we disregard the minor losses at the hose fittings and
bends, Eq. (4.19) simplifies to

$$\eta_C = 1 - \left(\frac{128\nu\rho q_c}{\Delta p_p \pi d^3}\right)\left(\frac{2L}{d}\right) \tag{4.23}$$

---

[9] Here, we are considering the hypothetical situation where the volumetric efficiency of the pump is 100%.
[10] We will use the same values for d, ρ and ν in the other examples of this section.

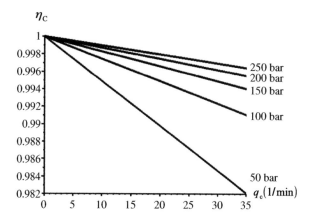

Figure 4.13 Conduit efficiency of the circuit shown in Figure 4.12 as a function of the flow, q_c, for some selected pressure differentials

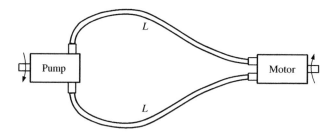

Figure 4.14 Pump and motor connected by long flexible hoses

Figure 4.15 shows the results for the circuit shown in Figure 4.14 with $L = 20$ m where we see that the pressure losses start to become significant, especially at lower pressures and higher flows.

Lastly, consider again the problem illustrated in Figure 4.14 with quick-disconnect couplings at the pump and motor input/output ports. It is reasonable to assume in this case that the local pressure drop at each hose fitting, Δp_L, can be approximately given by the curve in Figure 4.8, which was obtained for a similar inner diameter (1/2 in. = 12.7 mm). Moreover, we observe that the curve in Figure 4.8 can be approximated by the following equation:

$$\Delta p_L = \left(\frac{5}{92.5^2}\right) q_c^2 = \left[\left(5.84 \times 10^{-4}\right) q_c^2\right] \text{bar} = \left(58.4 q_c^2\right) \text{N/m}^2 \qquad (4.24)$$

where q_c is given in l/min.

Let us assume that the hose length, L, equals the total length $2a + 2b + 2c + 2L$ in Figure 4.12 and compare the metal tubes to the quick-disconnecting hose configuration this time. Equation (4.19) becomes

$$\eta_C = 1 - \left(\frac{128 v \rho q_c}{\Delta p_p \pi d^3}\right)\left(\frac{2a + 2b + 2c + 2L}{d}\right) - \frac{4\Delta p_L}{\Delta p_p} \qquad (4.25)$$

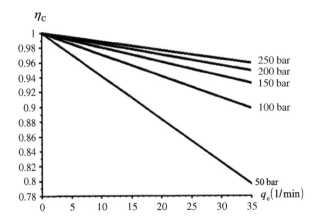

Figure 4.15 Conduit efficiency as a function of the flow, q_c, for some selected pressure differentials considering a 20 m long hose connection

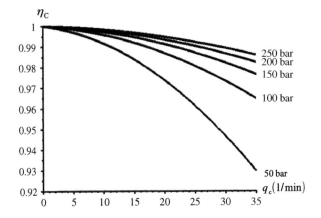

Figure 4.16 Conduit efficiency as a function of the flow, q_c, for some selected pressure differentials using quick-disconnecting hoses

Figure 4.16 is a graphical representation of Eq. (4.25) for $a = 0.1$ m, $b = 0.15$ m, $c = 0.05$ m and $L = 1.0$ m. We observe that the efficiency values differ very little from the values shown in Figure 4.13.

In summary, the use of quick-connecting hoses has slightly reduced the conduit efficiency when compared to the solid conduit solution (compare Figures 4.13 and 4.16). In addition, because of the quadratic dependence of the pressure losses with the flow, q_c, the efficiency drop is more accentuated in the hose circuit. However, the efficiency values are still high, which leads us to the conclusion that length is a predominant factor for the conduit efficiency (compare Figures 4.15 and 4.16).

4.3   Transmission Output

We saw that the output of a hydrostatic transmission is the power at the motor shaft, $T_m\omega_m$. However, we are usually interested in the power components individually, that is torque and angular speed. Therefore, it is useful to relate the motor output torque and speed, T_m and ω_m, to their pump (input) counterparts, T_p and ω_p.

We have seen in Section 4.1 that the output speed of a transmission is affected by the volumetric losses, whereas the output torque is influenced by the mechanical losses. These statements can be mathematically expressed as follows:

$$\begin{cases} \omega_m = \eta_{HT}^v R_T \omega_p \\[2mm] T_m = \dfrac{\eta_{HT}^m T_p}{R_T} \end{cases} \tag{4.26}$$

where η_{HT}^v and η_{HT}^m are the volumetric and mechanical efficiencies of the transmission, respectively, and R_T is the transmission ratio.

We proceed to give a formal proof for Eqs. (4.26). We begin with the output speed, ω_m. Consider a flow balance between the pump output, 1, and the motor input, 2, in Figure 4.3:

$$q_{po} - q_{cL} = q_{mi} \tag{4.27}$$

We have already seen that we can safely disregard the inter-conduit leakage, q_{cL}, in Eq. (4.27). Therefore, we can write that

$$q_{po} = q_{mi} \tag{4.28}$$

The pump output flow, q_{po}, can be written as the product of the maximum average flow, $D_p\omega_p$, and the volumetric efficiency of the pump, η_p^v. Because of the volumetric losses, the motor input flow, q_{mi}, must be bigger than $D_m\omega_m$, and can be written as the ratio between $D_m\omega_m$ and the volumetric efficiency of the motor, η_m^v.[11] Thus, Eq. (4.28) becomes

$$\eta_p^v D_p \omega_p = \frac{D_m\omega_m}{\eta_m^v} \tag{4.29}$$

Reorganizing the terms in Eq. (4.29), we obtain for the output speed, ω_m:

$$\omega_m = (\eta_p^v \eta_m^v)\left(\frac{D_p}{D_m}\right)\omega_p = \eta_{HT}^v R_T \omega_p \tag{4.30}$$

In Eq. (4.30), we have written the volumetric efficiency of the transmission, η_{HT}^v, as the product between the volumetric efficiencies of pump and motor individually. This is a fair

---

[11] We have already used this fact in the passage from Eq. (4.11) to Eq. (4.13).

assumption if we understand that leakages are not supposed to occur anywhere else but at the pump and the motor alone. The ratio between the pump and motor displacements, D_p/D_m, has already been identified with the maximum transmission ratio, R_T, in Eq. (1.21).[12]

We now proceed to prove the second part of Eqs. (4.26). In order to show that the output torque is given by the second expression in Eqs. (4.26), we must start from the definition of the overall transmission efficiency, η_{HT} (Eq. (4.6)), which relates the output and input powers:

$$T_m \omega_m = \eta_{HT} \omega_p T_p = (\eta_C \eta_p \eta_m) \omega_p T_p \tag{4.31}$$

Substituting ω_m, given by Eq. (4.30), into Eq. (4.31), we obtain

$$T_m = \left(\frac{\eta_C \eta_p \eta_m}{\eta_p^v \eta_m^v} \right) \left(\frac{T_p}{R_T} \right) \tag{4.32}$$

If we write the overall efficiencies of the pump and motor, η_p and η_m, as the products between their respective mechanical and volumetric efficiencies, we can rewrite Eq. (4.32) as

$$T_m = \left(\frac{\eta_C \eta_p^v \eta_m^v \eta_p^m \eta_m^m}{\eta_p^v \eta_m^v} \right) \left(\frac{T_p}{R_T} \right) = \frac{\eta_C \eta_p^m \eta_m^m T_p}{R_T} = \frac{\eta_{HT}^m T_p}{R_T} \tag{4.33}$$

where the mechanical efficiency of the transmission, η_{HT}^m, is given by the product of the mechanical efficiencies of the pump and the motor, η_p^m and η_m^m, and the conduit efficiency, η_C, respectively.

From Eq. (4.33), we conclude that the conduit efficiency, η_C, is in fact a mechanical efficiency and only influences the torque losses. This is expected since we have assumed the leakages between conduits, q_{cL}, as zero (see Eqs. (4.27) and (4.28)).

4.4 Steady-State Design Applications

In this section, we put in practice all that we have seen so far with two representative problems of steady-state transmission design. Two case studies will be considered. In Case Study 1, the output torque remains constant while the motor speed varies from zero to a maximum. For this first example, we have chosen a fixed-displacement motor/variable-displacement pump transmission, following the directions given in Section 4.1.1. In Case Study 2, the motor speed is kept constant while the pump speed is allowed to vary in time. In this particular situation, we will see that the best choice is a variable-displacement motor coupled with a fixed-displacement pump. These two examples are quite representative of many types of hydrostatic transmission and therefore will be presented here in details.

---

[12] Note that the volumetric losses were not present in Eq. (1.21). In fact, if we compare Eqs. (1.1) and (4.30), we see that in the presence of volumetric losses, the transmission ratio is modified and becomes equal to $\eta_{HT}^v R_T$.

4.4.1 Case Study 1. Fixed-Displacement Motor and Variable-Displacement Pump

Consider the elevator shown in Figure 4.17, where a hydrostatic transmission has been used together with a pair of gears to control the elevation speed. The fixed weight, W, is pulled up by a steel cable, which is looped around a sheave, whose radius is r. Since the radius and the weight are constants, a constant torque, T_m, is required from the hydraulic motor. This is an example of a variable-power transmission, because in order to accelerate or slow down the weight, the power input from the prime mover must vary accordingly (Figure 1.38).

To simplify matters, we consider that the pump and motor are connected by short hydraulic lines and therefore disregard the conduit losses in the circuit. As a result, the pressure differentials at the motor and the pump are the same and equal to Δp, as indicated. In addition, given that the motor displacement, D_m, and the torque, T_m, are constants, the pressure differential Δp will also be constant. Therefore, from Eqs. (3.4), we have that

$$\Delta p = \frac{T_m}{D_m} \tag{4.34}$$

The maximum possible value of the transmission ratio, $R_{T(max)}$, obtained by assuming that the circuit has no volumetric losses, will be given by

$$R_{T(max)} = \frac{D_{p(max)}}{D_m} \tag{4.35}$$

where $D_{p(max)}$ is the maximum pump displacement (remember that D_m is constant).

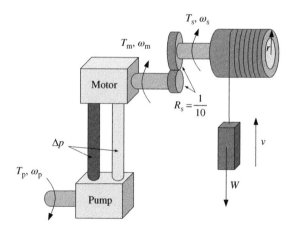

Figure 4.17 Hydrostatic transmission applied to an elevator

We know, however, that the actual transmission ratio depends on the volumetric efficiency of the transmission, η_{HT}^v. Therefore, we can only use Eq. (4.35) for an initial choice of the pump/motor size, as will be seen in the sequence.

4.4.1.1 Preliminary Pump and Motor Choices

Consider that the load, W, in Figure 4.17, weighs 6750 N, and that it must be lifted at a speed, v, varying between 0 and 1.5 m/s. Consider also that the sheave radius, r, is 0.1 m and the prime mover connected to the pump rotates at a constant speed of 2000 rpm (209.44 rad/s). Because of the elevated speed reduction required by the elevator, a couple of gears have been added between the motor shaft and the sheave with a transmission ratio, $R_s = 1{:}10$.

If we disregard the power losses at the gears, the required torque and maximum speed at the motor, T_m and ω_m, will be given by

$$\begin{cases} T_m = \dfrac{Wr}{R_s} = \dfrac{6750\ \text{N} \times 0.1\ \text{m}}{10} = 67.5\ \text{N m} \\[4mm] \omega_m = R_s \left(\dfrac{v}{r}\right) = 10 \times \dfrac{1.5\ \text{m/s}}{0.1\ \text{m}} = 150\ \text{rad/s} \cong 1433\ \text{rpm} \end{cases} \tag{4.36}$$

Once we know the torque at the motor, the pressure differential, Δp, can be calculated from Eq. (4.34):

$$\Delta p = \frac{67.5}{D_m} \tag{4.37}$$

where D_m should given in m^3/rad so that Δp can be expressed in N/m^2.

In Eq. (4.37), we observe that the motor displacement cannot be smaller than a minimum value that is determined by the maximum pressure differential allowed for the circuit. Another decisive factor for choosing motor and pump displacements, D_p and D_m, is the required transmission ratio. In this particular example, the motor speed must change continuously from 0 to 1433 rpm. Therefore, the maximum value of the transmission ratio, in the absence of volumetric loses, is

$$R_{T(max)} = \frac{1433}{2000} = 0.717 \tag{4.38}$$

Tables 4.2 and 4.3 have been taken from Table 3.3 and show a list of pump and motor displacements available from the manufacturer catalogues [8, 9]. The termination 'V' at the end of the model names indicates 'variable displacement', 'F' means 'fixed displacement' and 'T' is used for tandem units.

From Tables 4.2 and 4.3, we see that the combination of the M25PV variable-displacement pump and the M25MF fixed-displacement motor is in principle sufficient to cover the

Table 4.2 Danfoss series 40 pumps

Model	M25PV	M35PV	M44PV	M46PV	M25PT	M35PT	M44PT
D_p (cm³/rev)	24.6	35.0	43.5	46.0	24.6×2	35.0×2	43.5×2

Table 4.3 Danfoss series 40 motors

Model	M25MF	M35MF	M44MF	M35MV	M44MV	M46MV
D_m (cm³/rev)	25.0	35.0	44.0	35.0	44	46.0

transmission ratio range (0–0.717). In fact, the maximum estimated transmission ratio, considering that no volumetric losses exist, will be

$$R_{T(max)} = \frac{24.6\,\text{cm}^3/\text{rev}}{25\,\text{cm}^3/\text{rev}} = 0.984 \tag{4.39}$$

Given that the required value of the highest transmission ratio is 0.717, we expect that it should be possible to use this pump and motor combination, and limit the pump displacement to a value smaller than 24.6 cm³/rev, that is

$$D_{p(max)} = 0.717 \times 25\,\text{cm}^3/\text{rev} = 17.925\,\text{cm}^3/\text{rev} \tag{4.40}$$

It is important to keep in mind that the values of $R_{T(max)}$ and $D_{p(max)}$, obtained from Eqs. (4.39) and (4.40), are only approximations as we have not considered the volumetric efficiency of the transmission, η_{HT}^v, in our calculations.

Finally, it is necessary to check if the chosen motor produces a pressure differential beyond the maximum pressure indicated by the pump/motor manufacturer.[13] For the chosen pump and motor, the maximum pressure allowed in a continuous operation is 210 bar (Table 3.3). In this case, we know that the pressure differential is constant and can be obtained for the fixed-displacement motor by substituting $D_m = 25\,\text{cm}^3/\text{rev}$ into Eq. (4.37):

$$\Delta p = \frac{67.5\,\text{N m}}{25\,\text{cm}^3/\text{rev}} = \frac{67.5\,\text{N m}}{25 \times 10^{-6}\,\text{m}^3/(2\pi\,\text{rad})} \cong 170 \times 10^5\,\text{N/m}^2 \cong 170\,\text{bar} \tag{4.41}$$

From Eq. (4.41), we see that the system pressure at steady-state operation remains under the maximum allowed limit, which shows that our pump and motor choices are correct so far. The next step is to verify whether the desired transmission ratio can be achieved when the volumetric losses are considered.

4.4.1.2 Transmission Ratio and Volumetric Efficiency

The maximum transmission ratio is actually smaller than the ratio obtained from Eq. (4.39). In fact, the actual transmission ratio depends on the volumetric efficiency of the transmission and is given by Eq. (4.30):

$$R_T = (\eta_p^v \eta_m^v)\left(\frac{D_p}{D_m}\right) \tag{4.42}$$

---

[13] We are assuming that the conduits and the other transmission elements have been designed to safely bear the maximum pressure values allowed for the pump and the motor.

The question remains as to whether the chosen pump and motor are still able to provide a transmission ratio between 0 and 0.717 when volumetric losses are considered (remember that volumetric losses reduce the motor speed and therefore the transmission ratio). In order to answer this question, we must be able to solve Eq. (4.42) for R_T, which is not an easy task as the volumetric efficiency of the motor, η_m^v, depends on the motor speed, ω_m, itself (see Eq. (3.74)). Given that we only know the value of the pump speed (which is constant by definition), we must make use of the transmission ratio, R_T, to determine ω_m (remember that $\omega_m = R_T\omega_p$). We are then left with a cyclic problem for which an iterative procedure seems to be the best approach. A possible solution is to recur to a *predictor–corrector* algorithm ⬇, which can be described by the following steps:

1. Assign a minimum value for the pump displacement, D_p.
2. Predict the transmission ratio, $R_T = D_p/D_m$.
3. Obtain the motor speed, ω_m, using the predicted value of R_T ($\omega_m = \omega_p R_T$).
4. Use the rated values of the pump and motor speeds (4000 rpm, according to Table 3.3) to obtain the relative pump and motor speeds, ω_p^* and ω_m^*.
5. With the values of Δp^* ($\Delta p^* = 170/345 \cong 0.493$), ω_p^* and ω_m^*, calculate the volumetric efficiencies of the pump and motor using Eqs. (3.73), (3.74) and (3.94). Multiply them to obtain the volumetric efficiency of the transmission, η_{HT}^v.
6. Obtain the corrected value of the transmission ratio, $R_T^C = \eta_{HT}^v R_T$.
7. Obtain the corrected value of the motor speed, ω_m^C ($\omega_m^C = \omega_p R_T^C$).
8. With the corrected value of the motor speed, ω_m^C, calculate the overall efficiencies of the pump, motor and transmission. At this step, every other important parameter, such as the output power, can be obtained as a function of ω_m^C.
9. Increase the value of the pump displacement, D_p, by a small increment and return to Step 2 (i.e. make $D_p = D_p + \Delta D_p$). Care must be taken in the choice of the increment ΔD_p, as D_p cannot become zero during the iterative process.[14]

Figure 4.18 compares the corrected output speed, ω_m^C, considering an actual transmission ($\eta_{HT}^v < 100\%$) with the non-corrected output speed, ω_m, obtained for a 100% efficient transmission ($\eta_{HT}^v = 100\%$). We observe that despite the effect of the volumetric losses, the maximum output speed is still higher than the maximum required speed (1433 rpm), which means that, as far as the speed reduction is concerned, our pump/motor choice is quite acceptable. Note the dead zone for pump displacements smaller than (approximately) 1.25 cm³/rev, where no output speed is perceived.

Figure 4.19 shows the variation of the pump and motor overall efficiencies with the pump displacement factoring in the actual motor speed, shown in Figure 4.18. The hydrostatic transmission efficiency, estimated by the product of the individual efficiencies, has also been plotted. Again, note the dead zone, where the efficiency is zero for small values of the displacement, D_p.

The dead zone in Figure 4.19 occurs for pump displacements smaller than approximately 1.25 cm³/rev. The explanation for this can be found in the volumetric losses that take place at pump (see Figure 3.47(a)). When the pump displacement becomes sufficiently small, all

---

[14] Note that $D_p = 0$ makes Eqs. (3.73) and (3.74) undefined.

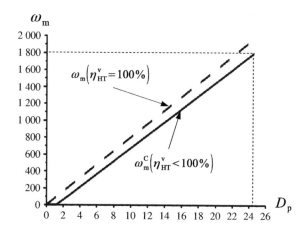

Figure 4.18 Effect of the volumetric losses on the output speed (rpm) at 170 bar (D_p in cm³/rev)

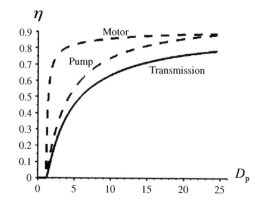

Figure 4.19 Overall efficiencies of the pump, motor and transmission at 170 bar as a function of the pump displacement, D_p (cm³/rev)

the pump flow ends up getting lost through internal and external leakages, and as a result, the volumetric efficiency of the pump becomes zero. Thus, for this particular transmission, the minimum pump displacement must be greater than 1.25 cm³/rev for an output to be produced at the motor.

4.4.1.3 Transmission Input and Output

The output of the transmission shown in Figure 4.17 is composed of the motor speed, ω_m, and torque, T_m. We have already shown the motor speed as a function of the pump displacement, D_p, in Figure 4.18. The torque, T_m, is constant by definition.

Concerning the transmission input, we observe that because of the transmission losses, the input power will necessarily be greater than the power at the motor shaft. This power has not yet been determined and must be given special attention, given that the prime mover choice depends on it. Because power is composed of torque and speed and we are assuming that the pump speed is constant (2000 rpm), we must determine the input torque from the information we have. Assuming that we have already calculated the mechanical efficiency of the transmission in Step 8 of the algorithm presented in the last section, we can start from Eq. (4.6) and obtain the following equation for the input torque, T_p:

$$T_p = T_m \left(\frac{\omega_m}{\eta_{HT} \omega_p} \right) = T_m \left(\frac{\eta_{HT}^v D_p \omega_p}{\eta_{HT} D_m \omega_p} \right) = T_m \left(\frac{D_p}{\eta_{HT}^m D_m} \right) \tag{4.43}$$

Substituting $T_m = 67.5$ Nm into Eq. (4.43), we obtain

$$T_p(Nm) = \frac{67.5 D_p}{\eta_{HT}^m D_m} \tag{4.44}$$

Figure 4.20 shows the variation of the input torque with the pump displacement. The input power, $P_p = T_p \omega_p$, can be obtained from Eq. (4.43) as follows:

$$P_p = \frac{T_m \omega_m}{\eta_{HT}} \tag{4.45}$$

It is important to note that for Eq. (4.45) to be consistent, the angular speed, ω_m, must be given in radians per unit of time. Therefore, for ω_m given in rpm and T_m given in N m, the following expression must be used:

$$P_p(W) = \frac{T_m \omega_m \pi}{30 \eta_{HT}} \tag{4.46}$$

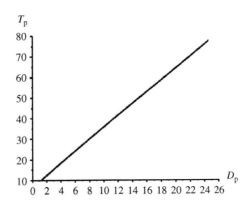

Figure 4.20   Input torque, T_p (Nm), as a function of the pump displacement, D_p (cm³/rev)

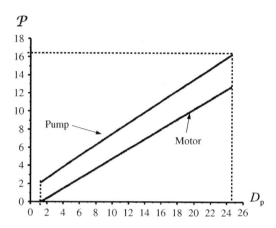

Figure 4.21 Power at the pump and motor, $\mathcal{P}$ (kW), as a function of the pump displacement, D_p (cm³/rev)

In this example, $T_m = 67.5$ N m and $\omega_p = 2000$ rpm. Thus, we may write Eq. (4.46) as

$$\mathcal{P}_p(W) = \frac{67.5\pi\omega_m}{30\eta_{HT}} \cong 7.07\frac{\omega_m}{\eta_{HT}} = 7.07\left(\frac{D_p}{D_m}\right)\frac{\eta_{HT}^v\omega_p}{\eta_{HT}} \cong 14137\left(\frac{D_p}{D_m\eta_{HT}^m}\right) \qquad (4.47)$$

Figure 4.21 shows the variation of the input power at the pump and the output power at the motor with the pump displacement, D_p.

We have seen in this case study that the pressure differential between the lines remains constant because of the constant torque at the motor shaft. However, we know that pump and motor efficiencies depend on both speed and pressure (see Eqs. (3.73), (3.74), (3.103) and (3.104)). Given that the pressure differential is directly related to the torque at the motor, we might wonder whether we are working in the best efficiency region in this case. In other words, would there be any particular values of the load, W, and speed, v, for which the transmission would operate in the highest efficiency zone? To answer this question, consider Figure 4.22 where we show the overall efficiencies of the transmission for three pressure differentials, each corresponding to a different torque at the motor. We see that the transmission efficiency drops drastically for low pressures. However, for pressures between 170 and 210 bar[15] we do not perceive any practical difference. The same happens for the prime mover speed, where we see that the efficiency does not change much for $\omega_p \geq 2000$ rpm while showing a great reduction at 50 rpm. Since our problem requires a 2000 rpm prime mover speed and a pressure differential of 170 bar, we conclude that we are indeed operating in the best performance region.

In this section, we have carried out a preliminary study on a variable-displacement pump and fixed-displacement motor transmissions, considering the steady-state regime of

---

[15] Maximum continuous pressure, according to Table 3.3.

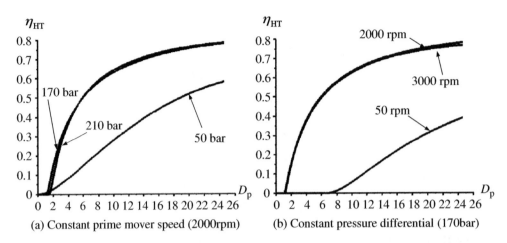

Figure 4.22   Transmission efficiency for (a) different pressure differentials and a constant prime mover speed and (b) different prime mover speeds and a constant pressure differential

operation. Despite the fact that this type of transmission is, by far, the most commonly used in practical applications, there will be occasions where a fixed-displacement pump and variable-displacement motor configuration are the most appropriate. This situation will be explored in the following case study.

4.4.2   Case Study 2. Fixed-Displacement Pump and Variable-Displacement Motor

Sometimes, the use of a variable-displacement motor is necessary in a hydrostatic transmission. Suppose that a situation requires a very high transmission ratio, that is, we want the output shaft to run considerably faster than the input shaft. Here, the pump would have to be much bigger than the motor so that the ratio D_p/D_m could prove satisfactory[16] (note that the displacement has to do with the actual size of the pump/motor). Consider the transmission shown in Figure 4.23, which represents a typical case where a high transmission ratio is required. Attached to the pump is a power source, which provides a variable input power at a relatively low-speed range. The motor must rotate at a constant (higher) speed all the time during operation.[17] In addition, assume that pump and motor are placed far away from one another and are connected by a pair of ½ in. inner diameter hoses through quick-disconnect couplings (see Figures 4.7 and 4.8).

---

[16] Take, for example the pumps and motors displayed in Tables 4.1 and 4.2. The maximum ratio between the pump and motor displacements is obtained for a M46PT variable-displacement pump (greatest displacement available) and a M25MF fixed-displacement motor (smallest displacement available), that is $(46 \times 2)/25 = 3.68$.

[17] A typical example of such requirements occurs in wind electricity generation, where the turbine shaft (input) rotates at very low speeds in comparison to the generator shaft (output). We will say more about this application later in Chapter 8.

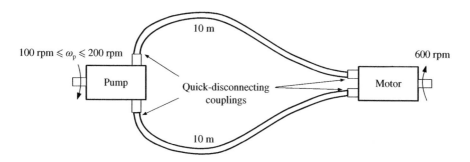

Figure 4.23   Variable input speed/constant output speed transmission

In this example, we consider that the power input at the pump shaft, P_p, is directly proportional to the pump speed, ω_p, and is given by the following expression[18]:

$$P_p(W) = 20\omega_p \tag{4.48}$$

where ω_p is expressed in rpm.

From Eq. (4.48), we obtain the input torque, T_p:

$$T_p = \frac{30P_p}{\pi\omega_p} = \frac{600}{\pi} \cong 191 \text{ Nm} \tag{4.49}$$

Figure 4.24 shows the speed and torque at the pump shaft as a function of the input power.

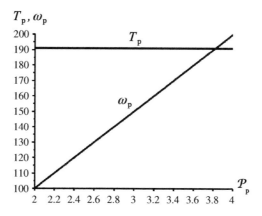

Figure 4.24   Input torque, T_p(Nm), and speed, ω_p(rpm), as a function of the input power, P_p(kW)

---

[18] One situation where power increases with speed occurs in wind turbines [11], although in such a case, the relation between power and speed is usually more complex than the linear relation given by Eq. (4.48).

4.4.2.1   Preliminary Pump and Motor Choices

We can make a preliminary choice of the pump and motor for the transmission based on the maximum pressure differential in the circuit and the transmission ratio, R_T. Therefore, beginning with Eqs. (4.49) and (3.3), after some unit adjustments, we have

$$\Delta p(\text{bar}) = \frac{20\pi T_p}{D_p} \cong \frac{11994.8}{D_p} \tag{4.50}$$

where D_p is given in cm^3/rev.

From Eq. (4.50), we see that the pressure differential is inversely proportional to the pump displacement. The maximum pressure differential allowed for a continuous operation is 210 bar, as indicated in Table 3.3. By making $\Delta p = 210$ bar in Eq. (4.50), we conclude that the minimum pump displacement is approximately 57.12 cm^3/rev in this case. As a result, the nearest model that can be safely used is the M35PT tandem pump, whose (maximum) displacement is 70 cm^3/rev (Table 4.2), and for which a corresponding pressure differential of approximately 171.35 bar is obtained. Note that such pressure differential is neither too low, as to compromise the transmission efficiency,[19] nor too high when compared to the 210 bar limit. We will therefore use this pump model in this example.

In order to choose the motor, we first observe that the value of the transmission ratio must fall within the following range (see Figure 4.23):

$$\frac{600}{200} \le R_T \le \frac{600}{100} \tag{4.51}$$

We may ignore the volumetric losses at this stage and write the transmission ratio as the quotient between the pump and the motor displacements. Therefore, considering the chosen M35PT pump at its maximum displacement, we have

$$R_T = \frac{D_{p(\text{max})}}{D_m} = \frac{70\,\text{cm}^3/\text{rev}}{D_m} \tag{4.52}$$

From Eqs. (4.51) and (4.52), we conclude that the motor displacement should vary between 11.7 and 23.3 cm^3/rev in order to satisfy the required transmission ratio range. The smallest variable-displacement motor in Table 4.3 that can be used in this case is the M35MF model, whose maximum displacement is 35 cm^3/rev. Thus, our preliminary pump/motor choice for this example is complete. We now need to study the effects of the volumetric losses on the transmission ratio.

4.4.2.2   Transmission Ratio and Volumetric Efficiency

We saw in Case Study 1 that in order to obtain the actual transmission ratio, we need to have the volumetric efficiency first. Since in this example it is the motor displacement that changes, it is meaningful to express the volumetric efficiency as a function of the motor

---

[19] The mechanical efficiency of the motor is particularly reduced at low-pressure differentials (see Eq. (3.101)).

displacement. An iterative procedure, similar to the one we used for solving Eq. (4.42), can be applied. The following algorithm ↓, very similar to the one used in Case Study 1, can be employed to calculate the variation of the volumetric efficiency of the transmission and the modified transmission ratio as a function of the motor displacement:

1. Start by choosing an appropriate value for the initial motor displacement,[20] D_m. For example, a good choice is to start with $D_m = 5\,\mathrm{cm^3/rev}$.
2. With the value of D_m, estimate the transmission ratio $R_T = D_p/D_m$.
3. Divide the motor speed, ω_m, by the transmission ratio, R_T, to estimate the pump speed, ω_p.
4. Substitute D_p into Eq. (4.50) to obtain the pressure differential, Δp.
5. Use the rated values of the pump and motor speeds (3600 rpm for both the M35PT pump and the M35MF motor, according to Table 3.3) to obtain the relative pump and motor speeds, ω_p^* and ω_m^*.
6. With the values of Δp^* ($\Delta p^* = \Delta p/345$), ω_p^* and ω_m^*, calculate the volumetric efficiencies of the pump and motor using Eqs. (3.73), (3.74) and (3.94). Multiply them to obtain the volumetric efficiency of the transmission, η_{HT}^v.
7. Obtain the corrected value of the transmission ratio, $R_T^C = \eta_{HT}^v R_T$.
8. Obtain the corrected value of the pump speed, ω_p^C ($\omega_p^C = \omega_m/R_T^C$).
9. With the corrected value of the pump speed, ω_p^C, calculate the efficiencies η_C, η_p, η_m and η_{HT}. At this step, every other important parameter, such as the output power, can be obtained as a function of ω_p^C.
10. Increase the value of the motor displacement, D_m, by a small increment and return to Step 2 (i.e. make $D_m = D_m + \Delta D_m$). Then proceed until the maximum motor displacement (35 cm³/rev) is reached.

Figure 4.25 shows the variation of the pump speed with the motor displacement factoring in a hypothetical lossless transmission ($\eta_{HT}^v = 100\%$) and the real case scenario where $\eta_{HT}^v < 100\%$. Note that the motor displacement must now vary between (approximately) 7.1 and 18.7 cm³/rev to maintain the correct speed range.

4.4.2.3 Transmission Output

The output power of the transmission, P_m, relates to the power input at the pump, P_p, through the following expression:

$$P_m = \eta_C \eta_p \eta_m P_p = \eta_{HT} P_p \tag{4.53}$$

The conduit efficiency, η_C, in Eq. (4.53) is a function of the flow inside the hydraulic lines, q_c. It is reasonable to assume that q_c equals the pump output flow, in this case given by

$$q_c = \eta_p^v \omega_p D_p = \eta_p^v \omega_p R_T D_m \tag{4.54}$$

---

[20] Note that because of the volumetric losses, the displacement range will be different from the originally estimated interval (between 11.7 and 23.3 cm³/rev).

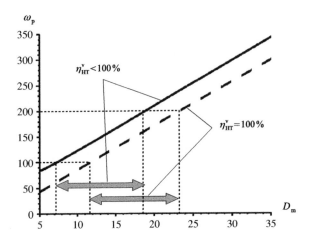

Figure 4.25 Input speed, ω_p (rpm) as a function of the motor displacement, D_m (cm³/rev) for different volumetric efficiencies

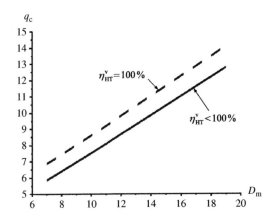

Figure 4.26 Conduit flow, q_c (l/min), as a function of the motor displacement, D_m (cm³/rev), for different volumetric efficiencies

For an inner diameter of ½ in. (12.7 mm) and an ISO VG 32 fluid ($\nu = 32 \times 10^{-6}\,\mathrm{m}^2/\mathrm{s}$), the turbulence limit becomes (see Eq. (2.52))

$$q_c = 3.47 \times 12.7 \cong 44\,1/\mathrm{min} \tag{4.55}$$

Figure 4.26 shows the variation of the conduit flow, q_c, with the motor displacement, D_m, according to Eq. (4.54) for the actual case, with the volumetric losses at the pump ($\eta_p^v < 100\%$) and the hypothetical scenario where no leakages occur at the pump ($\eta_p^v = 100\%$) being considered. In both cases, the flow remains below 44 l/min, which indicates that a laminar regime is maintained during operation.

Given that the flow inside the conduits is laminar, we can use an expression similar to Eq. (4.25) to calculate the transmission efficiency[21]:

$$\eta_C = 1 - \left(\frac{128\nu\rho q_c}{\Delta p_p \pi d^3}\right)\left(\frac{2L}{d}\right) - \frac{4\Delta p_L}{\Delta p_p} \tag{4.56}$$

where the term $2a + 2b + 2c + 2L$ in the numerator of Eq. (4.25) has been substituted by $2L$ (total hose length in Figure 4.23, i.e. $2L = 20$ m). The pressure differential, Δp_p, is given by Eq. (4.50) and the localized pressure drops at the quick-disconnect couplings, Δp_L, are given by Eq. (4.24).

Figure 4.27 shows the variation of the conduit efficiency, η_C, and the overall efficiencies, η_p, η_m and η_{HT}, as a function of the motor displacement. Note that the overall efficiency of the motor is greatly reduced at small displacements.

Figure 4.28 compares the transmission efficiency, η_{HT}, obtained by disregarding the conduit efficiency, η_C, in Eq. (4.53), with the actual curve shown in Figure 4.27, where we observe that the conduit losses do not have much influence on the transmission efficiency, especially at lower motor displacements.

To have a better understanding of why the overall efficiency of the transmission is so dependent on the motor displacement (Figure 4.27), we have plotted the mechanical and volumetric efficiencies of the motor individually in Figure 4.29. We clearly see that the mechanical efficiency of the motor is the most affected. The reason for that can be understood if we remember the force balance on a swashplate motor, shown in Figure 3.24. We see that the torque component of the driving force, F, decreases with the swashplate angle, α. Thus, at

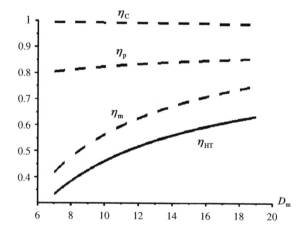

Figure 4.27 Transmission efficiencies η_C, η_p, η_m and η_{HT} as a function of the motor displacement, D_m (cm³/rev)

---

[21] Remember that Eq. (4.25) was originally written for the transmission illustrated in Figure 4.14, which, apart from the difference in the hose lengths, is identical to the transmission shown in Figure 4.23.

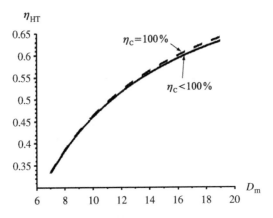

Figure 4.28 Variation of the overall efficiency of the transmission, η_{HT}, with the motor displacement, D_m (cm³/rev) for different conduit efficiencies

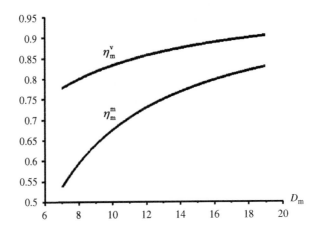

Figure 4.29 Volumetric and mechanical efficiencies of the motor, η^v_m and η^m_m, as a function of the displacement, D_m (cm³/rev)

small motor displacements, most of the incoming energy is spent on increasing the pressure between the pistons and the swashplate, and as a consequence, the output torque and the mechanical efficiency are considerably reduced.

Figure 4.30 shows the output power as a function of the motor displacement (see Eqs. (4.53), (4.46), and (4.30)). The hypothetical case where the transmission efficiency is 100% is also shown for comparison.

4.5   External Leakages and Charge Circuit

In Chapter 1, we learned that the charge pump is responsible for replenishing the external leakages in the circuit as well as feeding other auxiliary systems in the transmission such as

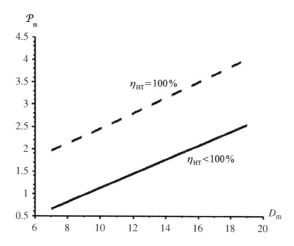

Figure 4.30   Power output, P_{m} (kW), as a function of the motor displacement, D_{m} (cm^3/rev) for different transmission efficiencies

the pump/motor servo-control. Thus, a proper charge pump selection would have to consider all these particularities. For example, an expression for the charge pump flow, q_{r}, similar to the one below, has been proposed in Ref. [12]:

$$q_{\mathrm{r}} = q_{\mathrm{L}} + q_{\mathrm{CT}} + q_{\beta} + q_{\mathrm{A}} \qquad (4.57)$$

where

- q_{L} represents the total leakages at the pump and the motor.
- q_{CT} is the flow needed for the pump/motor servo-control. For most applications, it can be assumed that $1.9\,\mathrm{l/min} < q_{\mathrm{CT}} < 7.6\,\mathrm{l/min}$ [12].
- q_{β} represents the volumetric losses due to compressibility effects.
- q_{A} is the flow required by other auxiliary devices.

The calculation of each term in Eq. (4.57) is beyond the scope of this book. However, with the information we have so far, we are able to estimate the first term, q_{L}, corresponding to the pump/motor leakages. Thus, in what follows, we illustrate the determination of the charge pump displacement, D_{cp}, by considering only the term, q_{L}, in Eq. (4.57). The student should have in mind that an accurate determination of D_{cp} depends on the specific configuration of each circuit, and every case must be dealt with separately. For the time being, we assume that we are dealing with the circuit shown in Figure 4.3, with no other auxiliary devices and no need for a servo-control at the pump or motor ($q_{\mathrm{A}} = 0$ and $q_{\mathrm{CT}} = 0$ in Eq. (4.57)). Still assuming that the fluid is incompressible, we write that $q_{\beta} = 0$.

If we define q_{pLe} as the external pump leakages and q_{mLe} as the external motor leakages, the displacement required from the charge pump, D_{cp}, can be approximated by the following equation:

$$D_{\mathrm{cp}} = \frac{q_{\mathrm{pLe}} + q_{\mathrm{mLe}}}{\omega_{\mathrm{p}}} \qquad (4.58)$$

It is not easy to determine the external leakages q_{pLe} and q_{mLe} based on information from the pump and the motor catalogues (remember that the volumetric efficiency curves do not separate external and internal leakage flows). One reasonable solution is to assume that all the leakages in the circuit are external. Such assumption will result in selecting an over-sized charge pump, which is acceptable because it will allow us to err on the side of caution. We then obtain the following expression for the charge pump displacement:

$$D_{cp} \cong \frac{q_{pL} + q_{mL}}{\omega_p} = \frac{D_p\omega_p(1 - \eta_p^v) + D_m\omega_m\left(\dfrac{1}{\eta_m^v} - 1\right)}{\omega_p} \tag{4.59}$$

where q_{pL} and q_{mL} were obtained from Eqs. (3.15) and (3.32).

Equation (4.59) can be rewritten as

$$D_{cp} = D_p(1 - \eta_p^v) + \left(\frac{D_m\omega_m}{\omega_p}\right)\left(\frac{1}{\eta_m^v} - 1\right) \tag{4.60}$$

Suppose that we want to select a charge pump for the elevator shown in Case Study 1. Since we are dealing with a variable-displacement pump/fixed-displacement motor, it is important to plot the variation of the charge pump displacement, D_{cp}, as a function of the main pump displacement, D_p. Figure 4.31 shows the results ⬇.

From the curve in Figure 4.31, we see that the maximum value of the charge pump displacement is around $2.2 \, \text{cm}^3/\text{rev}$. This means that if we pick a charge pump with a $2.5 \, \text{cm}^3/\text{rev}$ displacement, we can in principle cover the whole operational range.

It is interesting to register the different results that would have been obtained if we had considered a constant value for the volumetric efficiencies in Eq. (4.60) instead of using the actual variation of η_p^v and η_m^v in our calculations. Suppose that we had used a constant value

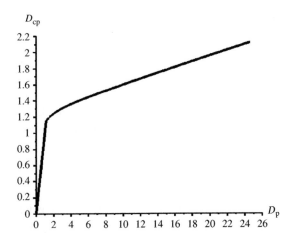

Figure 4.31 Charge pump displacement, D_{cp} (cm^3/rev) as a function of the main pump displacement, D_p (cm^3/rev), for the elevator in Case Study 1

of 90% for the volumetric efficiencies of the pump and the motor.[22] In this case, Eq. (4.60) would have become

$$D_{cp} = D_p(1 - 0.9) + \left(\frac{D_m\omega_m}{\omega_p}\right)\left(\frac{1}{0.9} - 1\right) \tag{4.61}$$

Given that the highest value of the charge pump displacement, D_{cp}, occurs at the maximum pump displacement, D_p, we can substitute D_p, D_m, ω_p and ω_m with their respective values – 24.6 cm³/rev, 25.0 cm³/rev, 2000 rpm and 1433 rpm – into Eq. (4.61) to obtain

$$D_{cp} = 24.6 \times (1 - 0.9) + \left(\frac{25 \times 1433}{2000}\right)\left(\frac{1}{0.9} - 1\right) = 4.45\,cm^3/rev \tag{4.62}$$

We observe that the charge pump obtained with the constant efficiency assumption is approximately twice the displacement calculated from the actual efficiency charts.

Because of the complexities involved in the correct determination of the charge pump displacement, some manufacturers make use of simplified 'rules of thumb' based on their experience. For instance, in Ref. [9], it is suggested that the charge pump displacement should equal 10% of the sum of the pump and motor displacements for the most common situations, that is

$$D_{cp} = \frac{D_p + D_m}{10} = \frac{24.6 + 25.0}{10} \cong 5.0\,cm^3/rev \tag{4.63}$$

Comparing Eqs. (4.62) and (4.63), we observe that the manufacturer recommendation coincides with the use of a constant volumetric efficiency of 90% in Eq. (4.61). This is a safer approach, considering that other important elements like the charge pump servo-control have not been included in the calculations.

Lastly, we note that a small tank (reservoir) connected to the pump and motor cases (Figure 4.32) is needed for the charge circuit. The tank size depends on the charge circuit flow, q_r (Eq. (4.57)). However, typical recommended values for the volume in closed-circuit transmissions correspond to ½ of the numerical value of the charge pump flow per minute [9]. For example, by this rule, a charge pump flow of 10 l/min would require a reservoir with a capacity for 5 litres. An inline filter must also be provided and can be placed before the charge pump, as indicated in the figure. The heat exchanger and the loop-flushing module, which also appear in the figure, will be addressed in the following section.

4.6 Heat Losses and Cooling

4.6.1 Sizing of the Heat Exchanger

The power lost into heat, Q, is directly associated with the mechanical efficiency of the transmission, η_{HT}^m, and can be estimated by

$$Q = (1 - \eta_{HT}^m)P_p \tag{4.64}$$

---

[22] Sometimes, seen as a 'typical value' for hydrostatic transmissions [13].

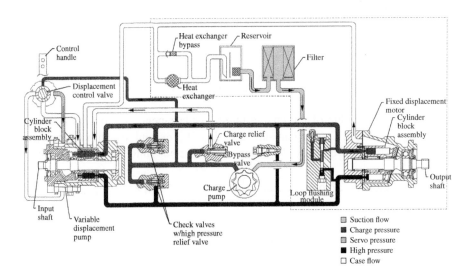

Figure 4.32   Typical hydrostatic transmission (courtesy of Danfoss Power Solutions)

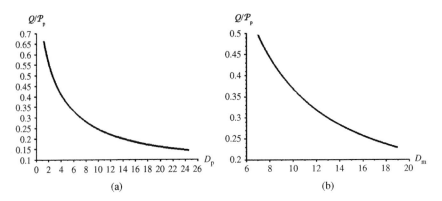

Figure 4.33   Ratio between the heat losses and the power input relative to (a) Case Study 1 and (b) Case Study 2 (D_p and D_m in cm³/rev)

Figure 4.33 (a) shows the ratio Q/P_p versus the pump displacement, D_p, relative to the situation depicted in Case Study 1, where the conduit efficiency, η_C, was considered 100%. Figure 4.33(b) shows the curve relative to Case Study 2, where Q/P_p is represented as a function of the motor displacement, D_m ⬇. We observe that the heat losses can become as high as 70% of the input power. In the situation shown in Figure 4.33(a), the highest ratio Q/P_p corresponds to lowest transmission ratio, where most of the input energy is turned into heat at the pump. On the other hand, the curve shown in Figure 4.33(b) reaches the maximum value at the highest transmission ratio, where the mechanical losses at the motor are dominant.

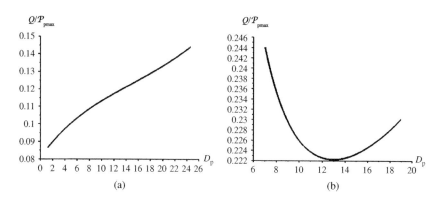

Figure 4.34   Ratio between the heat losses and the maximum power input relative to (a) Case Study 1 and (b) Case Study 2 (D_p and D_m in cm^3/rev)

Figure 4.34 shows the ratio between the heat losses and the *maximum* input power relatively to Case Studies 1 and 2 (approximately 16.2 and 4 kW, respectively) ⬇. Note that this is a different representation when compared to Figure 4.33, where the input power changed with the pump/motor displacement. In Case Study 1, we observe that the heat exchanger should be designed for a capacity of at least equal to 15% of the maximum input power of the transmission, that is 2.43 kW. Such demand increases for the situation presented in Case Study 2, where approximately 25% of the maximum input power is needed at small motor displacements, that is 1.0 kW. In both cases, the heat losses reach a maximum at the highest value of D_p/D_m.

Other aspects must be accounted for when sizing the correct heat exchanger for the transmission; these aspects include the existence of a loop-flushing module, the environment temperature and the conduits material. The subject is rather complex and will not be expanded upon in this book. However, apart from the calculation method shown in this section, some practical guidance has been given by hydrostatic transmission manufacturers for a quick selection. For instance, it has been reported that for constant-displacement pump/variable-displacement motor transmissions, it is usually safe to supply a heat exchanger with at least 25% of the maximum input power available [14]. Another manufacturer has recommended sizing the heat exchanger to operate at 1/3 of the maximum transmitted power [15].

4.6.2   Loop Flushing

Sometimes, a loop-flushing module is used in the circuit to forcedly drain out some fluid from the low-pressure line into the reservoir [16, 17, 9]. A loop-flushing valve removes hot fluid from the low-pressure side of the transmission for additional cooling and filtering, increasing the lifetime of the transmission components and keeping the oil viscosity within acceptable levels. Given that the oil is filtered before returning to the main circuit through the charge pump, eventual impurities, not detectable in the leaking flows, can be removed

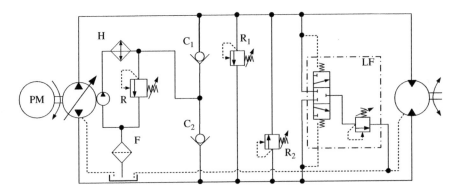

Figure 4.35 Hydrostatic transmission equipped with a flushing valve module (LF)

from the hydraulic fluid. Figure 4.35 shows a hydrostatic transmission with a conventional flushing module (LF) incorporated.

The operation of the loop-flushing module shown in Figure 4.35 can be described as follows: imagine that the high-pressure line is the upper conduit. As soon as the pressure gets high enough to overcome the spring adjustment of the three-way loop-flushing valve,[23] the spool moves downwards, causing the lower conduit to be connected to the tank through the relief valve placed within the loop-flushing module. The loop-flushing relief valve is responsible for limiting the flushing flow to match the design settings. Note that the presence of a loop-flushing module will interfere in the sizing of the charge pump, because the flush flow, q_{LF}, should now be added to the flow, q_r, in Eq. (4.57). In the configuration shown in Figure 4.35, the directional flush valve is set to open when the pressure in the circuit rises above the average (e.g. in eventual peaks) so that chunks of hydraulic fluid are sporadically flushed into the tank for cooling and filtering.

Exercises

(1) In Figure 4.3, the pressure at point 4 has been considered the lowest in the circuit. Assuming a clockwise flow, as shown in the figure, does this fact depend on the pressure set for the charge pump? Explain.

(2) What is the effect of a higher charge circuit pressure on the efficiency of the transmission shown in Figure 4.3?
Tip: Consider writing the charge pump power, P_r, in Eq. (4.7) as a function of the charge pump flow and pressure.

(3) In a hypothetical non-leaking hydrostatic transmission composed of a 100% efficient pump and a 100% efficient motor, both pump and motor are connected by two equal-length hoses. The resisting torque at the motor shaft and the motor displacement are 50 N m and 25 cm³/rev, respectively, and the pump supplies a flow of 30 l/min at

---

[23] Other means of activating the flush-looping valve and even other types of valves can be used (see, e.g. Ref. [18]).

140 bar. The transmission uses a 60 cSt hydraulic fluid whose density is 870 kg/m³ and the inner diameter of the hoses is 15 mm. Disregard the power needed to drive the charge pump and obtain an expression for the transmission efficiency as a function of the hose length, L.

(4) Consider the open-circuit transmission shown in Figure 1.44 and show that the pressure at the motor input can be written as the product between the pressure at the pump output and the conduit efficiency, that is $p_{mi} = \eta_C p_{po}$.

(5) Show that the angular speed, ω_m, in Eq. (4.45) must be given in radians per units of time for the equation to be physically consistent.

(6) In a variable-displacement pump and variable-displacement motor transmission, the pump and motor displacements are continuously adjusted in time according to Figure 4.36. Disregard the transmission losses and plot the motor speed, ω_m, for a constant pump speed, $\omega_p = 2000$ rpm.

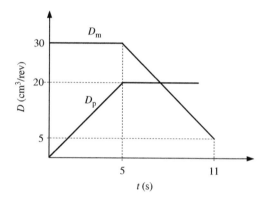

Figure 4.36 Pump and motor displacement variation in time

References

[1] Czynski M (2008) Energy efficiency of hydrostatic transmission-Comparing results of laboratory and simulation tests. Scientific Problems of Machines Operation and Maintenance, 2(154): 59–70.
[2] Fox RW, McDonald AT, Pritchard PJ (2004) Introduction to fluid mechanics, 6th Ed. John Wiley & Sons, USA.
[3] Parker-Hannifin (2011) Connector and tubing solutions-technical guide, UK.
[4] RYCO (2013) Accessories catalogue, RYCO – Hydraulic hose, couplings & fittings. http://www.ryco.com.au/tl_files/RYCO/PDF/catalogue/5%20Accessories.pdf. Accessed 26 October 2013.
[5] Parker-Hannifin (2013) General technical – Catalogue 4300, USA.
[6] CRANE (1986) Flow of fluids through valves, fittings and pipe – Technical Paper No. 410M. Crane Company, UK.
[7] Eaton (2011) Eaton-Weatherhead hose assembly master catalog W-HYOV-MC002-E3. Eaton Corporation, USA.
[8] Sauer-Danfoss (2010) Series 40 Axial Piston Motors Technical Information, S20L0636 Rev FD – October, USA.
[9] Sauer-Danfoss (2010) Series 40 Axial Piston Pumps Technical Information, S20L0635 Rev EJ – October, USA.
[10] HY-LOK (2002) 37° Flared SAE J514 Tube fittings with O-ring for fractional and metric, Catalog No. H – 240FF.

[11] RWE-npower (2012) Wind turbine power calculations: RWE npower renewables. http://www.raeng.org.uk/education/diploma/maths/pdf/exemplars_advanced/23_Wind_Turbine.pdf. Accessed 22 March 2012.

[12] Sauer-Sundstrand (1997) Applications manual BLN-9885. Section 1: selection of driveline components, US.

[13] Cundiff JS (2002) Fluid power circuits and controls: fundamentals and applications. CRC Press, Boca Raton, USA.

[14] Sundstrand (1974) 15-Series hydrostatic transmissions: service manual: bulletin 9646, USA.

[15] Sauer-Danfoss (1997) Applications manual BLN-9886. Section 4: transmission circuit recommendations, USA.

[16] Rexroth (2007) Flushing and boost pressure valve: catalog: RA 95 512/07.02, USA.

[17] Sauer-Danfoss (2010) Loop flushing valve: technical information, USA.

[18] Meier SM (2002) Loop flushing circuit. US Patent, number US 6,430,923 B1, 13 August 2002.

5

Dynamic Analysis of Hydrostatic Transmissions

In the previous chapter, we studied the steady-state behaviour of hydrostatic transmissions. In this chapter, we explore the operation of hydrostatic transmissions during transients, where compressibility effects are mostly evident and the system dynamics (i.e. inertia and acceleration) are considered. We begin with an overview of the main reasons why it is important to study the transient behaviour of hydrostatic transmissions and actuators. We then proceed with the development of a suitable physical and mathematical model for hydrostatic transmissions, where some important simplifications are discussed. Two case studies are presented to explore an analytic solution and a numerical method for the simulations.

5.1 Introduction

Three major considerations were embedded in the steady-state analysis carried out in Chapter 4:

- The load's inertia was neglected.
- The pump flow was considered as constant in time.
- The fluid was treated as incompressible.

Although it is possible to make a preliminary design of a hydrostatic transmission using these suppositions, each of these assumptions needs to be abandoned if an in-depth analysis of a hydrostatic transmission or actuator is required. In this section, we gradually explore the effects of the load's inertia, the pump flow and the fluid compressibility on hydrostatic transmissions. We begin by studying the influence of the load's inertia on pressure during transients. We then see how the oscillatory pump flow influences the angular speed of the motor. Lastly, we comment on the effects of fluid compressibility on the transmission – a theme that will be developed further in this chapter. We have chosen to address each topic as a sequence of independent subjects although, as will be seen, they are naturally interrelated.

Hydrostatic Transmissions and Actuators: Operation, Modelling and Applications, First Edition.
Gustavo Koury Costa and Nariman Sepehri.
© 2015 John Wiley & Sons, Ltd. Published 2015 by John Wiley & Sons, Ltd.
Companion Website: www.wiley.com/go/costa/hydrostatic

5.1.1 Pressure Surges during Transients

In the past chapters, we showed that the pressure differential between the pump and motor ports depended on the external torque applied to their shafts (see Eqs. (1.20), (1.22), (3.3) and (3.4)). In Chapter 4, we used these equations to estimate the pressure differential at steady-state regime, as in Eq. (4.34), where the torque at the motor, T_m, was constant. Now, we focus on what happens during the transient phase, that is, between the moment that the motor is at rest and the moment it reaches the steady-state regime (or vice versa).

Consider the situation illustrated in Figure 5.1, where a hydraulic motor is connected to a load. The combined motor–load inertia is J_0 and an external torque, T_{ext}, is applied to the motor shaft. To keep things simple, we disregard the friction between the mechanical parts. In this case, the Second Law of Newton combined with Eq. (1.22) gives

$$T_m = \Delta p D_m = J_0 \frac{d\omega_m}{dt} + T_{ext} \tag{5.1}$$

where ω_m is the angular speed of the motor shaft.

Equation (5.1) can be written in a simplified way by using the 'upper dot notation' for the time derivative:

$$\Delta p = \frac{1}{D_m}(J_0 \dot{\omega}_m + T_{ext}) \tag{5.2}$$

where

$$\dot{\omega}_m = \frac{d\omega_m}{dt} \tag{5.3}$$

As mentioned earlier, at this stage, we assume that the motor speed varies until it reaches the steady-state regime. We can derive the steady-state equation by making $\dot{\omega}_m = 0$ in Eq. (5.2):

$$\Delta p = \frac{T_{ext}}{D_m} \tag{5.4}$$

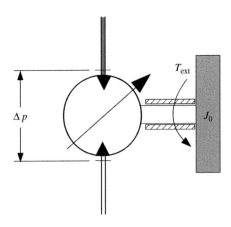

Figure 5.1 Hydraulic motor connected to a load

If we observe Eqs. (5.2) and (5.4), we note that during transients, a dynamic term containing the angular acceleration $\dot{\omega}$ appears. Such term can have a significant impact on the value of the pressure differential, Δp, causing it to rise up or fall down to prohibitive levels, depending on the values of $\dot{\omega}_m$, J_0 and D_m. A rise in the pressure (pressure overshoot) can be dangerous for the circuit as it may damage the hydraulic components or even cause structural failure. On the other hand, if the pressure falls down to very low levels (pressure undershoot), an instantaneous evaporation may occur, leading to a transient 'vaporous cavitation'.[1]

Figure 5.2 illustrates two typical situations in which the inertia of the load influences the pressure differential Δp. In the first situation (Figures 5.2(a) and (b)), the motor shaft is accelerating ($\dot{\omega}_m > 0$). Because of the load's inertia, a higher pressure differential is required to speed up the load from $\omega_m = 0$ to $\omega_m = \omega_s$, and a pressure overshoot occurs. In the second situation (Figures 5.2(c) and (d)), the motor shaft is initially at steady-state speed (ω_s). As the motor decelerates towards rest, the load's inertia tries to keep the shaft running forward, causing a momentary depressurization in the hydraulic lines (pressure undershoot). In both cases, there exists a point of maximum acceleration where the magnitude of the pressure differential reaches a peak.

As mentioned in the earlier chapters, there is a limit for transient pressure overshoots over which the risk of damaging the hydraulic components exists. For instance, we saw in Table 3.3 that the maximum pressure allowed for the Danfoss series 40 pumps and motors is 345 bar. Assuming that the same pressure is allowed for the transmission conduits in Figure 4.35, the relief valves R_1 and R_2 should be sized in such a way that the pressure differential between conduits does not exceed 345 bar at the maximum flow.[2]

Pressure relief valves need to be fast enough to open during a pressure overshoot. However, the response time of a relief valve depends on several factors related to its design and mechanical construction.[3] Reference [4] states that, 'depending on the design, the valve opens only after a certain stroke so within the response time the pressure can rise far beyond the setting of the valve ... ' This means that some relief valves may not be effective in alleviating the pressure spikes in the circuit.[4]

In choosing relief valves for the hydrostatic transmission, we must be aware that the response time is of paramount importance because the valve needs to open rapidly when pressure overshoots take place. Such a characteristic has to do with the valve construction, and the manufacturer should be able to provide the necessary information for a correct selection.

---

[1] The phenomenon of vaporous cavitation is very complex and will not be addressed in this book. The reader can consult Refs. [1, 2] for a deeper discussion on this theme.

[2] This means that at maximum displacement when all the pump flow gets blocked at the motor side (e.g. the motor stops for some reason), the relief valve must still keep the pressure inside the transmission within the admissible limit of 345 bar. In this case, all the pump flow will be diverted into the low-pressure side through the corresponding relief valve (R_1 or R_2, depending on which line holds the higher pressure in Figure 4.35). Given that the pressure at the valve input only reaches 345 bar at the maximum flow, the valve must start opening at a relatively lower pressure, known as the *cracking pressure* of the valve.

[3] The dynamics of relief valves will not be addressed in this book. The interested reader can consult Ref. [3] and the references therein for more information.

[4] Typical response times of direct operated relief valves range from 2 to 5 ms [4].

Relief valve response time has been subject to a lot of research, and some improvements have been suggested over the years.[5] Complementary solutions have also been developed to rapidly stabilize eventual pressure spikes in the circuit resulting from high-load accelerations, such as the so-called anti-shock relief valves that are currently available in the market [6, 7, 4]. A somewhat different solution to be applied to the specific case of motor start-ups makes use of a special type of normally open directional valve in the circuit. The said valve has been patented under the name of *pressure limit control acceleration valve* [8], although it is possible to find similar commercial designs being simply named *acceleration valves* [9].

Figure 5.3 illustrates the use of acceleration valves (valves AV_1 and AV_2) in a hydrostatic transmission. Note that the valves AV_1 and AV_2 are pilot actuated against internal springs, and the flows through the pilot orifices and through the valves themselves are reduced by fixed line restrictions.

Consider, for example, the operation of the valve AV_1. When the pump starts sending flow to the motor, there may be a pressure surge because of the sudden acceleration of the

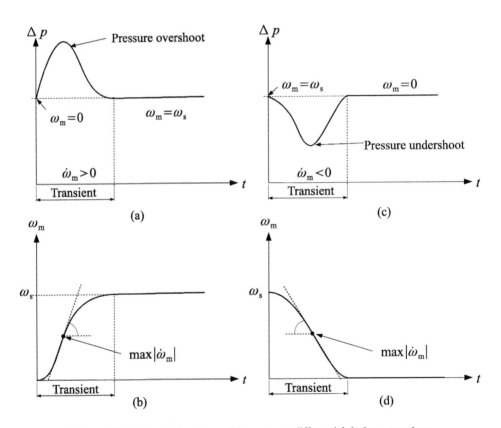

Figure 5.2 Typical behaviours of the pressure differential during a transient

---

[5] Much research has been carried out in the similar situation of water hydraulics where pressure surges are also very damaging (see Ref. [5]).

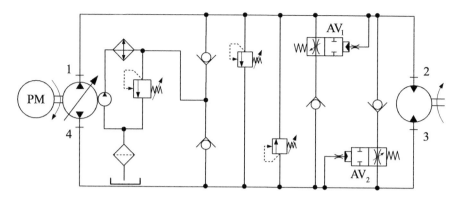

Figure 5.3   Acceleration valves in the circuit

load. At this time, the pressures at line 1–2 and at the pilot orifice of the valve AV_1 rise very fast. Because of the fixed restriction in the pilot line, the valve spool takes a while to respond to the pressure input, and the valve remains open long enough to cause the excess flow from the pressure overshoot to divert into the low-pressure line 3–4. Eventually, when the initial acceleration phase has passed and the pressure stabilizes in line 1–2, the valve spool is already displaced to its closed position, shutting the fluid passage between lines 1–2 and 3–4. The same operational principle is valid for the valve AV_2 when the flow direction is changed.

5.1.2   Mechanical Vibrations and Noise

We showed in Section 5.1.1 that pressure overshoots and undershoots can exist because of the dynamic term $J_0\dot{\omega}_m$ in Eq. (5.2). Now, we briefly investigate another cause of dynamic forces that occur in a hydrostatic transmission and in every hydraulic circuit: the oscillatory output flow coming from the pump. Here, we maintain the assumption that the fluid is incompressible and consider that there is no leakage in the conduit connecting the pump output to the motor input.

Consider, for example, an axial piston pump operating at an angular speed ω_p. It can be shown that in the absence of volumetric losses, the idealized displacement chamber[6] flow in a single piston, q_t^i, is given by [10]:

$$q_t^i = AR_m\omega_p\tan(\alpha)\sin(\theta_i) \tag{5.5}$$

where α, A and R_m are the swashplate angle, the piston area and the piston pitch radius, respectively (see Figure 3.25). The angle θ_i is the phase angle of the ith individual piston, as indicated in Figure 5.4, which shows the port plate of an axial piston pump (see Figure 3.23 for a similar representation).

---

[6] That is, the flow in and out of the displacement chamber.

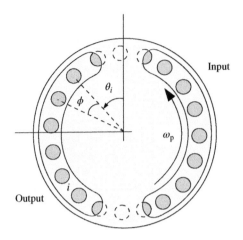

Figure 5.4 Phase angle of the *i*th piston in an axial piston pump

From Figure 5.4, we observe that the phase angle, θ_i, and the pump angular speed, ω_p, can be related through Eq. (3.13). Therefore, Eq. (5.5) can be rewritten as

$$q_t^i = AR_m\omega_p\tan(\alpha)\sin\left[\omega_p t + \left(\frac{2\pi}{N}\right)(i-1)\right] \tag{5.6}$$

Note that only the cylinders whose $q_t^i > 0$ should be considered when we are calculating the pump output flow, q_{pt}. Therefore, we can write that

$$q_{pt} = \sum_{i=1}^{N} q_{pt}^i \tag{5.7}$$

where q_{pt}^i is the individual flow contribution coming from the *i*th cylinder:

$$\begin{cases} q_{pt}^i = q_t^i \ \text{ if } q_t^i \geq 0 \\ q_{pt}^i = 0 \ \ \text{ if } q_t^i < 0 \end{cases} \tag{5.8}$$

Equations (5.6)–(5.8) give the ideal output flow of an axial piston pump with N cylinders. The angular speed of the motor, ω_m, can be approximately obtained by dividing the flow, q_{pt}, by the motor displacement, D_m, that is, $\omega_m = q_{pt}/D_m$. With this simplification, we can write the instantaneous angular acceleration of the motor shaft, $\dot{\omega}_m$, as (we leave the demonstration as an exercise to the student):

$$\dot{\omega}_m = \frac{d\omega_m}{dt} = \left[\frac{AR_m\omega_p^2\tan(\alpha)}{D_m}\right]\sum_{i=1}^{N}\cos\left[\omega_p t + \left(\frac{2\pi}{N}\right)(i-1)\right] \tag{5.9}$$

Remark

Because of the restriction imposed by Eqs. (5.8), the terms within the cosine brackets in Eq. (5.9) that should be considered in the summation are the ones whose corresponding sines are positive.

Substituting $\dot{\omega}_m$, given by Eq. (5.9), into Eq. (5.2) gives the pressure differential, Δp, as a function of time:

$$\Delta p = \left[\frac{A J_0 R_m \omega_p^2 \tan(\alpha)}{D_m^2} \right] \sum_{i=1}^{N} \cos\left[\omega_p t + \left(\frac{2\pi}{N} \right)(i-1) \right] + \frac{T_{ext}}{D_m} \qquad (5.10)$$

As a practical example, consider that $N = 7$, $A = 2\,\text{cm}^2$, $\alpha = 0.22$ rad ($12.61°$), $R_m = 4$ cm, $D_m = 25\,\text{cm}^3/\text{rev}$, $J_0 = 0.8\,\text{kg m}^2$, $\omega_p = 1000$ rpm and $T_{ext} = 67.5$ N m. Figure 5.5 shows $\omega_m(t)$ and $\Delta p(t)$ for this configuration.

In Figure 5.5, we observe that the pressure oscillates around the average value of 170 bar, which corresponds to an external torque of 67.5 N m, applied to the motor shaft

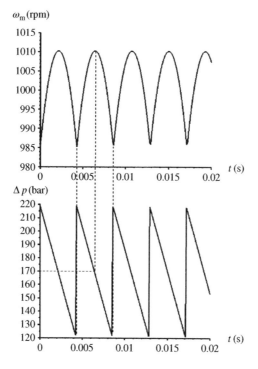

Figure 5.5 Instantaneous angular speed and pressure differential in a typical hydraulic motor connected to a hydrostatic pump

(see Eq. (4.41)). During the first half of every speed pulse, the motor acceleration is positive, which results in the instantaneous pressure differential being higher than this average value. On the other hand, on the second half of the speed pulse, the acceleration becomes negative and the instantaneous pressure differentials become lower than 170 bar. Therefore, each pulse of the angular speed contains two pressure spikes.

Pressure and flow oscillations ultimately translate into mechanical vibrations and noise, the latter being simply the acoustic propagation of the vibrations caused by the oscillating pressure acting on the circuit components.[7] In fact, mechanical vibrations in hydraulic circuits are sometimes referred to as noise indistinctively [11]. Other factors, such as fluid turbulence, can contribute to generation of noise in hydraulic circuits, but the flow pulsations coming from the pump are the most common noise source [11, 12]. Noise suppressors that help reduce the undesirable effects of pressure-induced oscillations can be found in the market (see Ref. [13]).

5.1.3   Natural Circuit Oscillations

In the last two sections, we considered that the fluid was incompressible. The immediate consequence of this assumption is that, in the absence of losses and disregarding the elasticity of the hydraulic lines and components, all the fluid energy coming from the pump will be transformed into mechanical energy in the end, that is, mechanical vibrations, noise and useful work at the motor shaft. However, if fluid compressibility is taken into consideration, the energy flow between the pump and the motor is significantly modified.

Consider, for example, the situation illustrated in Figure 5.6, where two cylinders are connected using a compressible fluid as the power transmission medium. The cylinder on the left plays the role of a pump, pushing the actuator on the right against a constant load F at a speed v_p. The actuator responds to the pump flow and moves at a speed v_c. The status of the actuator at times t_1, t_2 and t_3 is shown on the left and the evolution of the pump and actuator speeds v_p and v_c, together with the inner pressure, p, is shown in the curves on the right. A detailed explanation of the figure is given in the sequence.

An initial observation that can be made about Figure 5.6 is that if an incompressible fluid had been used in the circuit, there would be no difference between the curves $v_p(t)$ and $v_c(t)$ during operation. However, because of the fluid compressibility, the volume of fluid inside the cylinder at the end of the pump stroke, Ax_1, is smaller than the volume that would have been obtained using an incompressible fluid, Ax_p (see Figure 5.6 for $t = t_0$ and $t = t_1$). This means that part of the mechanical work input at the pump between $t = 0$ and $t = t_1$ does not transform into kinetic energy but is stored in the form of elastic[8] energy within the fluid mass, which has its volume reduced by $A\Delta x$. As a result, the cylinder speed v_1, at $t = t_1$, is necessarily smaller than the corresponding 'incompressible speed', v_m. The energy stored in the fluid is then sent back to the cylinder as the fluid expands between $t = t_1$ and $t = t_2$.

---

[7] It has been observed that mechanical vibrations caused by pressure oscillations in the circuit contribute to the failure of moving parts and metal tube assemblies in general [11].

[8] The term 'elastic' here simply means that the fluid is able to return to its original volume after the hydrostatic pressure returns to its original value. In a general sense, elasticity encompasses both the fluid compressibility and expansibility and is sometimes used in the definition of bulk modulus (e.g. 'Bulk modulus of elasticity' [14]).

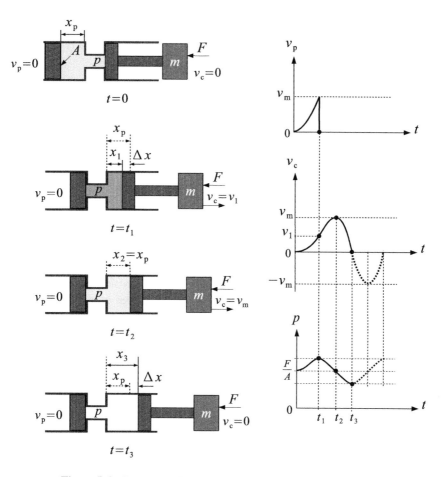

Figure 5.6   Pressure and speed in circuit with a compressible fluid

During this time interval, the mass m accelerates until the point where $v_c = v_m$. However, the acceleration is slowly reduced, changing signs at $t = t_2$. At this point, all the energy stored in the fluid returns to the load attached to the cylinder, which moves to the right at the speed v_m. Finally, the load decelerates from v_m to zero until it reaches the rightmost position, at time t_3.

In the absence of energy losses, the exchange between elastic energy and kinetic energy will continue to happen causing the cylinder piston to oscillate between the positions x_1 and x_3 (i.e. $x_p - \Delta x$ and $x_p + \Delta x$). The corresponding fluctuation in pressure is indicated by the dashed curve that appears as a continuation of the $p(t)$ curve after $t = t_3$. Note that the oscillatory pattern of the pressure does not depend on the input coming from the pump but is an inherent feature of the circuit itself.[9] Thus, in this sense, these oscillations can be regarded as *natural* as opposed to *forced* oscillations (e.g. the oscillations caused by the pump output flow in Section 5.1.2).

---

[9] If we study Figure 5.6 carefully, we see that the oscillations remain constant even after the pump input has ceased.

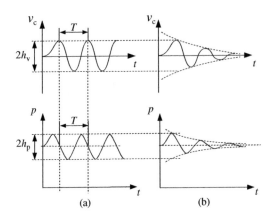

Figure 5.7   Speed and pressure versus time: (a) undamped mode and (b) damped mode

Natural oscillations are usually damped out after a while because of the energy losses in the circuit. It is also possible to add energy dissipation devices to the circuit in order to increase oscillation damping, such as the acceleration valves mentioned in Section 5.1.1. Figure 5.7 illustrates the damped and undamped $v_c(t)$ and $p(t)$ curves for the situation pictured in Figure 5.6.[10]

Two important characteristics of the pressure and speed curves, namely, the period (T) and the amplitude (h_v and h_p for the speed and pressure curves, respectively), are shown in Figure 5.7. The period is the time needed for one oscillation cycle to be completed and is the same for speed and pressure (Figure 5.7(a)). A more usual parameter is the natural frequency, defined as the number of cycles per unit time. The amplitude is the maximum (peak) value obtained in time and depends on the variable being considered, that is, speed or pressure. The immediate effect of the energy losses is to reduce the amplitude of the oscillations until these are completely eliminated, as shown in Figure 5.7(b). The natural frequency is also influenced by the energy losses but to a much lesser extent when compared to the amplitude.

Before we move on to the next section, let us summarize two important points concerning the frequency and amplitude of the natural oscillations:

1. The natural frequency is the same for pressure and speed, and depends on the physical characteristics of the hydraulic circuit, that is, it does not change if the initial input force and speed are modified.
2. The amplitude of the natural oscillations depends on the external input and the physical characteristics of the hydraulic system. For example, in the situation shown in Figure 5.6, the maximum pressure depends on the load mass and on the first derivative of the speed (acceleration) at $t = t_1$. In addition, the acceleration depends on the initial movement of the pump and the fluid compressibility.

---

[10] In more technical terms, the oscillations shown in Figure 5.7 are *underdamped*, basically meaning that the circuit still oscillates for a while before the steady-state regime is reached (see Section B.2.1 for details).

5.1.4   Resonance and Beating

In the previous section, we studied the case where a single input signal coming from the pump produced an oscillatory output at the cylinder. Consider, now, the situation where the input signal is repeated from time to time, as illustrated in Figure 5.8, which shows a particular case where the frequency of the input signal equals the natural frequency of the circuit. The oscillations are not damped in the illustration for a better visualization of the problem. In this figure, we observe that there is a repetitive energy input from the left cylinder (pump), which periodically adds to the already stored energy, causing the speed amplitude to rise up at every pump pulse (the piston on the left moves and then stops in an intermittent manner along its stroke). Note that if the periodic input lingered on infinitely, the output speed v_c would also grow to infinite values. This phenomenon is known as *resonance*, which is a term borrowed from the study of acoustical phenomena such as sound waves and music.

If the energy dissipation in the circuit is low, resonance can raise the output speed (and pressure) to very high levels. As seen in Figure 5.8, the pre-requisite for circuit resonance is that the external perturbation (e.g. the pump input) fulfils the following requirements:

(a) Is periodic, that is, it has a constant period of oscillation T.
(b) Repeats itself at the circuit's natural frequency.

If these two conditions are met, there is a chance[11] for resonance to take place if the oscillations are not rapidly damped. These conditions should therefore be avoided as much as possible. Another undesirable situation happens when the frequency of the external

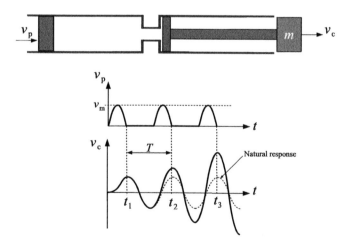

Figure 5.8   Periodic input signal and resonance

---

[11] A third condition is that the input circuit is in phase with the natural oscillations, for example, a positive input pulse must match a positive output speed pulse. This is usually the case when the oscillatory input itself initiates the natural circuit oscillations, as in Figure 5.8.

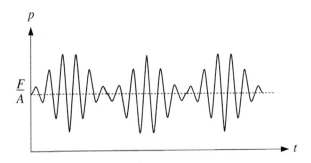

Figure 5.9 Beating

excitation is nearly equal to the natural frequency of the system. In such case, a phenomenon named *beating* occurs, where the amplitude of the oscillations comes to a maximum and then descends to zero periodically, as illustrated in Figure 5.9, when the pressure inside the circuit is represented as a function of time.

Conventional means of reducing resonance and beating in hydraulic circuits involve increasing the energy dissipation to favour oscillation damping.[12] We have already mentioned some of those techniques in Section 5.1.1 (e.g. the use of acceleration valves in the circuit). Other methods include the use of specially designed hoses [11] or the addition of passive mass–spring dampers [16, 17]. Active dampers, where an inverted pressure wave is constantly fed into the circuit, have also been proposed as a means of reducing fluid-borne noise [18]. Figure 5.10 illustrates the operation of active and passive dampers in a hydraulic circuit.

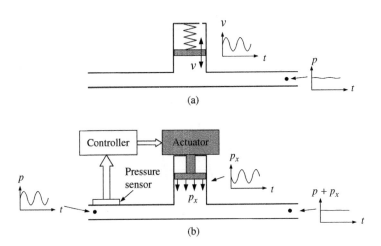

Figure 5.10 Schematic representation of a (a) passive mass–spring damper and an (b) active piezoelectric damper

---

[12] Damping is critical for a good operation of a hydraulic system. For instance, it has been reported that the absence of damping mechanisms caused the crash of a test fighter aircraft, as there was resonance in the aircraft's hydraulic system [15].

Note that the piezoelectric actuator in Figure 5.10 senses the actual pressure variation, $p(t)$, in the hydraulic line, and then produces a reversed wave, $p_x(t)$, so that both are cancelled out in the end (see Ref. [18] for a more detailed explanation).

The precise analytical study of pressure oscillations in a hydraulic circuit would be a formidable task given the number of variables involved [13]. Therefore, many rely on experimental vibrational analyses for detecting unusual circuit operations [19].

5.1.5 Summary

Here is a list of the most important points we have covered in this section:

1. If the inertia of the circuit components is considered, dynamic forces appear due to the speed variations in time. Such forces ultimately translate into pressure spikes (overshoots and undershoots) that can be damaging to the whole circuit if not properly attenuated.
2. The pulsating flow coming from the pump also results in the acceleration of moving parts and can produce dynamic forces and pressure spikes. The resulting vibration ultimately translates into audible noise.
3. Fluid compressibility tends to perpetuate the oscillatory pattern created by the dynamic forces mentioned in items 1 and 2. The fluid and the other elastic elements of the circuit act like springs by storing and releasing energy indefinitely in a frequency that is inherent to the circuit itself.
4. Once natural oscillations are established in the circuit, they interact with the external forces. If the external forces are themselves oscillatory with a frequency that matches the natural frequency of the circuit, resonance is likely to take place, and the amplitude of the speed and pressure oscillations can reach prohibitive levels with time.
5. Both natural and forced oscillations can be damped out through correctly sized energy dissipation devices.

5.2 Modelling and Simulation

In this section, we show the general steps to obtain a mathematical model of a hydrostatic transmission. Instead of trying to model the transmission in detail, together with the valves and the charge circuit as shown in Figure 5.3, we shall base our analysis on the simplified physical model shown in Figure 5.11, where the following nomenclature has been used[13]:

- q_{pLi}, q_{pLe}, q_{mLi} and q_{mLe} are the internal and external leakages at the pump and the motor, respectively. Despite the representation shown in the figure, we must remember that those leakages occur inside the pump and motor, from the high-pressure port into the tank (external) or into the low-pressure port (internal).
- q_r is the charge pump flow (we are assuming that there are no losses within the charge pump itself so that the input and output flows, q_{ri} and q_{ro}, as they appear in Figure 4.3, are the same, and both are equal to q_r).
- q_{cL} represents any other flow between the high and low-pressure conduits (e.g. the flow through the relief valves).

---

[13] The nomenclature adopted in Figure 5.11 follows the convention used in Figures 3.38 and 4.3.

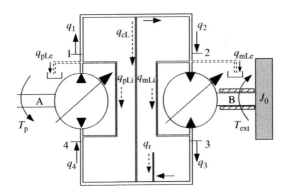

Figure 5.11   Simplified transmission model

- J_0 is the moment of inertia of the load.
- T_p is the input torque at the pump shaft.
- T_{ext} is the external torque at the motor shaft.

Once the transmission model has been properly set, we can make use of well-known physical laws to obtain a mathematical description of it. This will be done in the following section.

5.2.1   Basic Equations

In order to arrive at a suitable mathematical model for the transmission shown in Figure 5.11, we first perform an analysis of the mass flows along the circuit. Figure 5.12 shows an exploded view of the hydrostatic transmission, where the volumetric flows through the conduits, pump and motor are displayed in detail. To help the student visualize the local flows, we have divided the transmission into six different regions 1–6, which appear in the figure surrounded by dashed rectangles. Note the new flows q_{pt} and q_{mt} in regions 2–5. These are the pump and motor instantaneous flows obtained for an ideal situation where volumetric losses are not present, as explained in Sections 3.2.2 and 5.1.2.

Concerning Figure 5.12, we make two initial simplifying assumptions:

1. Pressure losses along the conduits are too low and can be ignored. This is a very reasonable supposition for short-length conduits as seen in Section 4.2.
2. The fluid density, ρ, is a function of pressure alone. This assumption, together with the first supposition, indicates that the density does not change along the high-pressure line (between the pump output and the motor input). The same is valid for the low-pressure line (between the motor output and the pump input). Again, this is true for short-line circuits, as explained in the paragraph following Eq. (2.45).

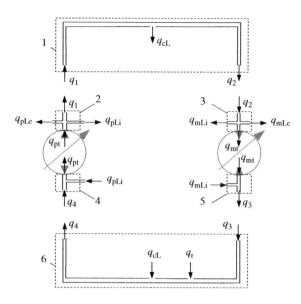

Figure 5.12 Volumetric circuit flows

Now, let p_{12}, p_{34}, ρ_{12} and ρ_{34} be the pressures and densities inside the conduits 1–2 and 3–4, respectively. Let us also denote the inner conduit volumes at regions 1–6 by $V_1, V_2, \ldots, V_6$. The mass balances within every region in Figure 5.12 can then be written as

$$
\begin{cases}
\text{Region 1}: \; \rho_{12}\left(q_1 - q_{cL} - q_2\right) = V_1 \dfrac{d\rho_{12}}{dt} \\[2mm]
\text{Region 2}: \; \rho_{12}(q_{pt} - q_1 - q_{pLe} - q_{pLi}) = V_2 \dfrac{d\rho_{12}}{dt} \\[2mm]
\text{Region 3}: \; \rho_{12}(q_2 - q_{mLi} - q_{mLe} - q_{mt}) = V_3 \dfrac{d\rho_{12}}{dt} \\[2mm]
\text{Region 4}: \; \rho_{34}(q_4 - q_{pt} + q_{pLi}) = V_4 \dfrac{d\rho_{34}}{dt} \\[2mm]
\text{Region 5}: \; \rho_{34}(q_{mt} - q_3 + q_{mLi}) = V_5 \dfrac{d\rho_{34}}{dt} \\[2mm]
\text{Region 6}: \; \rho_{34}(q_3 + q_r + q_{cL} - q_4) = V_6 \dfrac{d\rho_{34}}{dt}
\end{cases}
\tag{5.11}
$$

Equations (5.11) can be simplified if we assume that $V_2 \approx V_3 \approx V_4 \approx V_5 \approx 0$ in Figure 5.12. If we also consider that the inner volumes of the conduits 1–2 and 3–4 are the same, that is, $V_1 = V_6 = V$, we can rewrite Eqs. (5.11) as

$$
\begin{cases}
\text{Region 1}: \ q_1 - q_{cL} - q_2 = \dfrac{V}{\rho_h}\dfrac{d\rho_{12}}{dt} \\[2mm]
\text{Region 2}: \ q_{pt} - q_1 - q_{pLe} - q_{pLi} = 0 \\[2mm]
\text{Region 3}: \ q_2 - q_{mLi} - q_{mLe} - q_{mt} = 0 \\[2mm]
\text{Region 4}: \ q_4 - q_{pt} + q_{pLi} = 0 \\[2mm]
\text{Region 5}: \ q_{mt} - q_3 + q_{mLi} = 0 \\[2mm]
\text{Region 6}: \ q_3 + q_r + q_{cL} - q_4 = \dfrac{V}{\rho_1}\dfrac{d\rho_{34}}{dt}
\end{cases}
\tag{5.12}
$$

Equations (5.12) can be further reduced if we substitute the values of q_1 through q_4 obtained from the equations corresponding to regions 2–5 into the first and the last equations, corresponding to regions 1 and 6. We therefore have

$$
\begin{cases}
\dfrac{V}{\rho_{12}}\dfrac{d\rho_{12}}{dt} = \left(q_{pt} - q_{mt}\right) - \left(q_{pLe} + q_{pLi} + q_{mLi} + q_{mLe} + q_{cL}\right) \\[3mm]
\dfrac{V}{\rho_{34}}\dfrac{d\rho_{34}}{dt} = \left(q_{mt} - q_{pt}\right) + \left(q_{mLi} + q_{pLi} + q_r + q_{cL}\right)
\end{cases}
\tag{5.13}
$$

The density derivatives that appear in Eqs. (5.13) can be generally developed as

$$
\frac{d\rho}{dt} = \frac{dv^{-1}}{dt} = -\frac{1}{v^2}\frac{dv}{dp}\frac{dp}{dt} = -\frac{1}{v^2}\left(-\frac{v}{\beta_E}\right)\frac{dp}{dt} = \left(\frac{\rho}{\beta_E}\right)\frac{dp}{dt}
\tag{5.14}
$$

where $v = \rho^{-1}$ is the specific volume of the fluid and β_E is the effective bulk modulus.
We can then rewrite Eqs. (5.13) as

$$
\begin{cases}
\left(\dfrac{V}{\beta_E}\right)\dfrac{dp_{12}}{dt} = (q_{pt} - q_{mt}) - (q_{pLe} + q_{pLi} + q_{mLi} + q_{mLe} + q_{cL}) \\[3mm]
\left(\dfrac{V}{\beta_E}\right)\dfrac{dp_{34}}{dt} = (q_{mt} - q_{pt}) + (q_{mLi} + q_{pLi} + q_r + q_{cL})
\end{cases}
\tag{5.15}
$$

Equations (5.15) relate the pressures at both sides of the circuit to the internal flows. Another important expression relates the torque, T_m, at the motor shaft to its angular speed, ω_m [20]:

$$
T_m = J_0 \frac{d\omega_m}{dt} + b_v \omega_m + T_{ext}
\tag{5.16}
$$

where $b_v \omega_m$ is a general viscous friction term.[14]

---

[14] We will see more about friction in Chapter 7 when we study hydrostatic actuator modelling. In this chapter, we assume that $b_v = 0$ for simplicity.

Equations (5.15) and (5.16) constitute the basic expressions for modelling the transmission shown in Figure 5.11. Generally speaking, we may say that Eqs. (5.15) are related to the hydraulic energy within the circuit, whereas Eq. (5.16) has to do with the mechanical energy at the motor shaft. Given the number of variables involved, it is not possible to solve these equations as they are. Additional expressions must be added so that the number of variables can be reduced to match the number of equations. The complexity of the final mathematical model depends directly on these expressions and the simplifications that can be made.

Let us now give two particular examples of actual transmission modelling. In the first example (Case Study 1), the angular speed of the motor shaft and the pressure differential between conduits are obtained explicitly as a function of time; in the second example (Case Study 2), we approach the problem numerically.

5.2.2 Case Study 1. Purely Inertial Load with a Step Input

In this case study, we assume that the pump and motor flows, q_{pt} and q_{mt}, are given by the average expressions (3.3) and (3.4), that is, $q_{pt} = q_p = \omega_p D_p$ and $q_{mt} = q_m = \omega_m D_m$, respectively. In addition, we also consider the following:

(a) The torque at the motor shaft, T_m, is given by Eq. (5.1) with $T_{ext} = 0$ and $J_0 = 0.1\,\text{kg}\,\text{m}^2$ (purely inertial load).

(b) The pump speed, ω_p, becomes constant and equal to 1000 rpm right at the start of the transmission operation ($t \geq 0$).

(c) The transmission is configured as fixed-displacement pump/fixed-displacement motor with $D_p = 24.6\,\text{cm}^3/\text{rev}$ and $D_m = 25.0\,\text{cm}^3/\text{rev}$. The other pump and motor parameters are given in Table 3.3.

(d) The circuit has no valves between the high and low-pressure lines so that we can write $q_{cL} = 0$ in Eqs. (5.15).

(e) There are no external leakages in the circuit ($q_{pLe} = q_{mLe} = 0$), which also rules out the use of a charge circuit ($q_r = 0$).

(f) Pump and motor are connected by two identical steel conduits whose lengths and inner diameters measure 1 m and 11.7 mm, respectively. The corresponding inner volume, V, of each conduit is, therefore, $1.06 \times 10^{-4}\,\text{m}^3$.

(g) For the effective bulk modulus, β_E, we use a roughly averaged value between the two steel tube limits given by Eqs. (2.30), that is, $\beta_E = 16 \times 10^8\,\text{N}/\text{m}^2$.

With these assumptions, we can rewrite Eqs. (5.15) and (5.16) as

$$
\begin{cases}
\dfrac{dp_{12}}{dt} = \left(\dfrac{\beta_E}{V}\right)(\omega_p D_p - \omega_m D_m) - \left(\dfrac{\beta_E}{V}\right)(q_{pLi} + q_{mLi}) \\[4mm]
\dfrac{dp_{34}}{dt} = \left(\dfrac{\beta_E}{V}\right)(\omega_m D_m - \omega_p D_p) + \left(\dfrac{\beta_E}{V}\right)(q_{mLi} + q_{pLi}) \\[4mm]
\dfrac{d\omega_m}{dt} = \dfrac{\Delta p D_m}{J_0}
\end{cases}
\tag{5.17}
$$

where $\Delta p = p_{12} - p_{34}$ and $T_m = \Delta p D_m$.

If the values of V, β_E, D_p, D_m and ω_p are known, Eqs. (5.17) become a system of three equations and five variables: p_{12}, p_{34}, ω_m, q_{pLi} and q_{mLi}.

Equations (5.17) can be reduced into a single equation by performing some simple mathematical operations. First, we multiply the second equation through by -1 and then add it to the first equation:

$$\begin{cases} \dfrac{d(\Delta p)}{dt} = \left(\dfrac{2\beta_E}{V}\right)[\omega_p D_p - \omega_m D_m - (q_{pLi} + q_{mLi})] \\[3mm] \dfrac{d\omega_m}{dt} = \dfrac{\Delta p D_m}{J_0} \end{cases} \tag{5.18}$$

Then, we differentiate both sides of the second equation in Eq. (5.18):

$$\frac{d(\Delta p)}{dt} = \left(\frac{J_0}{D_m}\right)\frac{d^2\omega_m}{dt^2} \tag{5.19}$$

Substituting the derivative of the pressure differential, Δp, given by Eq. (5.19), into the first of the equations in Eq. (5.18), we obtain, after some mathematical manipulations:

$$\frac{d^2\omega_m}{dt^2} + \left(\frac{2D_m^2\beta_E}{J_0 V}\right)\omega_m = \left(\frac{2D_m D_p \beta_E}{J_0 V}\right)\omega_p - \left(\frac{2D_m\beta_E}{J_0 V}\right)(q_{pLi} + q_{mLi}) \tag{5.20}$$

We observe, in Eq. (5.20), that there are three variables to be determined, ω_m, q_{pLi} and q_{mLi}, that is, the equation is still not solvable. One possible way to circumvent this is to admit a linear relation between the internal leakages, $q_{pLi} + q_{mLi}$, and the pressure differential, Δp, as follows (see Ref. [20] for a similar approach):

$$q_{pLi} + q_{mLi} = K\Delta p \tag{5.21}$$

where K is an "internal leakage coefficient" that must be experimentally determined.

Substituting K, obtained from Eq. (5.21), into Eq. (5.20) and writing Δp as a function of the first derivative of the motor speed (see the last equation in Eq. (5.17)), we finally obtain

$$\frac{d^2\omega_m}{dt^2} + \left(\frac{2\beta_E K}{V}\right)\frac{d\omega_m}{dt} + \left(\frac{2D_m^2\beta_E}{J_0 V}\right)\omega_m = \left(\frac{2\beta_E D_m D_p}{J_0 V}\right)\omega_p \tag{5.22}$$

Observe that if the coefficients of ω_m and $\dot{\omega}_m$ are not functions of the angular speed, ω_m, or its derivatives, Eq. (5.22) will be a second-order linear differential equation. This is highly desirable, given that linear equations are much simpler to solve and usually admit analytical solutions. In this respect, we see that the coefficient K is decisive for the determination of the nature of Eq. (5.22). In fact, in Eq. (5.22), the inner conduit volume V, the effective bulk modulus β_E, the moment of inertia J_0 and the motor displacement D_m can be considered as constants.[15] Therefore, it remains to be seen whether the coefficient K can be taken as a constant as well. We deal with this matter in the following section.

---

[15] Strictly speaking, the bulk modulus is not constant (see Section 2.1.2) and the conduit volume expands when the system is pressurized. However, in order to keep things simple, we assume that β_E and V are constants in our model.

5.2.2.1 Internal Leakages and Volumetric Efficiencies

The first problem that we find when trying to obtain K from experimental pump and motor data is that we do not usually have specific information about the pump and motor internal leakages. Volumetric efficiency charts only give us information about the overall leakage that takes place at the pump and the motor. We may, however, use the efficiency charts to estimate K, by assuming that the external leakages are zero.[16] With this assumption, the internal leakages of the pump and the motor can be approximately obtained by making $q_{pL} = q_{pLi}$ and $q_{mL} = q_{mLi}$ in Eqs. (3.15) and (3.32), as follows:

$$\begin{cases} q_{pLi} = \left(1 - \eta_p^v\right) D_p \omega_p \\ q_{mLi} = \left(\dfrac{1}{\eta_m^v} - 1\right) D_m \omega_m \end{cases} \tag{5.23}$$

From Eqs. (5.21) and (5.23), we have

$$K = \frac{1}{\Delta p}\left[\left(1 - \eta_p^v\right) D_p \omega_p + \left(\frac{1}{\eta_m^v} - 1\right) D_m \omega_m\right] \tag{5.24}$$

Using Eq. (4.29), we can also write Eq. (5.24) as

$$K = \frac{1}{\Delta p}\left[\left(\frac{1}{\eta_p^v} - 1\right)\frac{D_m \omega_m}{\eta_m^v} + \left(\frac{1}{\eta_m^v} - 1\right) D_m \omega_m\right] \tag{5.25}$$

Similar to Case Studies 1 and 2 in Chapter 4, we need a predictor–corrector algorithm to determine K as a function of ω_m. Considering that the pump displacement varies from D_{pmin} to D_{pmax} and that we have chosen a constant pressure curve from where the efficiency values can be calculated, we may proceed as follows ↓:

1. Make $D_p = D_{pmin}$. For example, we can choose $D_{pmin} = 0.01$ to avoid an eventual division by zero.
2. Predict the transmission ratio, $R_T = D_p/D_m$.
3. Obtain the motor speed, ω_m, using the predicted value of R_T ($\omega_m = \omega_p R_T$).
4. Use the rated values of the pump and motor speeds to obtain the relative pump and motor speeds, ω_p^* and ω_m^*.
5. With the value of the chosen pressure differential (Δp), obtain Δp^*, ω_p^* and ω_m^* to calculate the volumetric efficiencies of the pump and motor, using Eqs. (3.73), (3.74) and (3.94). Multiply them to obtain the volumetric efficiency of the transmission, η_{HT}^v.
6. Obtain the corrected value of the transmission ratio, $R_T^C = \eta_{HT}^v R_T$.
7. Obtain the corrected value of the motor speed, ω_m^C ($\omega_m^C = \omega_p R_T^C$).

---

[16] Note that in Section 4.5, we assumed that the internal leakages were zero (Eq. (4.58)). In that case, we overestimated the volumetric losses in the circuit, which was favourable for a safe choice of the charge pump. Likewise, here we also overestimate the coefficient K by assuming that all the leakages are internal. We may then conclude that the actual circuit will be less dissipative than the circuit we are simulating.

8. With the corrected value of the motor speed, ω_m^C and the volumetric efficiencies of the pump and the motor, η_p^v and η_m^v, obtain K from Eq. (5.25).
9. Increase the value of the pump displacement, D_p, by a small increment and return to step 2 (i.e. make $D_p = D_p + \Delta D_p$).

Figure 5.13 shows the variation of K with the motor speed, ω_m, for some selected values of the pressure differential, Δp. From the figure, it is clear that the variation of K with ω_m is approximately linear for every pressure differential.[17]

Keeping in mind that some simplifications are already embedded in Eq. (5.25), we seek to find a meaningful constant value of K obtained from the curves in Figure 5.13, which can be used in Eq. (5.22). In this respect, we note that the difference between the minimum and the maximum values of K considering the whole Δp and ω_m range is approximately given by $\Delta K = 0.026 - 0.014 = 0.012$. The average value of K, on the other hand, is $\overline{K} = 0.020$. Therefore, we may write that

$$K = \overline{K} \pm \frac{\Delta K}{2} = 0.020 \pm 0.06 \tag{5.26}$$

Equation (5.26) shows that using an average value for K in Eq. (5.22) incurs in a $\pm 30\%$ error in relation to the maximum and minimum values, respectively. On the other hand, we have seen that K, as expressed by Eq. (5.25), is not precise and is in fact being overestimated. Therefore, given that there is no practical use in being too rigorous about the coefficient K in this case, we have chosen to use $K = \overline{K}$ in this chapter.

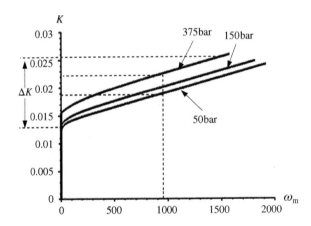

Figure 5.13 Variation of the coefficient K (l/(min bar)) with the motor speed, ω_m (rpm)

---

[17] Note that the variation of K with pressure is small. For instance, for $\omega_m = 1000$ rpm and Δp changing between 50 and 375 bar, K can be approximately taken as 0.021 ± 0.002 1/(min bar).

5.2.2.2 Natural Frequency and Damping Ratio

In order to find the natural frequency of oscillations and the damping ratio of the circuit, we need to solve Eq. (5.22) first. Let us begin by writing it in a more convenient way:

$$\frac{d^2\omega_m}{dt^2} + a\frac{d\omega_m}{dt} + b\omega_m = c \times 1(t) \tag{5.27}$$

where

$$a = \frac{2\beta_E\overline{K}}{V}, \quad b = \frac{2D_m^2\beta_E}{J_0V} \quad \text{and} \quad c = \frac{2\beta_E D_m D_p \omega_p}{J_0V} \tag{5.28}$$

The unit step function[18] $1(t)$ in Eq. (5.27) indicates that the pump speed suddenly changes from zero to the steady-state speed, ω_p, at $t = 0$. This can be practically accomplished by connecting the pump to the prime mover through a clutch, for example. We can therefore add the following initial conditions to Eq. (5.27):

$$\omega_m(0^-) = 0 \quad \text{and} \quad \dot\omega_m(0^-) = 0 \tag{5.29}$$

where 0^- indicates an instant of time immediately prior to $t = 0$ (see Figure B.2).

The problem described by Eqs. (5.27) and (5.29) is identical to the problem described by Eqs. (B.36), (B.37) and (B.43) in Appendix B, whose solution is given by Eq. (B.63).[19] We can therefore write the solution of Eqs. (5.27) and (5.29) as

$$\omega_m = \left(\frac{c}{b}\right)\left\{1 - e^{-\left(\frac{a}{2}\right)t}\left[\cos\left(\omega_\gamma t\right) + \left(\frac{a}{2\omega_\gamma}\right)\sin(\omega_\gamma t)\right]\right\} \tag{5.30}$$

where the damped natural frequency, ω_γ, is obtained from Eq. (B.21):

$$\omega_\gamma = \sqrt{b - \frac{a^2}{4}} = \sqrt{\frac{2D_m^2\beta_E}{J_0V} - \frac{\beta_E^2\overline{K}^2}{V^2}} \tag{5.31}$$

The damping ratio, ζ, is given by (see Eq. (B.20)):

$$\zeta = \frac{a}{2\sqrt{b}} = \sqrt{\frac{\overline{K}^2 J_0\beta_E}{2VD_m^2}} \tag{5.32}$$

Note that the damped natural frequency, ω_γ, can be immediately identified in Eq. (5.30), as it appears being multiplied by time within the sine and cosine functions. The same cannot

---

[18] See Eq. (B.34) for the definition of a unit step function.
[19] At this point, we recommend the reading of Section B.2 for those who are not familiar with the basics of ordinary differential equations.

be said about the damping ratio, ζ, which has been defined in Eq. (B.20) based on the homogeneous counterpart of Eq. (5.27), as explained in Section B.2.1. We also observe that

1. The damped natural frequency, ω_γ, and the damping ratio, ζ, are influenced by the physical properties β_E, J_0 and $\overline{K}$ and the geometric parameters V and D_m. The pump speed (problem input), ω_p, does not have any influence on ω_γ or ζ.
2. The maximum amplitude of the oscillations is directly proportional to the coefficient c in Eq. (5.30) and therefore is strongly influenced by the pump speed, ω_p (see Eqs. (5.28)).
3. For the limit when $t \to \infty$, Eq. (5.30) becomes identical to Eq. (1.21), obtained for the 100% efficient steady-state operation:

$$\lim_{t \to \infty}(\omega_m) = \frac{c}{b} = \frac{D_p \omega_p}{D_m} \tag{5.33}$$

Substituting the numerical values of ω_p, D_p, D_m, J_0, V, β_E and $\overline{K}$ into Eqs. (5.28) and then into Eqs. (5.31) and (5.32), after performing the necessary unit conversions, we obtain

$$\omega_\gamma = 47.41 \text{ rad/s} \quad \text{and} \quad \zeta = 0.73 \tag{5.34}$$

According to Ref. [21], in hydraulic circuits, a good damping ratio should range between 0.7 and 1. Based on this premise, we can say that in principle the energy dissipation in the circuit is satisfactory. However, a deeper analysis should still be made to carefully inspect the pressure differential and the motor speed.

5.2.2.3 Pressure and Speed Output

The output of the transmission described by Eq. (5.27) is the angular speed of the motor shaft, ω_m. On the other hand, the input is the motor speed, ω_p, which appears on the right side of the equation. The relation between the input and output is clear in the solution equation (5.30), where ω_m is directly written as a function of ω_p. Now, if we differentiate both sides of Eq. (5.30) with respect to time, we can use the second of Eqs. (5.18) to obtain the pressure differential, Δp. The result is given here without a formal development:

$$\Delta p = \frac{J_0 c}{b D_m}\left[\omega_\gamma + \frac{a^2}{4\omega_\gamma}\right] e^{-\left(\frac{a}{2}\right)t}\sin(\omega_\gamma t) \tag{5.35}$$

Figure 5.14 shows the motor speed and the pressure differential as functions of time, obtained from Eqs. (5.30) and (5.35), using the numerical data given at the beginning of this section. We see that the oscillations in pressure and speed are quickly damped within less than 0.2 s. Note that the pressure differential reaches 800 bar during the initial overshoot, a value that is far beyond the pump and motor limits (345 bar, according to Table 3.3).

It is interesting, at this stage, to plot the hypothetical curves $\omega_m(t)$ and $\Delta p(t)$ that would have been obtained for an undamped system, that is, with $\overline{K} = 0$ in Eqs. (5.27) and (5.28). In this case, the damping ratio, ζ, would become zero and the (undamped) natural frequency,

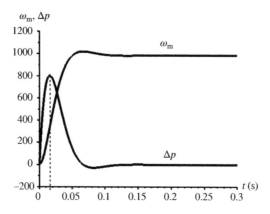

Figure 5.14   Pressure differential (bar) and motor speed (rpm) as functions of time

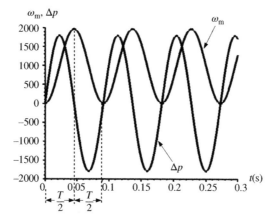

Figure 5.15   Pressure differential (bar) and motor speed (rpm) as functions of time (undamped case)

ω, would become 69.13 rad/s, which corresponds to $f = 11.05$ Hz. Because of the fact that no dissipation is present within the circuit, the pressure spikes reach their maximum in this situation. Note that this hypothetical undamped circuit can reveal some important limits, such as the maximum value of the pressure overshoots and the maximum natural frequency that can be achieved using this circuit configuration. Figure 5.15 shows the curves $\omega_m(t)$ and $\Delta p(t)$ obtained in this case.

From Figure 5.15, we see that the pressure overshoots happen within the time lapse $T/2$, where T is the oscillation period given by (see Eq. (B.23)):

$$T = \frac{2\pi}{\omega_\gamma} \tag{5.36}$$

Considering the undamped case, we have that $T/2 = \pi/(69.13) \cong 0.045$ s. Given that $\omega_\gamma < \omega$, we conclude that the undamped model gives a lower limit for the pressure spike interval (45 ms, in this example). Therefore, the relief valve needs to be able to respond in as fast as 22.5 ms to alleviate the pressure overshoots in time (note that the higher pressure value occurs within $T/4$ seconds after $t = 0$ in Figure 5.15).

5.2.2.4   Pressure Overshoot Attenuation

We saw in Figure 5.14 that the pressure differential reached a peak of 800 bar, at the highest acceleration point. It is imperative, therefore, to reduce the magnitude of this momentary pressure spike in order to satisfy the maximum circuit requirements. One of the possible means to accomplish this is through relief valves. Relief valves provide a momentary extra leakage between the high and low-pressure conduits, which increases the leakage coefficient, $\overline{K}$, and therefore the damping ratio, ζ. For instance, the coefficient, $\overline{K}$, which corresponds to the critical damping ratio, $\zeta = 1$, is given by (see Eq. (5.32)):

$$\overline{K} = \sqrt{\frac{2VD_m^2}{J_0\beta_E}} \tag{5.37}$$

If we substitute the numerical values of V, J_0, β_E and D_m into Eq. (5.37), we obtain $\overline{K} = 0.0275\ 1/(\text{min bar})$. Figure 5.16 shows the curves $\Delta p(t)$ and $\omega_m(t)$ for $\overline{K} = 0.0275$ (critically damped circuit, $\zeta = 1$) and $\overline{K} = 0.0825$ (overdamped circuit, $\zeta = 3$). Observe that the overdamped circuit has lowered the magnitude of the pressure overshoot significantly (from over 800 bar to less than 300 bar).

Another means of reducing pressure spikes is to provide a smoother variation of the pump speed at the transmission start-up. In fact, we know that magnitude of pressure overshoots and undershoots is directly associated with the maximum angular acceleration of the motor, $\dot{\omega}_m$, which in turn can be reduced by a smoother pump acceleration. Note that in this case study, there was an abrupt change in the pump speed at time $t = 0$ as ω_p changed from 0 to 1000 rpm instantaneously. Therefore, we can say that the right-hand side of Eq. (5.27) represents the 'worst-case scenario' in terms of pump acceleration. We deal with this subject in more details in Case Study 2.

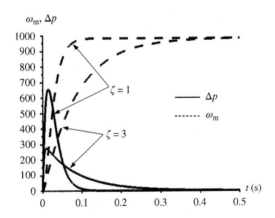

Figure 5.16   Effect of damping factor on the pressure differential (bar) and motor speed (rpm)

5.2.2.5 Final Considerations

If we observe Eqs. (5.30) and (5.35), we see that the transient outputs $\omega_m(t)$ and $\Delta p(t)$ are influenced by several transmission parameters. For instance, the moment of inertia of the motor and the load, J_0, affects the damped natural frequency, the damping ratio and the amplitude of the pressure overshoots. Figure 5.17 shows the curves $\omega_m(t)$ and $\Delta p(t)$ considering $J_0 = 1.0 \text{ kg m}^2$ (10 times the original value) while keeping the same values for the other parameters (ω_γ and ζ, obtained from Eqs. (5.31) and (5.32), are also indicated in the figure). We see that the higher moment of inertia has caused an increase in the magnitude of the pressure overshoot and a considerable increase in the damping ratio (2.3 against the former 0.73). The damped natural frequency, on the other hand, did not change much, decreasing from 47.41 to 45.32 rad/s.

Geometrical parameters such as the inner conduit volume and motor displacement also affect the transmission output. However, these are likely to remain unchanged once the transmission is running. On the other hand, the effective bulk modulus, β_E, can be easily altered by the presence of entrapped air in the system or by a temperature rise, for example, as pointed out in Section 2.1.2. For instance, it has been reported that the bulk modulus of hydraulic fluids can be reduced by 61% for a 37.8 °C temperature rise [22]. Figure 5.18 shows the pressure differential and speed during a transient considering a 50% reduction in the value of β_E, that is, $\beta_E = 8 \times 10^8 \text{ N/m}^2$. We see that the damping ratio has decreased from 0.73 to 0.51, which now falls below the recommended range of 0.7–1.0 [21]. The natural frequency also dropped from 47.41 to 41.91 rad/s.

5.2.3 Case Study 2. Variable Pump Flow

In Case Study 1, we obtained the transmission output (pressure differential and motor speed) of the simplified model shown in Figure 5.11, considering the particular situation where the pump flow was given by $\omega_p D_p$ and remained constant thorough the transmission operation. The resulting mathematical model was simple enough to be analytically solved and yet provided some important insight into the transmission operation.

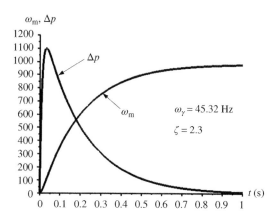

Figure 5.17 Influence of the load's inertia on the motor speed (rpm) and pressure differential between conduits (bar)

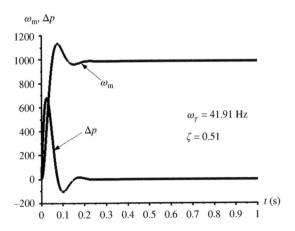

Figure 5.18 Pressure differential (bar) and motor speed (rpm) considering a 50% reduction in the effective bulk modulus

In this section, we study the case where the pump flow, q_{pt}, is slowly increased from zero to steady-state values. To make the situation as close to reality as possible, flow ripples will also be considered in the pump flow, which, in this example, will be given by Eqs. (5.6)–(5.8), repeated here as follows:

$$\begin{cases} q_t^i = AR_m \omega_p \tan(\alpha) \sin\left[\omega_p t + \left(\dfrac{2\pi}{N}\right)(i-1)\right] \\ q_{pt}^i = q_t^i \quad \text{if } q_t^i \geq 0 \\ q_{pt}^i = 0 \quad \text{if } q_t^i < 0 \\ q_{pt} = \displaystyle\sum_{i=1}^{N} q_{pt}^i \end{cases} \tag{5.38}$$

In Eqs. (5.38), both the pump speed, ω_p, and the swashplate angle, α, are allowed to change in time. In both cases, a variable pump flow is obtained.[20] For instance, we may choose to control the pump flow by changing the swashplate while keeping the pump speed constant, which is as follows:

$$\alpha(t) = \alpha_{max} e^{\left(-\frac{1}{ht}\right)} \quad \text{and} \quad \omega_p = \text{constant} \tag{5.39}$$

where h and α_{max} are constants.

Similarly, we can control the pump flow through changing the speed, ω_p, while keeping α constant:

$$\omega_p(t) = \omega_{max} e^{\left(-\frac{1}{ht}\right)} \quad \text{and} \quad \alpha = \text{constant} \tag{5.40}$$

where ω_{max} is constant.

---

[20] Using a variable-displacement pump (changing the swashplate angle) or a variable speed prime mover (changing the pump speed) apparently produces identical results for the pump output flow. Observe, however, that changing the pump speed also changes the output flow frequency, which can interact with the circuit's natural frequency, as shown in Section 5.1.4.

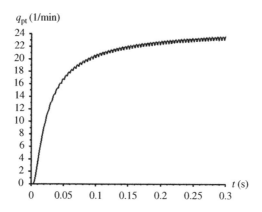

Figure 5.19   Pump flow versus time

As an example, Figure 5.19 shows the curve $q_{pt}(t)$ obtained from Eqs. (5.38) and (5.40), with $h = 50$, $N = 7$, $A = 2\,\mathrm{cm^2}$ and $R_m = 4$ cm, and with ω_p varying between 0 and $\omega_{max} = 1000$ rpm.

5.2.3.1   Pump Flow Frequency

Figure 5.20 shows the pump output between two consecutive moments in time for the pump operating at a constant speed, $\omega_p = 1000$ rpm. By simple observation, it is possible to obtain the following equations for the flow period, T_p, for $N > 1$, having the period of a single piston pump, T, as a reference:

$$\begin{cases} T_p = \dfrac{T}{2N}, & \text{if } N \text{ is odd} \\[4mm] T_p = \dfrac{T}{N}, & \text{if } N \text{ is even} \end{cases} \tag{5.41}$$

The corresponding flow frequency, f_p, can then be calculated as

$$\begin{cases} f_p = \dfrac{2N}{T} = \dfrac{\omega_p N}{\pi}, & \text{if } N \text{ is odd} \\[4mm] f_p = \dfrac{N}{T} = \dfrac{\omega_p N}{2\pi}, & \text{if } N \text{ is even} \end{cases} \tag{5.42}$$

where ω_p must be given in radians per unit of time.

As an example of the output flow frequency calculation, consider the situation where $N = 7$ and $\omega_p = 1000$ rpm (104.67 rad/s). Using the first equation in Eq. (5.42), we obtain

$$f_p = \frac{104.67 \times 7}{\pi} = 233.33 \text{ Hz} \tag{5.43}$$

The angular speed of the pump, which causes the output flow to oscillate in the circuit's natural frequency, can be calculated by substituting the pump flow frequency, f_p, in Eqs. (5.42),

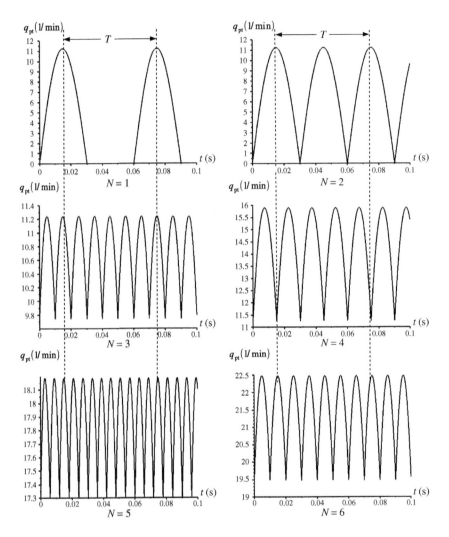

Figure 5.20   Output flow for a constant pump speed and a variable number of cylinders

by the damped natural frequency of the circuit $f = \omega_\gamma / 2\pi$, obtained from Eq. (5.31). It is easy to see that the corresponding pump speed, ω_p, will be given by

$$\begin{cases} \omega_p = \dfrac{\omega_\gamma}{2N}, & \text{if } N \text{ is odd} \\[2mm] \omega_p = \dfrac{\omega_\gamma}{N}, & \text{if } N \text{ is even} \end{cases} \qquad (5.44)$$

5.2.3.2 Mathematical Model

In this example, we assume that instantaneous motor speed, ω_m, is obtained from Eq. (3.4), that is, $\omega_m = q_{pt}/D_m$. We also consider that the motor displacement, D_m, is $25.0\,cm^3/rev$ and assume that Eq. (5.21) is valid with K being replaced by an average coefficient $\overline{K}$. With these considerations, Eqs. (5.15) and (5.16) reduce to

$$\begin{cases} \dfrac{d\Delta p}{dt} = \left(\dfrac{2\beta_E}{V}\right)(q_p - D_m\omega_m - \overline{K}\Delta p) \\[2mm] \dfrac{d\omega_m}{dt} = \dfrac{\Delta p D_m}{J_0} \end{cases} \qquad (5.45)$$

Equations (5.45) have two unknowns, ω_m and Δp, and therefore admit a general solution.

Similarly to Case Study 1, it is possible to write the system of first-order differential equations (5.45) as a single second-order ordinary differential equation (ODE). However, because of the solution method that will be used in this case study, it is convenient that we leave Eqs. (5.45) as they are. In fact, in this example, we employ a numerical method to find $\Delta p(t)$ and $\omega_m(t)$ instead of trying to obtain an analytical solution to the problem. Given that now we have two first-order ODEs, we need to specify the initial values for Δp and ω_m within Eqs. (5.45). In this aspect, we remember that the pump starts smoothly from rest and that the external torque at the motor is zero, which give

$$\omega_m(0) = 0 \quad \text{and} \quad \Delta p(0) = 0 \qquad (5.46)$$

Before we continue with the solution of Eqs. (5.45) and (5.46), let us introduce a classic numerical method for solving first-order systems of ODEs known as the Runge–Kutta method.[21]

5.2.3.3 The Runge–Kutta Method

Consider the following general system of first-order ODEs:

$$\begin{cases} \dfrac{dy_1}{dt} = f_1\left(y_1, y_2, \ldots, y_m, t\right) \\[2mm] \dfrac{dy_2}{dt} = f_2(y_1, y_2, \ldots, y_m, t) \\[1mm] \vdots \\[1mm] \dfrac{dy_m}{dt} = f_m(y_1, y_2, \ldots, y_m, t) \end{cases} \qquad (5.47)$$

---

[21] In honour of the German mathematicians Carl David Tolmé Runge (1856–1927) and Martin Wilhelm Kutta (1867–1944).

Equations (5.47) can be written in vector notation as

$$\frac{d\mathbf{y}}{dt} = f(\mathbf{y}, t) \tag{5.48}$$

where

$$\mathbf{y}^T = \begin{bmatrix} y_1 & y_2 & \cdots & y_m \end{bmatrix} \tag{5.49}$$

The Runge–Kutta method is a marching method. Marching methods require the knowledge of the solution vector, $\mathbf{y}$, at time $t_0 = 0$, from which the solution at any instant $t_n > t_0$ can be approximated. The algorithm is simple to be implemented and can be described as follows [23]:

1. From the known solution at time t_n ($\mathbf{y}_n$), compute the solution at time $t_{n=1}$ ($\mathbf{y}_{n+1}$) as

$$\mathbf{y}_{n+1} = \mathbf{y}_n + \frac{\Delta t}{6}(\mathbf{K}_1 + 2\mathbf{K}_2 + 2\mathbf{K}_3 + \mathbf{K}_4)$$

where ($\Delta t = t_{n+1} - t_n$). The coefficients $\mathbf{K}_1$ through $\mathbf{K}_4$ are given by

$$\begin{cases} \mathbf{K}_1 = f\left(\mathbf{y}_n, t_n\right) \\[2mm] \mathbf{K}_2 = f\left[\mathbf{y}_n + \Delta t\left(\dfrac{\mathbf{K}_1}{2}\right), \left(t_n + \dfrac{\Delta t}{2}\right)\right] \\[2mm] \mathbf{K}_3 = f\left[\mathbf{y}_n + \Delta t\left(\dfrac{\mathbf{K}_2}{2}\right), \left(t_n + \dfrac{\Delta t}{2}\right)\right] \\[2mm] \mathbf{K}_4 = f(\mathbf{y}_n + \Delta t\mathbf{K}_3, t_n + \Delta t) \end{cases}$$

2. Make $\mathbf{y}_n = \mathbf{y}_{n+1}$ and return to step 1.

The algorithm proceeds as long as it is needed. In the following section, we show the results obtained from the implementation of the Runge–Kutta method for the solution of Eqs. (5.45) and (5.46).[22]

5.2.3.4 Pressure and Speed Changes at Start-Up

Consider the case where the pump flow changes according to Eqs. (5.38) and (5.40). Note that the constant, h, in Eqs. (5.40), determines the speed in which the pump flow increases from zero to a maximum value. The limiting case, when $h \to \infty$, corresponds to a sudden variation of the pump speed, as assumed in Case Study 1.

---

[22] Note that we can write Eqs. (5.45) in vector form (Eq. (5.48)) by making $\mathbf{y}^T = \begin{bmatrix} \Delta p & \omega_m \end{bmatrix}$ and $[f(\mathbf{y}, t)]^T = \left[(2\beta_E/V)(q_p - D_m\omega_m - \overline{K}\Delta p) \quad \Delta p D_m/J_0\right]$.

Figure 5.21 shows the curves $\Delta p(t)$ and $\omega_m(t)$ for a fixed-displacement pump/fixed-displacement motor transmission, with ω_p changing between 0 and $\omega_{max} = 1000$ rpm. The pump swashplate angle was kept constant and was equal to 0.22 rad ($\cong 12.6°$). In addition, we have considered that $N = 7$, $A = 2\,\text{cm}^2$ and $R_m = 4$ cm in Eqs. (5.38). As for the other parameters, we have adopted the same values used in Case Study 1: $\beta_E = 16 \times 10^8\,\text{N/m}^2$, $V = 1.06 \times 10^{-4}\,\text{m}^3$, $J_0 = 0.1\,\text{kg}\,\text{m}^2$ and $\overline{K} = 0.020\,1/(\text{min}\,\text{bar})$.

It is interesting to compare the worst-case scenario shown in Figure 5.21 ($h = 50$) with the sudden motor acceleration shown in Figure 5.14, where the magnitude of the pressure differential, Δp, reached 800 bar, that is, approximately twice the value of the pressure spike obtained for $h = 50$ in Figure 5.21(a).

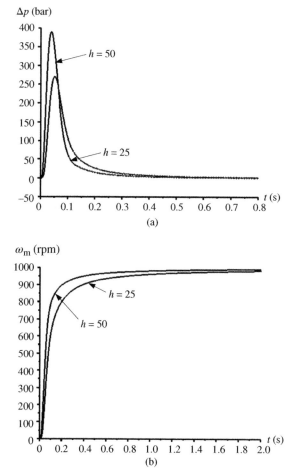

Figure 5.21   Pressure differential (a) and motor speed (b) for different values of the constant h

It can be shown that the same results shown in Figure 5.21 could have been obtained for a variable-displacement pump/fixed-displacement motor transmission, with the pump swashplate angle changing from zero to a maximum value (Eqs. (5.39)). In fact, the same average pump flow would have been produced in the end if we had followed that approach. However, there is a particular and important situation for which the results will differ. We deal with this subject in the following section.

5.2.3.5 Resonance and Beating

Consider that no load is attached to the motor shaft so that $J_0 = 0.0018$ kg m^2, corresponding to the moment of inertia of the M25MF motor in Table 3.3. This choice will reduce the damping ratio, ζ, and increase the damped natural frequency, ω_γ, as can be concluded from Eqs. (5.32) and (5.31), respectively. Consider also that we maintain the same values of β_E, V and $\overline{K}$, which we used in Case Study 1 and in the example given in Figure 5.21. It can be shown that the values of the damped natural frequency, ω_γ, and the damping ratio, ζ, are 512.82 rad/s and 0.098, respectively. Substituting this value of ω_γ into the first of Eqs. (5.44), we obtain the pump speed, ω_p, that produces an output flow whose frequency equals ω_γ for an axial piston pump with seven cylinders, that is, $\omega_p = 349.79$ rpm. Figure 5.22 shows the curves $\omega_m(t)$ obtained for two different situations:

- In Figure 5.22(a), the swashplate, α, is constant and equal to 0.22 rad. The pump speed, on the other hand, changes according to Eq. (5.40) with $h = 200$ and $\omega_{pmax} = 349.79$ rpm.
- In Figure 5.22(b), the pump speed is constant and twice the value of the resonant speed, that is, $\omega_p = 699.58$ rpm. To maintain the same output flow, the swashplate changes according to Eq. (5.39) with $\alpha_{max} = 0.11$ rad and $h = 200$.

We clearly see in Figure 5.22 that there is a difference in the ω_m behaviour at steady-state regime in each case. In Figure 5.22(b), the oscillations that we observe after $t = 0.2$ s are

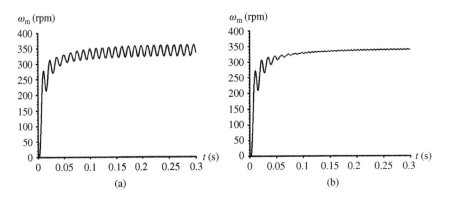

Figure 5.22   Variation of the motor speed with time: (a) resonant and (b) non-resonant case

basically those resulting from the flow ripples originating from the pump. In this case, the pump rotates at a speed that is twice the resonant speed and does not interact with the circuit's natural frequency. On the other hand, in Figure 5.22(a), the pump flow oscillates at the circuit's natural frequency, causing the amplitude of the motor fluctuations to become considerably higher. In fact, depending on the amount of energy dissipation in the circuit, the oscillation amplitude at the motor speed may grow very quickly. The limiting case is the situation in which no energy dissipation exists, which in our example corresponds to $\overline{K} = 0$.

Figure 5.23 shows two hypothetical situations where $\overline{K} = 0$. The angular speed of the pump for which resonance occurs is $\omega_p = 351.47$ rpm. In Figure 5.23(a), we consider the situation where the swashplate angle is constant and the pump speed changes from 0 to 351.47 rpm, with $h = 200$. In Figure 5.23(b), we make $\omega_{pmax} = 702.94$ rpm (twice the resonance speed) and vary the swashplate angle from 0 to 0.11 rad. Resonance occurs in the

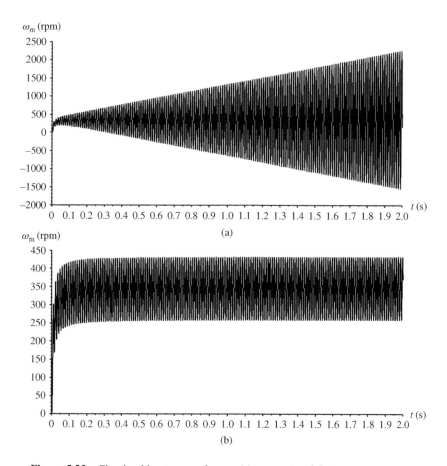

Figure 5.23   Circuit without energy losses: (a) resonant and (b) non-resonant case

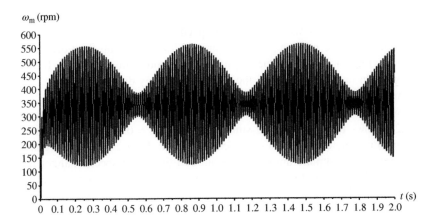

ω_m (rpm)

Figure 5.24 Beating phenomenon

situation illustrated in Figure 5.23(a), where we see the amplitude of the motor speed oscillations growing steadily in time.[23] The same does not happen in Figure 5.23(b), where the oscillations result from the fluctuating pump flow.

We end this chapter showing an example of the beating phenomenon. Figure 5.24 shows the situation where the pump speed changes from 0% to 98% of the resonance value ($\omega_p = 344.44$ rpm), while the swashplate angle remains constant. Again, we have considered $\overline{K} = 0$ in this example. Because of the proximity of the resonance speed, we observe a periodical rise in the motor speed amplitude.

From the results shown in Figures 5.22–5.24, we see that it is better to keep the pump running at a faster speed while controlling the motor through changing the pump displacement to avoid resonance in the circuit. In this aspect, the pump speed based on the undamped natural frequency, $\omega_p = 351.47$ rpm, can be seen as a minimal value under which resonance (or beating) may occur.

Exercises

(1) Give a formal demonstration of Eq. (5.5).

(2) Starting from Eq. (5.5), show that the average swashplate pump flow is given by

$$q_p = \frac{NAR_m\omega_p\tan(\alpha)}{\pi}$$

(3) Using the expression shown in the previous exercise, obtain the average output flow for the pump corresponding to Figure 5.5, where $N = 7$, $A = 2\,\mathrm{cm}^2$, $\alpha = 0.22$ rad ($12.61°$), $R_m = 4$ cm and $\omega_p = 1000$ rpm

---

[23] As expected, because of the timescale, the sinusoidal pattern of the curves $\omega_m(t)$ is not clearly distinguished.

(4) Starting from the expression for the average output flow in Exercise 2, demonstrate the first of Eqs. (3.42):

$$q_p = \left[\frac{\omega_p D_p^{max}}{\tan\left(\alpha_{max}\right)} \right] \tan(\alpha)$$

(5) Develop a mathematical model for the transmission shown in Figure 5.11 by considering the same assumptions made for Case Study 1, for the pump speed, ω_p, changing linearly in time ($\omega_p = kt$, where k is a constant), and the external torque, T_{ext}, being not equal to zero. Next, obtain an expression for the solution ω_m as a function of time.

(6) Show that in the absence of volumetric losses in the circuit, the mathematical model for Case Study 1 reduces to

$$\omega_m = \left(\frac{D_p \omega_p}{D_m} \right) [1 - \cos(\omega t)]$$

(7) Estimate the internal leakages at the pump and motor in l/min by considering a pressure differential of 170 bar and an average leakage coefficient, $\overline{K} = 0.020$ l/(min bar).

References

[1] Jiang D, Li S, Zeng W, Edge KA (2012) Modeling and simulation of low pressure oil-hydraulic pipeline transients. Computers and Fluids, 67: 79–86.

[2] Shu JJ (2003) Modelling vaporous cavitation on fluid transients. International Journal of Pressure Vessels and Piping, 80: 187–195.

[3] Handroos HM, Vilenius MJ (1990) The utilization of experimental data in modelling hydraulic single stage pressure control valves. Transactions of ASME, 112: 482–488.

[4] Robson P, Zähe B (2006) Anti shock relief valve, Technical paper. Sun Hydraulics UK.

[5] Zhang KQ, Karney BW, McPherson DL (2008) Pressure-relief valve selection and transient pressure control. Journal AWWA, 100(8): 62–69.

[6] Bosch-Rexroth (2012) Radial piston motor for slew drives MCR-X. Technical catalogue RE 15214/06.2012.

[7] Bucher Hydraulics (1999) Stacking anti-shock relief valve, Technical catalogue 400-P-300101-E-/01.99.

[8] Howeth D (1987) Pressure limiting acceleration control system and valve for hydraulic motors. US Patent, number 4,694,649 – 22 September 1987.

[9] Sundstrand (1974) 15-Series hydrostatic transmissions: service manual: bulletin 9646, USA.

[10] Manring ND (2000) The discharge flow ripple of an axial-piston swash-plate type hydrostatic pump. Journal of Dynamic Systems, Measurement, and Control, 122: 263–268.

[11] Hayes G, Lemond J (2013) Reducing noise in hydraulic systems. Parker-Hannifin Corporation, Parflex Division, USA.

[12] OSU (1974) Hydraulic system noise study. Annual report FPRC-4M2. Oklahoma State University, USA. Available at http://www.dtic.mil/dtic/tr/fulltext/u2/a011170.pdf. January 2014.

[13] Wilkes & McLean (2013) The hydraulic noise, shock, vibration and pulsation suppressor, Technical brochure. Wilkes & McLean Ltd, USA.

[14] Streeter VL (1962) Fluid mechanics, 3rd Ed., McGraw Hill, Japan.

[15] Sewall JL, Wineman DA, Herr RW (1973) An investigation of hydraulic-line resonance and its attenuation, NASA Technical Memorandum – NASA TM X-2787, USA. Available at http://ntrs.nasa.gov/archive/nasa/casi.ntrs.nasa.gov/19740004441_1974004441.pdf?origin=publication_detail. January 2014.

[16] Kela L (2010) Attenuating amplitude of pulsating pressure in a low-pressure hydraulic system by an adaptive Helmholtz resonator, PhD Thesis, University of Oulu, Finland.

[17] Mikota J (2001) A novel, compact pulsation compensator to reduce pressure pulsations in hydraulic systems. In Proceedings of ICANOV – International Conference on Acoustics, Noise and Vibration, Ottawa, Canada.

[18] Wang L (2008) Active control of fluid-borne noise, PhD Thesis, University of Bath, UK.

[19] Martin HR (2000) Vibrational analysis for fluid power systems, In: Totten GE (ed) Handbook of hydraulic fluid technology. Marcel Dekker Inc., New York.

[20] Stringer J (1976) Hydraulic system analysis: an introduction. John Wiley & Sons, USA

[21] Watton J (2009) Fundamentals of fluid power control. Cambridge University Press, UK.

[22] George HF, Barber A (2007) What is bulk modulus, and when is it important?. The Lubrizol Corp., Wickliffe, Ohio, USA. In Hydraulics & Pneumatics, http://hydraulicspneumatics.com. Accessed in January 2014.

[23] Kulakowski BT, Gardner JF, Shearer JL (2007) Dynamic modeling and control of engineering systems, 3rd Ed., Cambridge University Press, UK.

6

Hydrostatic Actuators

Chapters 4 and 5 focused on hydrostatic transmissions. In this chapter, we address the theme of hydrostatic actuators. First, we explore some key subjects that will be needed in the rest of the chapter, such as the operational quadrant description of hydraulic circuits and some basic energy management concepts. Next, we describe an array of typical hydrostatic actuator circuits in detail for both displacement-controlled actuators and electrohydrostatic actuators. We close the chapter by introducing the 'common pressure rail' (CPR) concept and its relation to hydraulic transformers.

6.1 Introductory Concepts

In this section, we deal with some introductory notions that are very important for the study of hydrostatic and electrohydrostatic actuators. We begin by introducing a common methodology to characterize the steady-state operation of a hydrostatic actuator in a graphical manner: the 'four-quadrant' representation.

6.1.1 Circuit Operational Quadrants

Let us start by establishing a convention for the force and speed signs in a hydraulic cylinder. In this book, we assume the following (see Figure 6.1(a)):

1. The cylinder speed, v, is positive when the cylinder extends, and negative otherwise.
2. The force, F, is positive if it is oriented against the velocity of the cylinder when it is extending.

Depending on the signs of the speed, v, and the force, F, the cylinder may operate in each one of the four quadrants, as shown in Figure 6.1(b). In the first quadrant, the cylinder is moving forward, pushing the load against a resistive force, F. In the third quadrant, the

Hydrostatic Transmissions and Actuators: Operation, Modelling and Applications, First Edition.
Gustavo Koury Costa and Nariman Sepehri.
© 2015 John Wiley & Sons, Ltd. Published 2015 by John Wiley & Sons, Ltd.
Companion Website: www.wiley.com/go/costa/hydrostatic

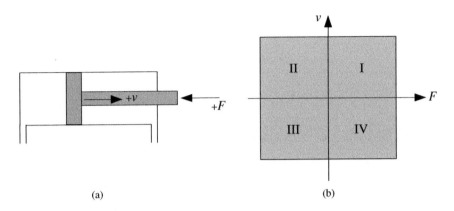

(a)   (b)

Figure 6.1   Sign convention (a) and quadrant operation (b) for a single-rod cylinder

cylinder is retracting, pulling the load against a force, F. Note that, in both cases, energy is being consumed by the pump in order to push and pull the load.

In quadrants II and IV, the load is helping the cylinder to move. In the second quadrant, the force F helps the cylinder move forward, whereas in the fourth quadrant, the force F helps the cylinder move backward.

Each quadrant in Figure 6.1(b) characterizes a mode of operation for the hydrostatic actuator. In this book, we adopt the nomenclature used in Ref. [1], which divides the operational modes as follows:

- Quadrants I and III – pumping mode.
- Quadrants II and IV – motoring mode.

The reason for labelling the operational modes in quadrants I and III as 'pumping' is obvious, since in those quadrants the pump supplies energy to the cylinder so that it can move the load forwards and backwards, respectively. On the other hand, it is possible to use the load energy in quadrants II and IV to reversely drive the pump, which in that case causes the pump to operate as a motor. Therefore, we generally use the term 'motoring mode' for quadrants II and IV.

6.1.2   Energy Management

Not all hydrostatic actuators show the same behaviour when operating in a particular quadrant. In fact, the operational characteristics in each quadrant depend on the manner in which the hydraulic circuit has been conceived. In spite of the fact that the cylinder receives energy from the load in quadrants II and IV, the way in which the actuator deals with that incoming energy depends on the hydraulic circuit design. In some circuits, the energy from the load is simply wasted, whereas in others, it can be reused (or *regenerated*). In this aspect, we observe that some circuits are able to regenerate most of the incoming energy, while others can only regenerate part of it and sometimes in only one of the two motoring quadrants. Therefore, we speak of energy regeneration if this energy is stored to be reused later

[2]. Likewise, we refer to regenerative actuators as the actuators whose circuits are able to regenerate energy from the load.[1]

> A regenerative actuator is one in which the energy received from the load at quadrants II and IV is stored to be reused later on.

Hydrostatic actuators do not make any use of throttling to control the cylinder speed. This constitutes a clear advantage over valve-controlled systems where there will always be metering losses to some extent, even in more elaborate circuits[2] [2].

6.1.3   Cylinder Stiffness

Sometimes, the cylinder is required to stop at a certain position under the effect of an external load. In such cases, it is necessary for the cylinder to remain steady so that it can hold the load in place. In other words, the cylinder must become 'stiff'. Cylinder stiffness can be obtained by properly blocking the flow at the input and output ports. In displacement-controlled actuators, this is usually done by setting the pump swashplate angle, α, to zero. The same is not observed in electrohydrostatic actuators, where the pump displacement is fixed.

6.1.4   Double-Rod and Single-Rod Actuators

We now discuss the main differences between a double-rod and a single-rod cylinder in a hydrostatic actuator. In order to make the text simpler, hereafter we will use *double-rod actuator* when discussing a hydrostatic actuator with a double-rod cylinder. Similarly, a hydrostatic actuator equipped with a single-rod cylinder will be called a *single-rod actuator*.

Double-rod actuators do not present major design difficulties and are very similar to hydrostatic transmissions in their construction to some extent. Figure 6.2 illustrates a basic hydrostatic actuator with a double-rod cylinder. The circuit itself does not need any add-ons other than an oil replacement unit and proper relief valves, not shown in the figure. The cylinder's speed and direction is completely controlled by the variable-displacement pump. Note that cylinder stiffness is guaranteed for axial piston pumps at zero displacement, but the same is not true for fixed-displacement pumps where the force applied to the piston rod can create an internal pressure that is high enough to turn the pump shaft backwards.

Single-rod actuators have some peculiarities that require more complex hydraulic circuits. For instance, consider the single-rod actuator shown in Figure 6.3. In Figure 6.3(a), the cylinder is moving to the right, and the fluid inside the circuit flows in a clockwise direction.

---

[1] Note that here we are referring to a concept different from the flow regeneration that is commonly referred to in valve-controlled regenerative hydraulic circuits.

[2] Valve-controlled actuators use a metering device (either a proportional valve or a flow control valve) to control the cylinder speed. If a constant flow pump is used, part of the flow is diverted into the tank through the relief valve to reduce the cylinder speed. More sophisticated circuits use a variable-displacement pump that adjusts its output flow according to the circuit needs, eliminating the unnecessary waste through the relief valve. However, the presence of flow control valves in the path between the pump and the cylinder still incurs in metering losses in any case.

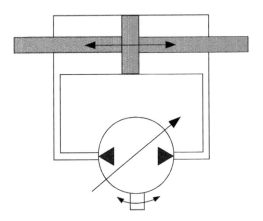

Figure 6.2   Pump-cylinder connection in a double-rod hydrostatic actuator

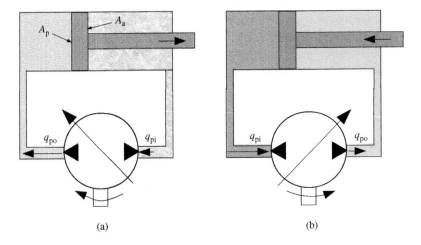

(a)   (b)

Figure 6.3   Unmatched flows in a single-rod hydrostatic actuator

In Figure 6.3(b), the pump reverses its direction in order to push the cylinder back to left. In both cases, the pump has a fixed output flow, q_{po}, which, in the absence of pump losses, should be equal to the input flow q_{pi}, for a sound operation. What we observe, however, is that the flow entering the pump, q_{pi}, is actually different from the pump output flow, q_{po}. Having Figure 6.3(a) in mind, we can write for q_{po} and q_{pi}:

$$\begin{cases} q_{pi} = A_a v \\ q_{po} = A_p v \end{cases} \tag{6.1}$$

where A_p and A_a are the areas of the piston and the ring (annulus) of the cylinder, respectively.

From Eq. (6.1), we obtain the relation between the output and input flows, q_{pi} and q_{po}:

$$q_{pi} = \left(\frac{A_a}{A_p}\right) q_{po} \tag{6.2}$$

Since $A_a < A_p$, in Eq. (6.2) we have that $q_{pi} < q_{po}$. The inverse occurs in the situation shown in Figure 6.3(b), where $q_{pi} > q_{po}$. In both cases, a problem needs to be solved. In the first case (Figure 6.3(a)), the flow entering the pump is insufficient to fill up the displacement chambers and we may have a 'pump starvation' phenomenon. In the second case (Figure 6.3(b)), the higher flow being forced into the pump causes the pressure inside the left conduit to rise. This pressure elevation creates a high opposing force on the piston side of the cylinder, causing it to stop moving. Note that the solution to the first problem is to inject more fluid into the right conduit, whereas in the second situation, fluid should be extracted from the left conduit. The obvious conclusion is that a pump cannot be simply connected to a double-rod cylinder, as shown in Figure 6.3. Some additions must be made to the hydraulic circuit in order to equalize the flows at the input and output ports of the pump.

6.2 Hydrostatic Actuator Circuits

In this section, we introduce a number of different circuits to be used with hydrostatic actuators. The largest part of the text is dedicated to single-rod actuators. Only one double-rod actuator circuit is presented in the end.

Single-rod actuators can be designed with one or two main pumps (single- or dual-pump actuators). We use 'main pumps' when discussing pumps that have a direct influence on the cylinder operation. If a pump exists only to help in the process of supplementing or extracting fluid from the circuit without directly interfering on the cylinder speed and force, we call it a secondary pump or a charge pump, depending on the specific role that it plays in the circuit. To keep things as simple as possible, pressure relief valves will not be represented in the circuits shown in this section.

6.2.1 Design 1. Dual-Pump, Open-Circuit, Displacement-Controlled Actuator

Figure 6.4 shows a circuit with two variable-displacement pumps connected to a common prime mover (PM) whose shaft rotates at an angular speed, ω_{pm} [3].

The displacements of the pumps in Figure 6.4 are individually adjusted during the cylinder expansion and retraction, and the relation between the flows, $q_{p1} = \omega_{pm}D_{p1}$ and $q_{p2} = \omega_{pm}D_{p2}$, is given by Eq. (6.2) by making $q_{po} = q_{p1}$ and $q_{pi} = q_{p2}$. As indicated in the figure, the displacements are different not only in intensity but also in direction, since the input and output ports of pump 2 are inverted in relation to the input and output ports of pump 1. Note that we have slightly modified the standard nomenclature in Figure 6.4 (Appendix A) for a better understanding of the circuit operation. We will proceed likewise in every circuit presented in this chapter.

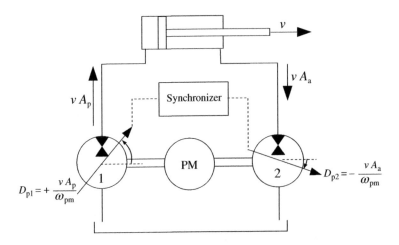

Figure 6.4   Dual-pump, open-circuit, displacement-controlled actuator

Substituting q_{pi} and q_{po} by $\omega_{pm}D_{p2}$ and $\omega_{pm}D_{p1}$ into Eq. (6.2), respectively, we obtain

$$D_{p2} = \left(\frac{A_a}{A_p}\right) D_{p1} \qquad (6.3)$$

where D_{p1} and D_{p2} are the displacements of the pumps 1 and 2.

In order to adjust both displacements, according to Eq. (6.3), a synchronizing device must be placed between the displacement controls of the two pumps, as shown in Figure 6.4. Given that the relation between the displacements is constant, the synchronizer may be a simple mechanical device connecting the swashplates of the pumps 1 and 2. However, as pump displacements are usually changed by means of electrovalves, an electric/electronic controller is often necessary.

Figure 6.5 shows the operation in the first and second quadrants when the cylinder is extending at a speed,[3] v. In the first quadrant, pump 1 delivers a flow equal to vA_p at the same time that pump 2, now in reverse operation, must be adjusted to a flow equal to vA_a. Because of the resisting force, F, the conduit connecting pump 1 to the cylinder becomes pressurized, indicated by a thicker line.

The situation in the second quadrant differs in two aspects. First, the pressurized line changes because of the resultant forces acting on the cylinder. Second, given that pump 2 now operates as a motor, the power required from the PM to keep a constant speed, v, at the cylinder is reduced. In fact, depending on the magnitude of the force F, there may not be the need of any extra power from the PM for the cylinder to extend. Note, however, that it is possible for the cylinder to accelerate to undesired speeds if the force F becomes sufficiently high. In such cases, reducing the displacements of both pumps according to Eq. (6.3) causes the pressure on the rod-side of the piston to rise. Consequently, the resultant force acting

---

[3] It is implicitly understood in this chapter that the magnitudes of the speed, v, and the force, F, do not need to remain the same in every quadrant of operation. We only maintain the same nomenclature at every operational quadrant for simplicity reasons.

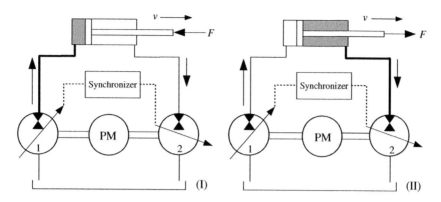

Figure 6.5 Operation in the first and second quadrants

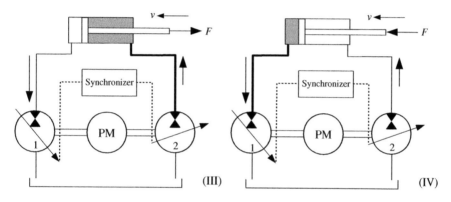

Figure 6.6 Operation in the third and fourth quadrants

on the cylinder is changed, making it possible for the speed to be controlled. Setting the displacements of both pumps to zero will block the input and output of the cylinder, causing it to stop.

Figure 6.6 shows the operation in the third and fourth quadrants where the pump displacements have been reversed because of the flow direction. Note the similarity between quadrants I and III and between quadrants II and IV. As mentioned earlier, it is only in quadrants II and IV that the cylinder is driven by the force F. Therefore, the same observations made for quadrant II concerning velocity control equally apply to quadrant IV.

6.2.2 Design 2. Dual-Pump, Closed-Circuit, Displacement-Controlled Actuator

Another type of dual-pump circuit is shown in Figure 6.7 [4]. In this circuit, the flow that comes from pump 2 is used to supplement the cylinder output flow, vA_a, so that the input and

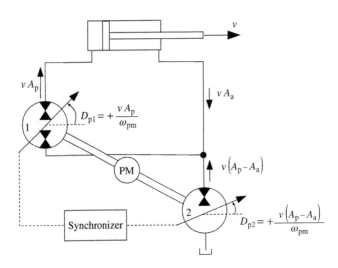

Figure 6.7   Dual-pump, closed-circuit, displacement-controlled actuator

output flows at pump 1 are matched during cylinder expansion. During cylinder retraction, pump 2 works reversely by draining flow from the rod-side into the tank. The cylinder speed, in this case, is directly controlled by adjusting the displacement of pump 1.

The relation between displacements is given by[4]

$$D_{p2} = \left(1 - \frac{A_a}{A_p} \right) D_{p1} \tag{6.4}$$

Figure 6.8 shows the operation in the first and second quadrants. In both quadrants, pump 2 supplies the missing flow into pump 1. In the second quadrant, however, pump 1 operates as a motor being driven by the flow coming out of the cylinder. If we compare this circuit to the open-circuit shown in Figure 6.5, we note that in this configuration, the driven pump (pump 2) needs to overcome a higher pressure differential between the input and output ports when compared to the driven pump in Design 1 (pump 1). Consequently, a greater part of the absorbed energy is used to keep the correct fluid flow within the circuit itself, which reduces the availability for energy regeneration. We observe that whenever it becomes necessary to slow the cylinder down, we simply need to reduce both pump displacements. Setting both displacements to zero stops the cylinder and holds it in place.

Figure 6.9 illustrates the operation in the third and fourth quadrants. Pump 1 now sends its flow, vA_p, to the rod-side port of the cylinder. However, given that the rod-side of the cylinder can only take a flow equal to vA_a, part of the flow coming from pump 1 must be diverted into the tank. This is carried out by pump 2, which extracts fluid from the circuit in both quadrants. With respect to the fourth quadrant, we can see that it differs from the second quadrant because no high pressure exists at the output port of pump 2 this time. As a result, the resisting torque at pump 2 is much lower, and less energy is required to keep the correct flows in the circuit.

---

[4] Observe that if a double-rod cylinder were used, we would have $A_p = A_a$ in Eq. (6.4) and D_{p2} would become zero, indicating that pump 2 could be removed from the circuit, as expected.

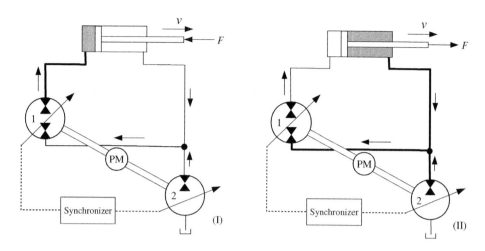

Figure 6.8 Operation in the first and second quadrants

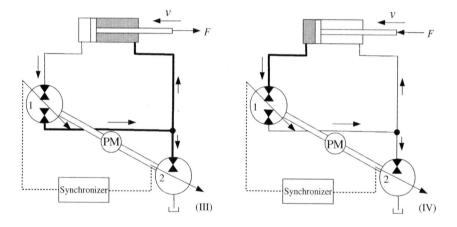

Figure 6.9 Operation in the third and fourth quadrants

6.2.3   Design 3. Dual-Pump Electrohydrostatic Actuator with Accumulators

Figure 6.10 shows an electrohydrostatic actuator originally designed to command the cylinders of a Boeing 787 flight simulator [5]. Two identical fixed-displacement pumps, each of them outputting a flow vA_a, are connected to the same electric motor and drive a differential cylinder whose relation between the areas of the piston and the ring is exactly 2:1. During cylinder extension (Figure 6.10(a)), the two pumps are connected in parallel and supply a flow equal to $2vA_a$ to the cap-side of the cylinder. Given that the area of the piston is twice the area of the ring, the total flow coming from the pumps is precisely the flow needed to extend the cylinder at a speed, v. On the other hand, the flow going out the rod-side of the cylinder exactly matches the input flow at pump 2. Therefore, there is no

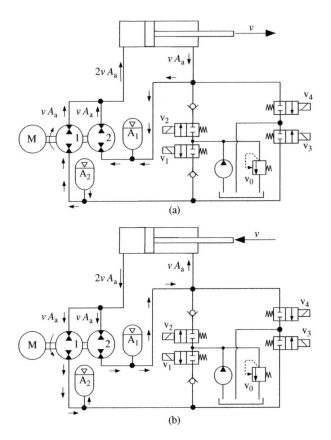

Figure 6.10 Dual-pump electrohydrostatic actuator with accumulators: (a) cylinder extension and (b) cylinder retraction

need to consume fluid from accumulator A_1, whose basic function is to keep the hydraulic line connected to the rod-side of the cylinder at a pre-adjusted pressure. Accumulator A_2 supplies the flow needed by pump 1, operating as a reservoir. When the cylinder retracts (Figure 6.10(b)), the flow coming from pump 2 is sufficient to push the cylinder to the left. Furthermore, the flow, $2vA_a$, coming from the cap-side is equally divided between the two pumps. Once again, accumulator A_2 plays the role of an oil reservoir as it is refilled by the flow coming from pump 1. Just like in the cylinder expansion, no flow is consumed[5] from accumulator A_1.

The circuit contains four 2×2 solenoid-operated valves, V_1–V_4, that can be used to load and unload the accumulators from a secondary charge circuit. Thus, valves V_1 and V_3 connect accumulator A_2 with either the charge pump or the tank, respectively. Similarly, valves V_2 and V_4 connect accumulator A_1 with the charge pump or the tank, as needed. Connecting accumulators A_1 and A_2 with the charge circuit is necessary to compensate for the volumetric losses of the circuit so that a minimum pressure, set by the relief valve, V_0,

---

[5] Despite the fact that no flow needs to be supplied by accumulator A_1, fluid can still flow into it, depending on the operational quadrant.

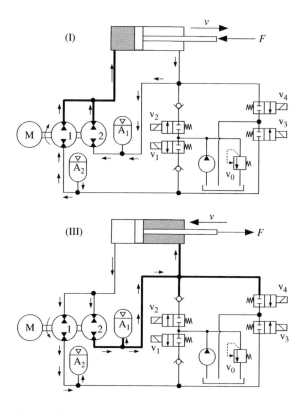

Figure 6.11 Operation in the first and the third quadrants

can be re-established at the low-pressure branches of the circuit. In addition, a means of discharging the accumulators is always necessary in any hydraulic circuit for safety reasons, which explains the inclusion of valves V_3 and V_4 in the circuit.

Figure 6.11 shows the circuit operation in the first and third quadrants. Note that as the pressure rises at the rod-side of the cylinder in the third quadrant, fluid is pushed into accumulator A_1, which acts as a shock absorber in the circuit. This shock absorbing feature, however, is not present in the first quadrant as none of the accumulators gets exposed to the high-pressure line.

Figure 6.12 shows the operation in the second and fourth quadrants. Once again, accumulator A_1 works as an energy absorber when the pressure on the rod-side line increases (quadrant II). Note that cylinder stiffness depends on the pressures at the rod and cap sides of the cylinder (established by accumulators A_1 and A_2, respectively), as well as on the torque produced by the electric motor.

6.2.4 Design 4. Circuit with an Inline Hydraulic Transformer

Figure 6.13 shows another configuration that can be used with single-rod actuators. In this design, the connection between the two pumps is made through a motor placed at the rod-side

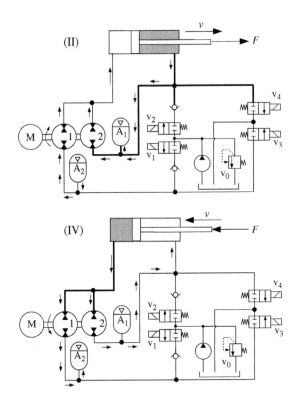

Figure 6.12 Operation in the second and fourth quadrants

line of the cylinder [6]. The set motor–pump, T, displayed as a unit inside the dash-dot rectangle, is known as *hydraulic transformer*.

In the situation illustrated in Figure 6.13(a), the cylinder is extending with a speed, v, and demands a flow, vA_p, from the main pump. Meanwhile, the flow going out of the rod-side of the cylinder, vA_a, turns the hydraulic motor, M, that is mechanically connected to a secondary pump, P. The flow coming from the secondary pump, $v(A_p - A_a)$, adds to the flow that comes out of the cylinder, vA_a, so a total flow, vA_p, reaches the input port of the main pump as required.

In Figure 6.13(b), the flow coming from the main pump into the cylinder equals vA_a. Because of the differential areas, the flow coming out of the cylinder is given by vA_p. Since the main pump outputs a flow vA_a ($vA_a < vA_p$), the secondary pump, P, needs to divert a flow $v(A_p - A_a)$ into the tank this time. This is accomplished by changing the position of the directional valve, V, as indicated in the figure. Note that since the pump and motor are mechanically connected, reversing the motor rotation also causes the secondary pump to reverse its flow direction when the cylinder speed changes sign.

Before analysing the four-quadrant operation in more details, let us draw our attention to the main component of the circuit: the hydraulic transformer, T. The name 'hydraulic transformer' has been identified as a 'unit composed of a pump and a motor in which the input flow energy at the motor, $q_m \Delta p_m$, is carried out to the pump as $q_p \Delta p_p$ [7]. Therefore,

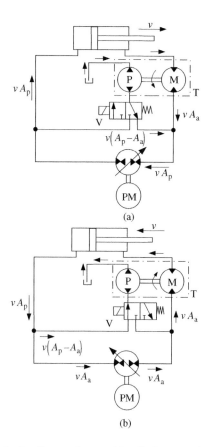

Figure 6.13   Circuit with a hydraulic transformer: (a) cylinder extension and (b) cylinder retraction

if we assume that no energy losses exist, we can write for the hydraulic transformer, T, in Figure 6.13:

$$\Delta p_p q_p = \Delta p_m q_m \tag{6.5}$$

where Δp_p and Δp_m are the pressure differentials at the pump and motor, respectively. Similarly, q_p and q_m represent the flows through the pump and the motor, respectively.

Equation (6.5) shows that either a flow amplification or reduction can be obtained through the use of a hydraulic transformer. The concept is identical to an electrical converter, in which current and voltage are varied to maintain a constant electric power. In the particular case of the circuit in Figure 6.13, we can develop Eq. (6.5) to obtain

$$\Delta p_p v(A_p - A_a) = \Delta p_m v A_a \tag{6.6}$$

From Eqs. (3.4) and (6.6), we have

$$D_p = D_m \left(\frac{T_p}{T_m}\right)\left(\frac{A_p}{A_a} - 1\right) \tag{6.7}$$

Since pump and motor are mechanically connected by the same shaft, $T_p = T_m$ in Eq. (6.7). Therefore, we can write that

$$D_p = D_m \left(\frac{A_p}{A_a} - 1 \right) \tag{6.8}$$

The interesting thing about this design is that no swashplate control is required to synchronize the displacements of the two pumps in the circuit. Moreover, only the main pump needs to have a variable displacement. However, it is true to say that there is now the need to control the directional valve, V, which was not present in the previous designs. Such valve control is quadrant dependent, as clearly illustrated in Figure 6.14, which shows the circuit operation in the first and second quadrants. We see, for example that the valve, V, needs to shift its position in the passage from the first to the second quadrant.

Figure 6.15 shows the operation in the third and fourth quadrants. Given that the motor M now rotates in the opposite direction in relation to quadrants I and II, the secondary pump directs the flow from the main line into the tank. Similar to quadrants I and II, we observe that a shift in the position of the directional valve is necessary between the third and fourth quadrants. We finally note that the cylinder speed can be easily controlled through changing the displacement of the main pump.

Although this actuator design does not make use of a synchronizer, as was the case in Designs 1 and 2, it is not practical because it is not possible to compensate for the external volumetric losses. For example, if a certain amount of fluid gets lost through leakages, the flow, $v(A_p - A_a)$, as supplied by the secondary pump becomes insufficient to meet the actuator needs. Given that no external control is exerted on the hydraulic transformer, there are no means of correcting for flow mismatches at runtime. A possible solution would be to replace the fixed-displacement pump in the hydraulic transformer by a variable-displacement pump, as shown in Figure 6.16. Note that this change will introduce the need to control the displacement of the secondary pump. In addition, such control will obviously depend on runtime information about the volumetric losses (or the corresponding pressure variations) in the

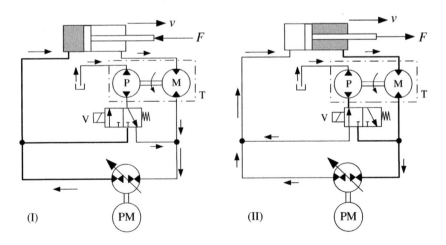

Figure 6.14 Operation in the first and second quadrants

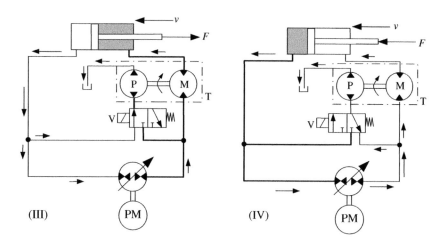

Figure 6.15 Operation in the second and fourth quadrants

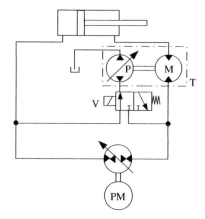

Figure 6.16 Circuit correction to account for external leakages (use of a variable-displacement pump in the hydraulic transformer)

circuit. Another major drawback is the fact that the pump–motor in the hydraulic transformer is generally 'bulky, heavy and complex' [8], which makes this design less appealing when compared to other solutions.

6.2.5 Design 5. Single-Pump Circuit with a Directional Valve

The actuator shown in Figure 6.17 is based on the design proposed in Ref. [9]. In this design, each side of the cylinder is alternately connected to the reservoir through the electrovalve, V_1. No charge pump is present, and the flows to and from the tank are possible because of the pressure differentials created in the circuit. Note that, similar to Design 4, this design also demands an external control of the electrovalve.

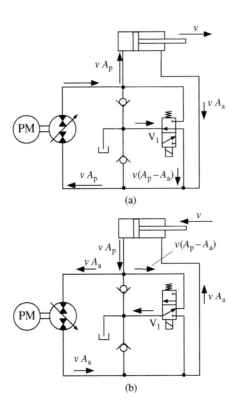

Figure 6.17 Single-pump circuit with a directional valve: (a) cylinder extension and (b) cylinder retraction

Figure 6.17(a) shows one situation when the cylinder is extending. In the figure, we observe that the solenoid of the valve V_1 remains inactive, connecting the rod-side of the cylinder to the tank. As a consequence, the lower pressure produced at the cylinder output due to the area differential creates the driving force that makes the fluid flow from the tank into the circuit. Similarly, Figure 6.17(b) shows the cylinder retracting. The solenoid of the valve V_1 is now activated, causing the excess fluid to be diverted into the tank. As will be seen shortly, the position of the directional valve, V_1, depends on the operational quadrant and not on the cylinder speed alone.

Figure 6.18 shows the operation in the first and third quadrants. Note that the commutation of the directional valve, V_1, coincides with the inversion of the pump flow. Another interesting feature of the circuit can be observed in the first quadrant where one of the check valves opens, making the pathway from the tank into the rod-side line less restricted. This is important to prevent cavitation because the pressure at the rod-side of the cylinder may reach considerably low levels. The same problem does not happen in the third quadrant, where the difference between the pressure on the cap-side and the tank is positive. In this case, increasing the restriction between the main line and the tank results in a higher pressure on the piston side, which, in turn, contributes to decelerating the cylinder.

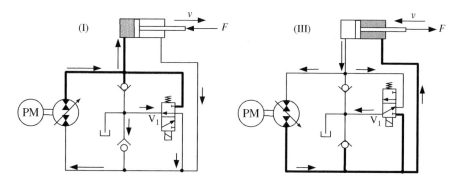

Figure 6.18   Operation in the first and third quadrants

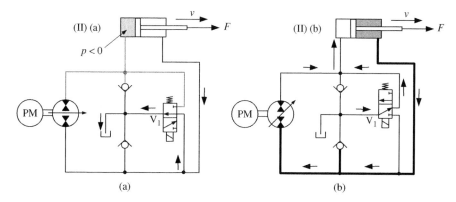

Figure 6.19   Operation in the second quadrant: (a) incorrect circuit configuration and (b) correct circuit configuration

Based on Figure 6.18, we could be tempted to keep the valve V_1 deactivated in the second operational quadrant, where the cylinder speed is positive. However, this is not a good solution, as can be seen in Figure 6.19(a), where we attempt to hold the cylinder in place by setting the pump displacement to zero. In this case, we see that a negative gauge pressure is created on the cap-side of the cylinder by the pulling force, F. The figure illustrates the case where this force is sufficiently high to accelerate the cylinder to the right, causing the fluid in the cap-side chamber to evaporate. We also observe that no control can be exerted on the cylinder speed while the valve V_1 connects the rod-side of the cylinder to the tank. Figure 6.19(b) corrects this problem by activating the directional valve, V_1. As a result, a high pressure is produced on the rod-side of the cylinder, which allows for a suitable speed control of the actuator through changing the pump displacement. The load energy is then transferred to the pump, which now operates as a motor.

A similar situation happens in the fourth quadrant, as can be seen in Figure 6.20. In Figure 6.20(a), we show the extreme case where we try to stop the cylinder by setting the pump displacement to zero. Although we do not observe any negative pressure in the

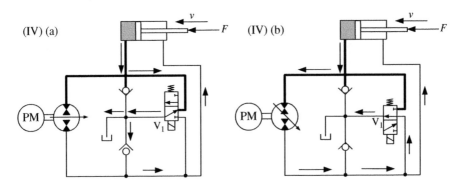

Figure 6.20   Operation in the fourth quadrant: (a) incorrect circuit configuration and (b) correct circuit configuration

circuit this time, no speed control exists and the cylinder accelerates to the left anyway. In Figure 6.20(b), this problem is corrected by shifting the valve position. The pump then operates as a motor and the cylinder speed can be controlled through the pump displacement.

6.2.6   Design 6. Single-Pump Circuit with Pilot-Operated Check Valves

The following design (Figure 6.21) uses two pilot-operated check valves and a charge circuit as a means of compensating for the uneven flows caused by the differential area [1]. The concept is somewhat similar to the hydrostatic transmission shown in Figure 1.31, where the charge pump compensated for the volumetric losses. Here, the charge pump compensates not only for the volumetric losses but also for the smaller flow coming from the rod-side of the cylinder.[6]

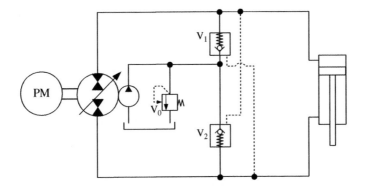

Figure 6.21   Single-pump circuit with pilot-operated check valves

---

[6] Depending on the magnitude of the flow needed at the rod-side of the cylinder, the charge circuit may be disregarded and a direct connection to the tank might be used instead, as in Design 5.

The operation and the circuit flows are better visualized through a quadrant-by-quadrant analysis. Figure 6.22 shows the operation in the first quadrant. During extension, the cylinder demands a flow vA_p from the pump. The flow coming out of the rod-side of the cylinder, vA_a, is then joined to the charge pump flow, $v(A_p - A_a)$, coming through check valve V_2, so that the same flow, vA_p, goes into the pump again. The excess flow from the charge pump (not shown in the figure) is deviated to the tank through the relief valve, V_0, which is also responsible for keeping the charge circuit pressure at a pre-determined level. Note that it is the pressure differential between the cap-side and the rod-side of the cylinder that opens valve V_2 while keeping valve V_1 closed.

Figure 6.23 shows the operation in the third quadrant. The pump now supplies the flow vA_a to the rod-side of the cylinder, which in turn pushes out a flow vA_p at the cap-side port. Part of this flow deviates into the tank through the check valve, V_1, and the relief valve, V_0, so that the pump can receive a flow vA_a at the input. Note that the charge pump flow is totally directed to the tank this time.

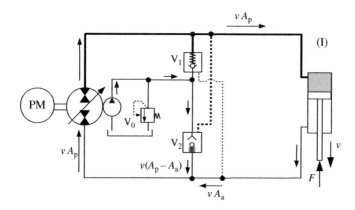

Figure 6.22   Operation in the first quadrant

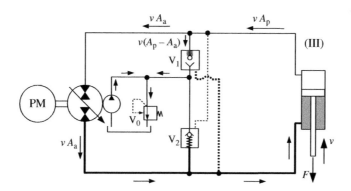

Figure 6.23   Operation in the third quadrant

In the second quadrant (Figure 6.24), the flow coming from the pump, vA_a, is insufficient to fill in the cap-side chamber of the cylinder, which demands a flow vA_p. However, the charge pump sends a flow $v(A_p - A_a)$, which adds to the pump flow, vA_a, so that the correct oil supply enters the cap-side of the cylinder. Note that the pump is now being driven by the external load, F, and therefore operates as a motor. Another important observation is that the cylinder speed, v, can be easily controlled through changing the pump displacement in case of an undesired acceleration.

Operation in the fourth quadrant is shown in Figure 6.25. By the action of the force F, the flow vA_p is forced into the pump, now operating as a motor. Therefore, part of the flow coming out of the pump must deviate into the tank through the check valve, V_2, and the relief valve V_0 before reaching the rod-side of the cylinder. The charge pump again discharges into the tank at a pressure determined by the valve V_0. We observe that the cylinder speed can be easily controlled by changing the pump displacement.

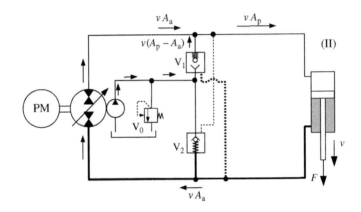

Figure 6.24   Operation in the second quadrant

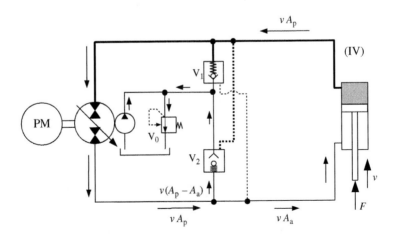

Figure 6.25   Operation in the fourth quadrant

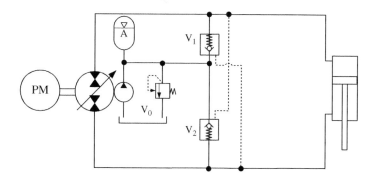

Figure 6.26   Single-pump circuit with pilot-operated check valves and an accumulator

A slight modification can be made in the circuit of Figure 6.21, which will significantly improve the energetic efficiency of the actuator. Note that the charge pump must supply the missing flow into the circuit in quadrants I and II. However, depending on the flow demands of the circuit, the charge pump can become considerably big, in this case. This comes at the cost of higher energy loss through the relief valve V_0 in quadrants III and IV when the charge pump flow is not being used and is instead being totally diverted into the reservoir. One possible solution is to connect an accumulator to the charge circuit, ensuring it is properly sized to supply the flow needed in quadrants I and II, as illustrated in Figure 6.26.

Figures 6.27(a) and (b) show the operation in the first and second quadrants of the circuit illustrated in Figure 6.26. In both situations, the accumulator discharges into the main circuit. Depending on the accumulator capacity, the size of the charge pump can be reduced to a minimum. In fact, considering the areas of the piston and the ring, A_p and A_a, and the cylinder stroke, L, the accumulator must be sized to hold a fluid volume at least equal to $(A_p - A_a)L$ at the pressure set by the valve, V_0.

Operation in the third and fourth quadrants is illustrated in Figures 6.28(a) and (b). Note that the accumulator is charged during operation.

A significant feature of the design shown in Figure 6.26 is that it may be easily modified so that more than one displacement-controlled actuator can share the same charge pump,[7] as shown in Figure 6.29. Valves V_5 and V_6 are incorporated to hold the load in position in case of an eventual emergency [10]. Single-pump circuits with pilot-operated check valves similar to those shown in Figures 6.26 and 6.29 have been extensively studied and documented in Refs. [10, 1, 11, 12].

6.2.7   Design 7. Single-Pump Circuit with Inline Check Valves

The actuator shown in Figure 6.30 is based on the design presented in Ref. [13]. It shares many of the same characteristics as the actuator shown in Figure 6.26, but because of the way in which the check valves are placed in the circuit, the stiffness of the cylinder is improved

---

[7] In this case, the accumulator and the charge circuit have to be sized to supply the flow needed by all the cylinders in the circuit.

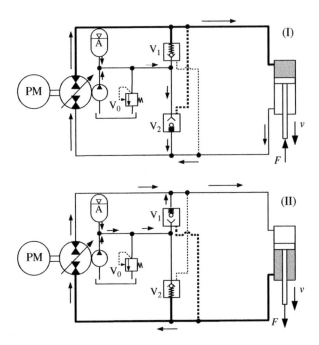

Figure 6.27 Operation in the first and second quadrants (accumulator discharge)

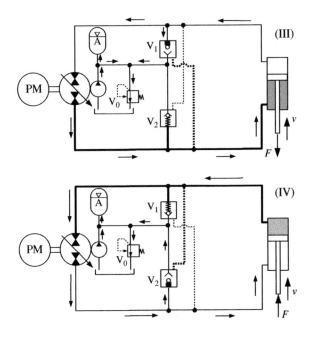

Figure 6.28 Operation in the third and fourth quadrants (accumulator charge)

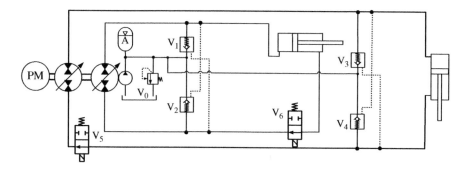

Figure 6.29 Two cylinders sharing one charge pump

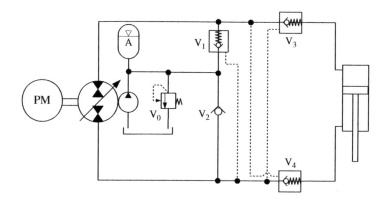

Figure 6.30 Single-pump circuit with inline check valves

when the swashplate of the pump is set to a neutral position. Therefore, this circuit is suitable to be used with electrohydrostatic actuators as well as with displacement-controlled actuators.

Operation in the first and third quadrants is illustrated in Figure 6.31 where we see that the check valves V_3 and V_4 remain open in both cases. The check valve V_2, on the other hand, opens spontaneously by the negative pressure differential created between the rod-side line and the tank (first quadrant). Note that accumulator A is charged in the third quadrant and discharged in the first quadrant.

Figure 6.32 shows the operation in the second quadrant. We see that the pump flow entering the cap-side of the cylinder, vA_p, naturally opens valve V_3. On the other hand, the pressure on the cap-side, p_c, still needs to grow to the point of being able to open valve V_4 before the cylinder starts moving. In this case, the check valve V_4 acts as a counterbalance valve, preventing the cylinder from accelerating by the action of the force, F. As a consequence, during operation, a pressure, p_r, is created at the rod-side of the cylinder.

Due to the flow differential between the input and output of the cylinder, a smaller pressure builds up at the pump input, which causes valve V_2 to open and give way to the flow coming

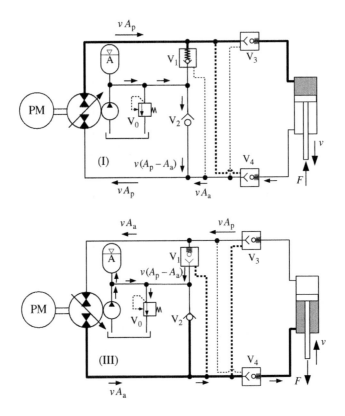

Figure 6.31   Operation in the first and third quadrants

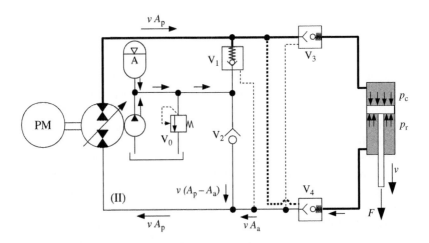

Figure 6.32   Operation in the second quadrant

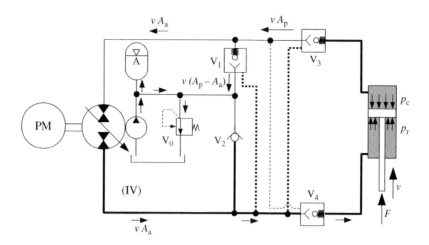

Figure 6.33   Operation in the fourth quadrant

from the charge circuit into the main line. Note that because of the inline check valves V_3 and V_4, the load is held in position even if the PM gets disconnected from the pump.

A similar situation happens in the fourth quadrant, as can be seen in Figure 6.33. The counterbalancing action of the check valve V_3 creates a higher pressure, p_c, at the cap-side of the cylinder, which prevents the load from accelerating upwards. The pump now sends in a flow vA_a into the rod-side chamber, which elevates the pressure at the rod-side line to p_r. When this pressure becomes sufficiently high to open valve V_3, the cylinder moves upwards, sending a flow, vA_p, into the pump. Since $vA_p > vA_a$, it is necessary to deviate part of the fluid into the reservoir. This is done through valve V_1, which remains open by the pressure, p_r, at the rod-side of the cylinder. Note that the accumulator, A, is charged when the circuit operates in the fourth quadrant.

6.2.8   Design 8. Energy Storage Circuit

At this point, after studying some hydrostatic actuator designs, one might be wondering what could be done to store the energy received from the load at quadrants II and IV. The design to be introduced here was conceptualized in Ref. [13] and presents one way in which this can be accomplished. The basic idea is to add an 'energy storage circuit' to the existing actuator. Figure 6.34 shows an example of an energy storage circuit [2, 13].

In the circuit shown in Figure 6.34, we identify the pump/motor P_S, the relief valve R_S, the high-pressure accumulator (HA) the directional valves V_A and V_B, and the two clutches C_M and C_P, which are intended to connect the shaft of the pump P_S to the PM (clutch C_M) and to the main pump of the hydrostatic actuator (clutch C_P).

Four modes of operation can be identified in the energy storage circuit. Let us name them storage, regeneration, discharge and bypass. Figure 6.35 shows the first three possibilities. Note that in these three modes of operation, we keep the PM clutch, C_M, disengaged so that

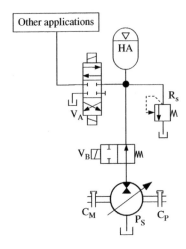

Figure 6.34   Energy storage circuit

all the energy flow happens between the actuator and the energy storage circuit. During storage (Figure 6.35(a)), energy is input through the clutch, C_P, and the pump/motor, P_S, running as a pump, loads the accumulator, HA, to a pressure determined by the adjustment of the relief valve, R_S. During this operation, the directional valve, V_A, must remain closed (centred position). The stored energy can be sent back to another hydraulic device (Figure 6.35(b)) or to the main pump itself (c). In both cases, the energy that had been stored is subsequently reused (regenerated). In order to allow the stored energy to be used by other applications, it is necessary to change the positions of valves V_A and V_B,[8] as shown in Figure 6.35(b). Note that it is also necessary to switch the swashplate angle of the pump P_S to neutral. On the other hand, by centring valve V_A and returning valve V_B to its open position, we allow the accumulator to discharge through the pump P_S, in which case the stored energy can be reused by the circuit through the clutch C_P (Figure 6.35(c)). Lastly, we observe that it is always possible to discharge the accumulator directly into the tank (discharge mode), as shown in Figure 6.35(d).

Whenever the energy storage circuit is not being used, we can set it to bypass mode, as shown in Figure 6.36. The figure represents the case where the accumulator HA has already been previously loaded. Therefore, it is necessary to set the directional valve V_A to its centred position while closing valve V_B at the same time to prevent the accumulator from discharging through the pump P_S. The pump swashplate must also be set to neutral at this stage. In fact, this setup simulates what would happen if the prime mover had been directly connected to the actuator, ignoring the energy storage circuit in between.

Figure 6.37 shows how to couple the energy storage circuit to the displacement-controlled actuator shown in Figure 6.26. The resulting circuit is now regenerative based on the reasons explained earlier.

The four-quadrant operation of the main circuit shown in Figure 6.37 has been explained before in details (see Figures 6.27 and 6.28). Therefore, we can assume that most of the

---

[8] Valve V_B has been placed in the circuit to help prevent the pressurized fluid from leaking back through the pump P_S into the tank.

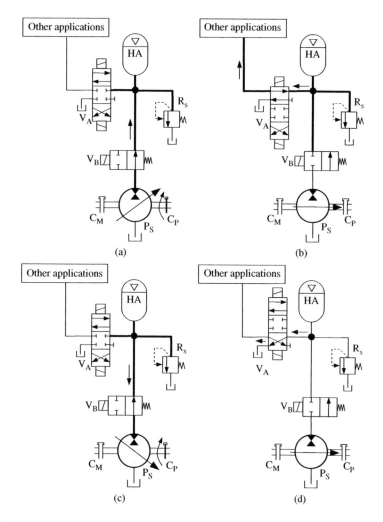

Figure 6.35 Operation of the energy storage circuit: (a) storage, (b) and (c) regeneration, (d) discharge

circuit operation is understood at this stage, and analyse the behaviour of the energy storage circuit in each quadrant instead.

Figure 6.38 illustrates the operation in the first and third quadrants (pumping mode). Because no energy is stored at this stage, the energy storage circuit can be set to bypass position, with the two clutches, C_M and C_P, engaged (note that we are assuming that the accumulator HA is charged as in Figure 6.36).

Figure 6.39 shows the operation in the second and fourth quadrants. In the figure, we have chosen to represent the prime mover running in both quadrants while disconnected[9] from the actuator through the clutch C_M. In the second quadrant, the load drives the main pump in a

---

[9] This is not always the case, given that in some circumstances the prime mover can be turned off during operation in the second and fourth quadrants.

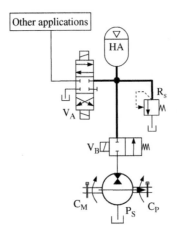

Figure 6.36   Bypass mode operation

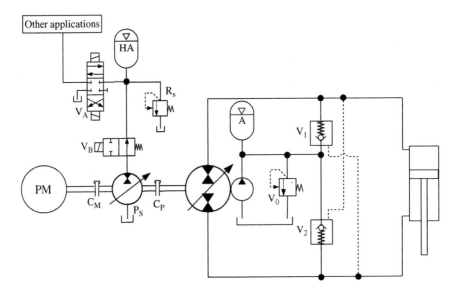

Figure 6.37   Regenerative displacement-controlled actuator with pilot-operated check valves

clockwise direction. The main pump gets connected to the pump P_S through the clutch C_P so that the accumulator HA can be charged. The same happens in the fourth quadrant, the only difference being that the swashplate angle of the main pump is now inverted because of the change in the flow direction. In both quadrants, speed control of the actuator is still attained through changing the displacement of the main pump. Note also that the accumulator HA helps stop the cylinder when it accelerates. In this aspect, the pressure set on the relief valve, R_S, determines the maximum energy to be regenerated and the corresponding braking action of the energy storage circuit on the cylinder.

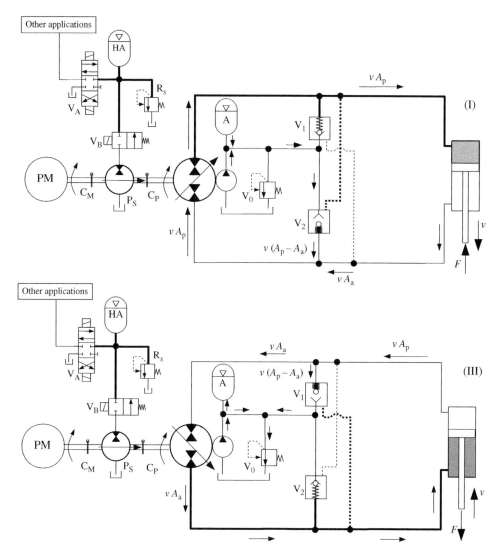

Figure 6.38 Operation in the first and third quadrants

The energy stored in the HA can be reused in the same circuit as illustrated in Figure 6.40, which shows how this can be done during pumping mode (first and third quadrants). Note that the swashplate angle of the pump P_S must become negative to maintain a clockwise rotation in both quadrants.

It is possible to couple the energy storage unit to other actuator designs as long as they allow for energy regeneration at the motoring quadrants. For instance, we could transform the circuit shown in Design 5 (Figure 6.17) into a regenerative circuit as shown in Figure 6.41, which illustrates the operation in quadrants II and IV. The energy storage process is very similar to the circuit shown in Figure 6.37 and needs no further explanation.

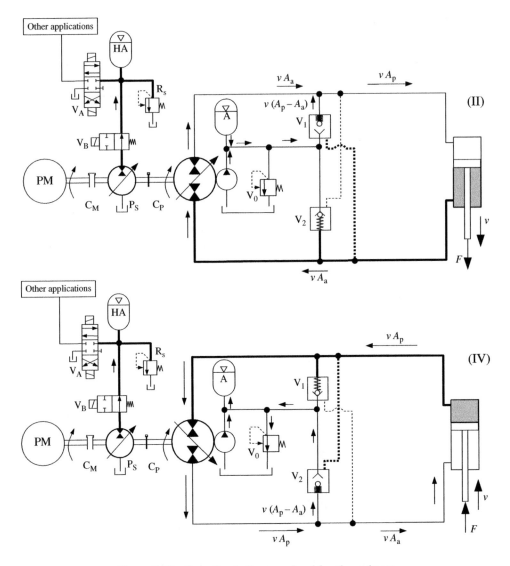

Figure 6.39 Operation in the second and fourth quadrants

Another interesting feature of the energy storage circuit is illustrated in Figure 6.42, where the cylinder needs to be kept in place and still exert a constant force, F, on a solid obstacle[10] (think of a hydraulic press, for example). The same situation could have been simulated with the circuit in Figure 6.37, and the conclusion in both cases would be that no energy needs to be supplied by the PM to hold the cylinder in place (as a matter of fact, the PM can even stop running as shown in the figure).

---

[10] The maximum force is determined by the regulation of the relief valve, R_S.

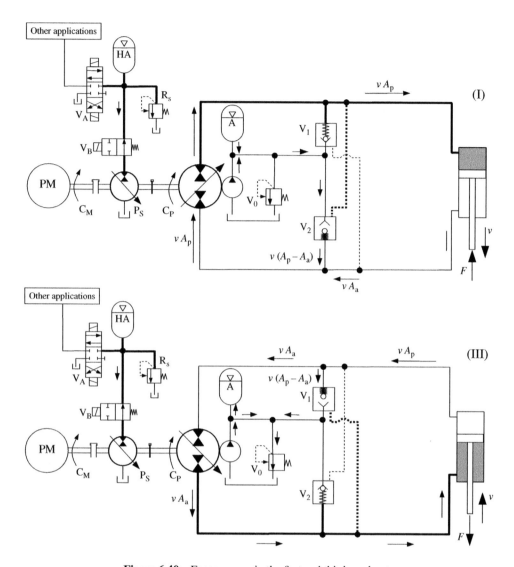

Figure 6.40 Energy reuse in the first and third quadrants

6.2.9 Design 9. Double-Rod Actuator

The design of double-rod actuators is relatively easy to be conceptualized when compared to single-rod actuators. Therefore, we present only one circuit in this section as illustrated in Figure 6.43. Note that the charge pump can be considerably smaller given that the charge circuit only needs to replenish the external leakages.[11] Two check valves, V_1 and V_2,

---

[11] In fact, the charge circuit may be replaced by a direct connection to the tank or to a hydraulic accumulator.

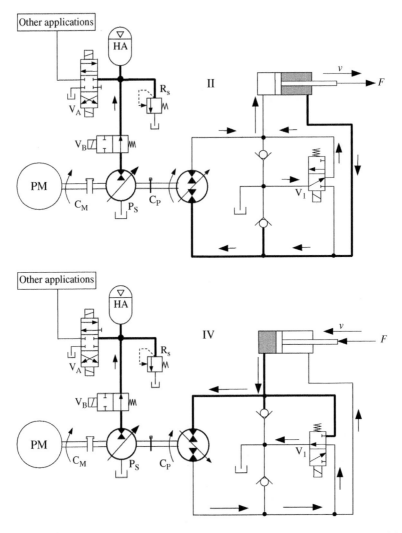

Figure 6.41 Operation in the second and fourth quadrants of a regenerative actuator with a directional valve

alternatingly connect the low-pressure branch of the circuit with the charge pump, which has its output pressure determined by the relief valve, V_0.

Figure 6.44 shows the operation in the first quadrant. The small dashed arrows represent a very small flow (for external leakage replenishment). The pressure at the junction, N, is determined by the relief valve, V_0, so that the low-pressure branch of the circuit is always exposed to the pressure set by the charge circuit. This is very important because we therefore guarantee that the lowest pressure in the circuit is kept above a safe minimal level.

Operation in the second quadrant is illustrated in Figure 6.45. We observe that the only difference between the first and the second quadrants is that the charge pump in the second quadrant connects to the upper branch of circuit through valve V_1.

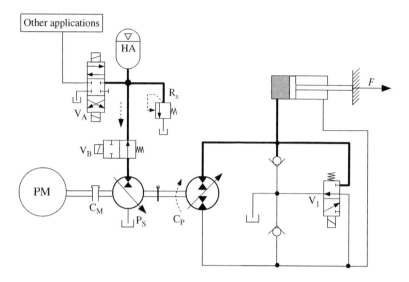

Figure 6.42   Holding a force F against an obstacle during the first quadrant

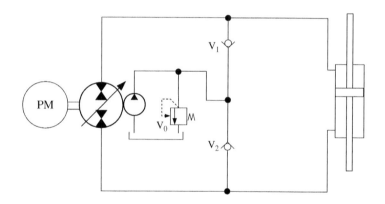

Figure 6.43   Double-rod actuator

Double-rod actuators are symmetric by nature. Therefore, there is no need to show the operation in the third and fourth quadrants here since it would only be a repetition of the first and second quadrants with the cylinder going in the opposite direction.

6.3   Common Pressure Rail and Hydraulic Transformers

We saw in Figure 6.29 that the concepts used for hydrostatic actuators can easily extend to multiple cylinders. However, each cylinder is still controlled by one pump/motor, and in the end, the same prime mover will be used to drive a series of independent displacement-controlled actuators. In general, multiple pumps are costly [8], but they constitute the only means of controlling multiple cylinders using hydrostatic actuators. In

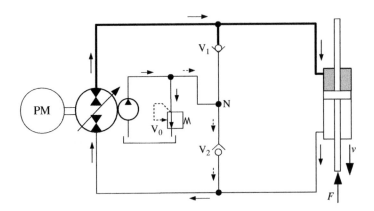

Figure 6.44   Operation in the first quadrant

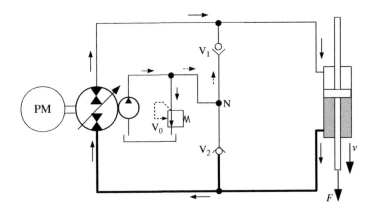

Figure 6.45   Operation in the second quadrant

fact, the pairing of a pump to a cylinder is implicit in the definition we gave of hydrostatic actuators in Section 1.6.2.

One way to circumvent the problem of multiple pumps is to use a common pressure rail (CPR). The CPR concept is better understood with a pneumatics analogy, where we connect each circuit to a line (or rail) with a constant pressure. The other line is the atmosphere, associated with the exhausts of the valves. For instance, Figure 6.46 illustrates the CPR principle in pneumatic circuits. One of the lines is connected to the pressure reservoir, which is constantly fed by the compressor. We can assume that in a sufficiently large reservoir the internal pressure, p_r, is constant. The other line connects to the atmospheric pressure,[12] that is, zero gauge. As a result, a constant pressure differential is available for every cylinder connected to these two lines, which guarantees that similar speeds and forces are developed in each actuator.

---

[12] This representation of a line at atmospheric pressure is not actually standard in pneumatic circuits. It is only used here for didactical reasons.

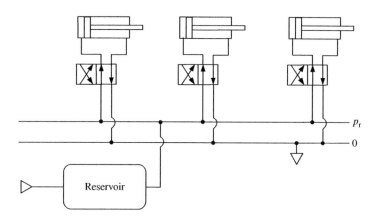

Figure 6.46 Common pressure rail concept in a pneumatic circuit

Figure 6.47 shows how a CPR can be built in a hydraulic circuit. In the figure, we observe two pressure rails, R_1 and R_2. The first one is connected to high-pressure accumulator (HA) while the second one is connected to the low-pressure accumulator (LA). Two hydrostatic pumps, P_1 and P_2, together with two relief valves V_1 and V_2, and a directional valve V_3 are responsible for keeping the lines R_1 and R_2 at their corresponding pressure levels. Similar to the pneumatic case, loads are connected in parallel to the pressure rails, and as fluid consumption takes place, the pressure in rail R_1 drops down. In that case, the directional valve, V_3, remains open, connecting the pump P_1 to the corresponding rail, R_1. As soon as the pressure at rail R_1 reaches a pre-determined level, the valve V_3 changes its position, discharging the pump, P_1, to the tank.

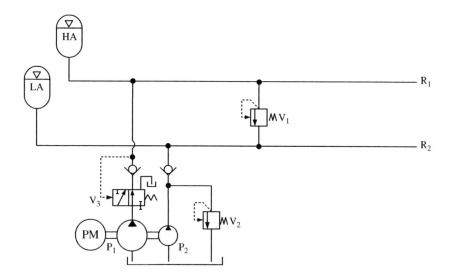

Figure 6.47 Common pressure rail

CPRs have the advantage of separating the loads from the energy source [8] and must be designed to the highest pressure required by the loads if valve-controlled actuators are connected to the rails, as in Figure 6.46. If a future expansion of the grid is expected, this fact must be taken into consideration. Note that the pressure differential between the two rails is what promotes the individual flows at the consuming points. This is the reason why CPR systems have been assigned the term 'imposed pressure' [8], as opposed to conventional hydraulic circuits in which the flows are imposed by the pump.[13]

High-pressure demands at individual consuming points along the grid can make CPRs very inefficient if valve control is to be used. This happens particularly when low-pressure applications are paralleled with high-pressure applications. In this case, throttling losses must be introduced to lower the pressure down to operational levels at the low-pressure consuming points. As an example, consider the two-cylinder grid shown in Figure 6.48 where two identical cylinders, A and B, are connected to the pressure rails through proportional directional valves. In the circuit, the pressure at line R_1 was chosen accordingly to the maximum load, $10F$ (pressure losses in the circuit have been ignored).

With respect to Figure 6.48, we observe that in order to lower the pressure at the cap-side of the cylinder A, it was necessary to partially displace the spool of the valve V_1. Therefore, the pressure drop from $10p$ to p, needed to reduce the cylinder force from $10F$ to F, came at the cost of heat dissipation at valve V_1. This requirement to greatly reduce the pressure makes the circuit energetically inefficient in comparison to the usual design where each cylinder is paired to a hydraulic pump. One possible attempt to solve this problem is to replace the directional valves by hydraulic transformers (see Section 6.2.4, for a brief explanation about hydraulic transformers). The idea is that a hydraulic transformer would eliminate throttling losses, making it possible for different loads to be connected to the rails efficiently.

The use of hydraulic transformers connecting the cylinders to the rails can be better understood with the help of Figure 6.49. In the figure, a special type of hydraulic transformer with

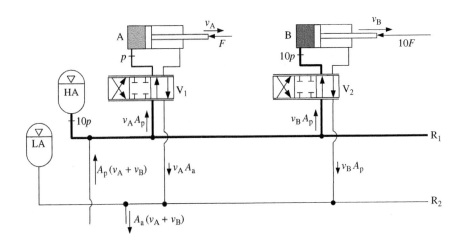

Figure 6.48 Common pressure rail with valve-controlled actuators

---

[13] Remember that in conventional hydraulics, the pressure inside the circuit increases because of the resistance to the flow coming from the pump.

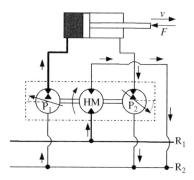

Figure 6.49   Operation in the first quadrant of a cylinder connected to the pressure rails by means of a dual-pump hydraulic transformer

two variable-displacement pump/motors, P_1 and P_2, is used. The pumps are connected to a reversible hydraulic motor, HM, through a common shaft. The situation is identical to the one shown in Section 6.2.1 (see Figure 6.4), with the hydraulic motor, HM playing the role of the prime mover.

There is no need to explain the four-quadrant operation of the circuit in Figure 6.49, since it has been explained before in Section 6.2.1. However, a couple of facts are worth noting:

- There can be either a pressure amplification or reduction depending on the pumps' displacements, whereas in valve-controlled cylinders, only reductions are possible. This feature allows us to use a smaller pressure at the rail R_1 and still move heavier loads.
- It is possible to regenerate energy coming from the load by reversing the rotation of the motor HM, therefore transferring fluid from the low-pressure rail, R_2, into the high-pressure rail, R_1.

An alternative design uses a single-pump transformer together with a directional valve [14], as shown in Figure 6.50. Note that the directional valve is not supposed to introduce significant throttling losses in the circuit, in this case.

The cylinder in Figure 6.50 can operate in the second and fourth quadrants in regenerative mode, transferring energy from the load into the high-pressure rail, as seen in Figure 6.51. Because the sign of the swashplate angle of the pump, P, does not change, the motor HM rotates in the other direction under the influence of the force, F. Fluid is then forcedly transferred from the low-pressure rail, R_2, into the high-pressure rail, R_1, in both quadrants, which results in energy from the load being stored in the HA (see Figure 6.48). Note that a correct control of the directional valve, V, is crucial for a sound operation of the circuit.

Despite the advantages of using hydraulic transformers as a means for controlling the cylinder, they are not practical for two reasons:

- Hydraulic transformers are bulky and difficult to be placed at the consuming points.
- Hydraulic transformers, being composed of pumps and motors, have the potential for considerable mechanical and volumetric losses so that they may end up being less efficient than the valve control itself [8].

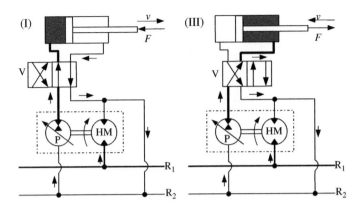

Figure 6.50 Cylinder control with a conventional (two-quadrant) hydraulic transformer operating in the first and third quadrants

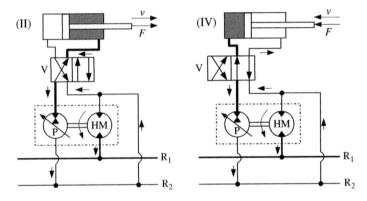

Figure 6.51 Cylinder control with a conventional hydraulic transformer operating in the second and fourth quadrants

Recently, a new type of hydraulic transformer designed by the Dutch company Innas B.V. has gained attention for promising a high efficiency associated with a small size. The Innas Hydraulic Transformers (IHTs) are based on the floating cup principle (see Section 3.3.2.4, Chapter 3), and according to Ref. [8] are 'small and light weighted' when compared to conventional hydraulic transformers. Moreover, research has been carried out on models that can also substitute the dual-pump transformer shown in Figure 6.49 so that the cylinder may operate in four quadrants without the need of an extra directional valve [15]. Figure 6.52(a) was published in Ref. [8] and shows a 20 kW IHT prototype, using a pencil as a referential to give an idea of the actual size of the transformer. Figure 6.52(b) shows the symbol adopted for this IHT in comparison with the symbol of a conventional hydraulic transformer. Finally, Figure 6.52(c) shows the connection with a hydraulic cylinder in a CPR circuit [14].

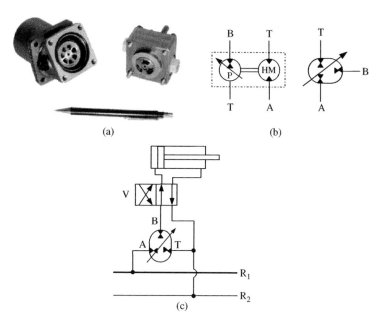

(a)   (b)

(c)

Figure 6.52   (a) Innas IHT prototype (courtesy of Innas B.V.); (b) IHT symbol in comparison with the conventional hydraulic transformer symbol; (c) cylinder control in a CPR

Exercises

(1)   How does pump 2 interfere with the cylinder speed in the circuit shown in Figure 6.7? And how does it influence the force exerted by the cylinder rod? Based on your answers, do you think that this design has been correctly named a 'dual-pump actuator' or should it be classified as a 'single-pump actuator', according to the definition given at the beginning of Section 6.2?

(2)   What are the risks of misadjusting the relative displacements of the two pumps in Designs 1 and 2?

(3)   Could the fixed-displacement pumps in Design 3 have different displacement values? Discuss the matter.

(4)   Would it be possible to use a constant speed PM and two variable-displacement pumps in the circuit shown in Figure 6.10 in order to make it a displacement-controlled actuator? Discuss the matter.

(5)   In the circuit of Figure 6.42, consider that the displacements of the energy storage pump, P_S, and the main pump, are D_s and D_p, respectively. Consider also that the pressure at the tank is 0 bar gauge. Write an expression for the force F as a function of the piston area, A_p, and the pressure, p_s, at the input of the relief valve, R_s.

References

[1] Williamson C, Ivantysynova M (2010) Stability and motion control of inertial loads with displacement controlled hydraulic actuators. Proceedings of the 6th FPNI – PhD Symposium – West Lafayette, USA, pp. 499–514.

[2] Wendel GR (2000) Regenerative hydraulic systems for increased efficiency. Proceedings of the National Conference on Fluid Power – NCFP, April 4–6, pp. 199–206, USA.

[3] Hydraulics and Pneumatics (2011) Control of single-rod cylinders with two pumps. http://www.hydraulicspneumatics.com/200/Issue/Article/False/66450/Issue. Accessed 29 June 2011.

[4] Hydraulics and Pneumatics (2011) Control of single-rod cylinders with two pumps http://www.hydraulicspneumatics.com/200/Issue/Article/False/67493/Issue. Accessed 29 June 2011.

[5] Cleasby KG, Plummer AR (2008) A novel high efficiency electro-hydrostatic flight simulator motion system. Proceedings of the Fluid Power and Motion Control – FPMC 2008, Bath, UK, pp 437–449.

[6] Alaydi JY (2008) Mathematical modeling for pump controlled system of hydraulic drive unit of single bucket excavator digging mechanism. Jordan Journal of Mechanical and Industrial Engineering, 2–3: 157–162.

[7] Marktheidenfeld RS (2005) Hydraulic transformer. US Patent, number 6,887,045 B2, 3 May 2005.

[8] Vael G, Achten P, Fu Z (2000) The Innas hydraulic transformer the key to the hydrostatic common pressure rail. SAE Technical Paper 2000-01-2561.

[9] Hewett AJ (1994) Hydraulic circuit flow control. US Patent, number 5,329,767, 19 July 1994.

[10] Rahmfeld R, Ivantysynova M (2001) Displacement controlled linear actuator with differential cylinder-a way to save primary energy in mobile machines. 5th International Conference on Fluid Power Transmission and Control – ICFP 2001, pp. 296–301, Hangzhou, China.

[11] Williamson C, Zimmerman J, Ivantysynova M (2008) Efficiency study of an excavator hydraulic system based on displacement-controlled actuators. Bath ASME Symposium on Fluid Power and Motion Control – FPMC2008, pp. 291–307.

[12] Zimmerman JD, Ivantysynova M (2010) Reduction of engine and cooling power by displacement control. Proceedings of the 6th FPNI – PhD Symposium – West Lafayette, USA, pp. 339–352.

[13] Wendel GR (2002) Hydraulic system configurations for improved efficiency. Proceedings of the 49th National Conference on Fluid Power – NCFP, pp. 567–573, USA.

[14] Inderelst M, Sgro S, Murrenhoff H (2010) Energy recuperation in working hydraulics of excavators. Fluid Power and Motion Control, pp. 551–562.

[15] Achten P, Van Den Brink T, Potma J, Schellekens M, Vael G (2009) A four-quadrant hydraulic transformer for hybrid vehicles. The 11th Scandinavian International Conference on Fluid Power, SICFP'09, 2–4 June 2009, Linköping, Sweden.

7

Dynamic Analysis of Hydrostatic Actuators

Chapter 5 introduced the dynamic analysis of hydrostatic transmissions. In this chapter, we use the same modelling principles for hydrostatic actuators. As a result, the basic aspects of dynamic analyses as were explained in Section 5.1 can also be applied here.

This chapter is outlined as follows. First, we shall develop a mathematical model for a specific single-rod circuit from Chapter 6. The circuit must be carefully chosen as we would like to use the same model on a particular case where the cylinder has a double rod as well. After developing the basic equations, we perform a more detailed analysis of the hydraulic cylinder and the circuit valves individually. Finally, we illustrate the theory with one representative case study.

7.1   Introduction

Consider the simplified version of the displacement-controlled actuator illustrated in Figure 6.21 and shown again in Figure 7.1, where the charge circuit has been replaced by a direct connection to the tank. Consider also the following assumptions:

1. The axial piston pump has a constant speed and a variable displacement and is allowed to operate over centre.
2. No volumetric or mechanical losses occur at the pump.
3. Cylinder leakages and conduit losses are very small and can be disregarded.
4. The two piloted check valves, V_1 and V_2, are identical.

In what follows, we develop a mathematical model for the actuator shown in Figure 7.1. We begin by looking at a general mass flow balance in the circuit, together with the force

Hydrostatic Transmissions and Actuators: Operation, Modelling and Applications, First Edition.
Gustavo Koury Costa and Nariman Sepehri.
© 2015 John Wiley & Sons, Ltd. Published 2015 by John Wiley & Sons, Ltd.
Companion Website: www.wiley.com/go/costa/hydrostatic

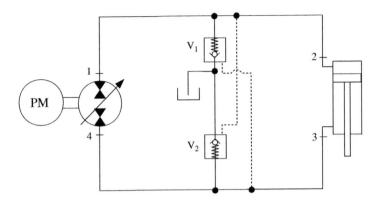

Figure 7.1 Single-rod displacement-controlled actuator

balance at the cylinder and then proceed with a more detailed analysis of the cylinder and the valves.

7.2 Mathematical Model

The circuit shown in Figure 7.1 is simple enough to serve as the starting point for our analysis. In fact, only one additional assumption – also used in Chapter 5 – will be necessary, that is, we consider the fluid density, ρ, to be a function of the pressure, p, alone. If we recall the assumption that there are no conduit losses in the circuit, the density will not change along the lines connecting the pump to the cylinder. It is therefore pertinent to define $\rho_{12}, p_{12}, \rho_{34}$ and p_{34} as the fluid densities and pressures in lines 1–2 and 3–4, respectively.

Similar to what was done in Section 5.2, we first develop the basic equations using the conservation of mass along the circuit and the second law of Newton at the cylinder.

7.2.1 Basic Equations

Figure 7.2 shows a general view of the volumetric flows inside the hydrostatic actuator displayed in Figure 7.1. Similar to what was done in Figure 5.12, we have divided the actuator into six different control volumes, defined by the dashed rectangles 1–6. The following nomenclature has been used in the figure:

- q_{pt} is the instantaneous pump flow obtained for an ideal situation where volumetric losses are not present (see Sections 3.2.2 and 5.1.2).
- q_{V_1} and q_{V_2} are the flows through the pilot-operated check valves V_1 and V_2. Note that depending on the quadrant of operation, these flows may change their direction or even become zero (see Figures 6.22–6.25).
- A_p and A_a are the areas of the piston and the ring of the cylinder.

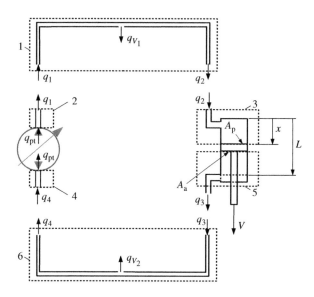

Figure 7.2   Volumetric circuit flows

- x and v are the instantaneous position and speed of the cylinder rod, respectively.
- L is the piston stroke (maximum value of x).

Let $V_1 \ldots V_6$ be the inner volumes of the circuit parts within the control regions 1–6 in Figure 7.2. The following mass balances can be written for each of these regions:

$$\begin{cases} \text{Region 1: } \rho_{12}\left(q_1 - q_{v_1} - q_2\right) = V_1\dfrac{d\rho_{12}}{dt} \\[2mm] \text{Region 2: } \rho_{12}(q_{pt} - q_1) = V_2\dfrac{d\rho_{12}}{dt} \\[2mm] \text{Region 3: } \rho_{12}q_2 = \dfrac{d(V_3\rho_{12})}{dt} \\[2mm] \text{Region 4: } \rho_{34}(q_4 - q_{pt}) = V_4\dfrac{d\rho_{34}}{dt} \\[2mm] \text{Region 5: } -\rho_{34}q_3 = \dfrac{d(V_5\rho_{34})}{dt} \\[2mm] \text{Region 6: } \rho_{34}(q_3 - q_{v_2} - q_4) = V_6\dfrac{d\rho_{34}}{dt} \end{cases} \qquad (7.1)$$

Equations (7.1) can be further simplified if we observe that the volumes V_2 and V_4 are very small and can therefore be disregarded. In addition, the volumes V_3 and V_5 can be written as functions of the cylinder areas, A_p and A_a, and the cylinder displacement, x, as follows:

$$\begin{cases} V_3 = A_p x \\ V_5 = A_a (L - x) \end{cases} \tag{7.2}$$

With V_3 and V_5 given by Eqs. (7.2), we can expand the derivatives within parentheses in Eqs. (7.1):

$$\begin{cases} \dfrac{d\left(V_3\rho_{12}\right)}{dt} = \rho_{12}\dfrac{dV_3}{dt} + V_3\dfrac{d\rho_{12}}{dt} = \rho_{12}A_p v + A_p x\dfrac{d\rho_{12}}{dt} \\ \dfrac{d(V_5\rho_{34})}{dt} = \rho_{34}\dfrac{dV_5}{dt} + V_5\dfrac{d\rho_{34}}{dt} = -\rho_{34}A_a v + A_a(L - x)\dfrac{d\rho_{34}}{dt} \end{cases} \tag{7.3}$$

where

$$v = \frac{dx}{dt} \tag{7.4}$$

Substituting Eqs. (7.3) into Eqs. (7.1) and making $V_2 = V_4 = 0$, we obtain

$$\begin{cases} \text{Region 1: } \rho_{12}\left(q_1 - q_{V_1} - q_2\right) = V_1\dfrac{d\rho_{12}}{dt} \\[2mm] \text{Region 2: } q_1 = q_{pt} \\[2mm] \text{Region 3: } \rho_{12}q_2 = \rho_{12}A_p v + A_p x\dfrac{d\rho_{12}}{dt} \\[2mm] \text{Region 4: } q_4 = q_{pt} \\[2mm] \text{Region 5: } -\rho_{34}q_3 = -\rho_{34}A_a v + A_a(L - x)\dfrac{d\rho_{34}}{dt} \\[2mm] \text{Region 6: } \rho_{34}(q_3 - q_{V_2} - q_4) = V_6\dfrac{d\rho_{34}}{dt} \end{cases} \tag{7.5}$$

Equations (7.5) can be further simplified to

$$\begin{cases} \left(\dfrac{V_1 + A_p x}{\rho_{12}}\right)\dfrac{d\rho_{12}}{dt} = q_{pt} - A_p v - q_{V_1} \\[3mm] \left[\dfrac{V_6 + A_a (L - x)}{\rho_{34}}\right]\dfrac{d\rho_{34}}{dt} = A_a v - q_{pt} - q_{V_2} \end{cases} \tag{7.6}$$

Making use of the general relation between density and pressure given by Eq. (5.14), after some mathematical manipulations, we obtain

$$\begin{cases} \dfrac{dp_{12}}{dt} = \beta_E\left(\dfrac{q_{pt} - A_p v - q_{V_1}}{V_1 + A_p x}\right) \\[3mm] \dfrac{dp_{34}}{dt} = \beta_E\left[\dfrac{A_a v - q_{pt} - q_{V_2}}{V_6 + A_a (L - x)}\right] \end{cases} \tag{7.7}$$

The conduits within regions 1 and 6 in Figure 7.2 are usually identical. In such a case, $V_1 = V_6 = V$ in Eqs. (7.7), which finally gives

$$
\begin{cases}
\dfrac{dp_{12}}{dt} = \beta_E \left(\dfrac{q_{pt} - A_p v - q_{V_1}}{V + A_p x} \right) \\[4mm]
\dfrac{dp_{34}}{dt} = \beta_E \left[\dfrac{A_a v - q_{pt} - q_{V_2}}{V + A_a (L - x)} \right]
\end{cases}
\tag{7.8}
$$

7.2.1.1   Force Balance

Similar to what was done in Eq. (5.16), a force balance at the actuator can be carried out so that an additional equation can be derived. Figure 7.3 shows the single-rod cylinder moving against an inertial load and an external force, F. The friction force, F_f, that develops between the piston and the case is also shown in the figure.

Assuming that the sum of the masses of the rod, the piston and the load in Figure 7.2, is m, we can perform a force balance on the set piston–rod–load to obtain

$$
\frac{dv}{dt} = \frac{(p_{12} A_p - p_{34} A_a) - (F + F_f)}{m}
\tag{7.9}
$$

7.2.1.2   Steady-State and Transient Models

The mathematical model for the actuator shown in Figure 7.1 can be obtained from Eqs. (7.4), (7.8) and (7.9), which are grouped together in Eqs. (7.10):

$$
\begin{cases}
\dfrac{dp_{12}}{dt} = \beta_E \left(\dfrac{q_{pt} - A_p v - q_{V_1}}{V + A_p x} \right) \\[4mm]
\dfrac{dp_{34}}{dt} = \beta_E \left[\dfrac{A_a v - q_{pt} - q_{V_2}}{V + A_a (L - x)} \right] \\[4mm]
\dfrac{dv}{dt} = \dfrac{(p_{12} A_p - p_{34} A_a) - (F + F_f)}{m} \\[4mm]
\dfrac{dx}{dt} = v
\end{cases}
\tag{7.10}
$$

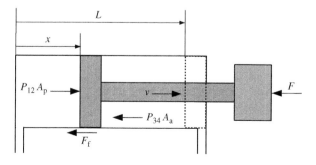

Figure 7.3   Forces acting on a single-rod hydraulic cylinder

Assuming that the pump flow, q_{pt}, is known, we observe that the four differential equations (7.10) contain seven variables: x, v, q_{v_1}, q_{v_2}, p_{12}, p_{34} and F_f. Therefore, we need three other equations to be able to find a solution to the transient mathematical model. The first equation comes from writing the friction force, F_f, as a function of the cylinder speed, v. The other two equations can be obtained through relating the check-valve flows, q_{v_1} and q_{v_2}, to the pressure differential, $(p_{12} - p_{34})$. The complexity of these relations depends on the models used for the valve flows and the cylinder friction. Note that in the particular case where the cylinder speed, v, and the pressures p_{12} and p_{34} are constants (steady-state operation), Eqs. (7.10) simplify to

$$\begin{cases} p_{12} = \dfrac{F + F_f}{A_p} + p_{34} \left(\dfrac{A_a}{A_p} \right) \\ q_p - A_p v - q_{v_1} = 0 \\ A_a v - q_p - q_{v_2} = 0 \end{cases} \tag{7.11}$$

where the time-dependent pump speed, q_{pt}, has been replaced by the constant (average) value, q_p (see Eqs. (3.3)).

7.2.2 Cylinder Friction

Finding an accurate mathematical model for the friction, F_f, in Eqs. (7.10) and (7.11) is not easy, given the number of variables involved. Friction has been thoroughly studied over the years and many friction models have been developed [1–3]. Given the introductory nature of this book, we limit ourselves to presenting some relatively simple steady-state-based models[1] that are sufficient for our purposes. We begin with the Coulomb[2] model, which can be applied for the dry friction between two moving bodies.

7.2.2.1 Coulomb Model

The Coulomb model assumes that the friction force, F_f, in Figure 7.3, does not depend on the magnitude of the speed, v, and is given by the following equation, which is valid for $v \neq 0$:

$$F_f = F_C \, \text{sgn}(v) \tag{7.12}$$

where F_C is known as the Coulomb friction [2], and the *signum* function, sgn(v), is generally defined for a given variable, x, as

$$\text{sgn}(x) = \frac{x}{|x|} \tag{7.13}$$

---

[1] In all models that follow, we assume that the piston speed, v, is either constant or zero.
[2] In honour of the French physicist Charles Augustin de Coulomb (1736–1806).

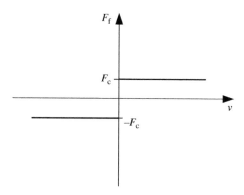

Figure 7.4   Coulomb friction as a function of the cylinder speed

Note that the same sign convention for the velocity and force used in Figure 6.1 is valid for Eq. (7.12), and that the friction, F_f, is not defined for $v = 0$.

Figure 7.4 shows a graphical representation of the Coulomb friction as a function of the cylinder speed.

7.2.2.2   Stiction Model

The Coulomb model can only be used when there is relative motion between two dry surfaces; in cases when no relative motion exists between the parts in contact, a static friction (or "stiction") develops. It has been experimentally observed that the force needed to overcome the static friction between two bodies in contact is actually bigger than the Coulomb friction, given by Eq. (7.12). Thus, let F_R be the resultant force acting on the piston at $v = 0$. It has been shown that the static friction, F_f, in this case, is given by [3]:

$$\begin{cases} F_f = F_R, \text{ when } v = 0 \text{ and } |F_R| < F_S \\ F_f = F_S \operatorname{sgn}(F_R), \text{ when } v = 0 \text{ and } |F_R| \geq F_S \end{cases} \quad (7.14)$$

where F_S is a threshold value ($F_S > F_C$) and F_R can be obtained from Figure 7.3 as

$$F_R = p_{12}A_p - p_{34}A_a - F \quad (7.15)$$

The static friction can be better visualized with the help of Figure 7.5, which shows the variation of the friction force, F_f, with the resultant force, F_R. As the force, F_R, grows, the piston remains in static equilibrium until the maximum value, F_S, is reached. At this point,

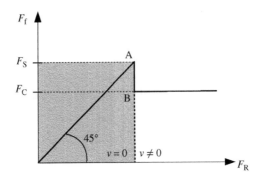

Figure 7.5   Static and slip friction between dry surfaces

the piston breaks into motion and the friction force, F_f, drops from F_S to the Coulomb friction, F_C, along the vertical line A–B.

The stiction model given by Eqs. (7.14) and represented by Figure 7.4 has been credited to Morin[3] [3]. It can be inferred from the model that friction cannot be considered as a function of the speed alone, being undefined at $v = 0$. On the other hand, for $v \neq 0$, the Coulomb model states that friction is constant and does not depend on the magnitude of the relative speed, v.

7.2.2.3   Viscous Friction

In a hydraulic cylinder, the viscosity of the oil film between the piston and the case adds a component to the Coulomb friction called *viscous friction*. The viscous friction can be obtained through the study of the laminar flow between two moving surfaces, in a similar way to what was done in the development of the resisting viscous torque, $(K_\omega \mu)\omega$, in Eq. (2.100). In fact, if we consider the Couette flow between the two plates, as shown in Figure 2.1, we can easily obtain an equation of the following kind for the case of a 'wet' piston-case friction:

$$F_f = F_v v \tag{7.16}$$

where F_v is the viscous friction coefficient, usually defined as constant.[4]

Figure 7.6 shows a graphical representation of the viscous friction as a function of the cylinder speed. Note that F_f is now continuous at $v = 0$ (compare Figures 7.4 and 7.6).

In an actual hydraulic cylinder, the three kinds of friction studied so far will generally take place, that is, the stiction (when $v = 0$) and the Coulomb and viscous friction (when $v \neq 0$). Therefore, a combination of these three models is expected to fit the experimental data more closely. We shall see how this can be done in the following section.

---

[3] Arthur Jules Morin. French physicist (1795–1880).
[4] It has been observed that the coefficient F_v also depends on the oil viscosity, which ultimately varies with the speed, v. Therefore, the assumption of a constant F_v in Eq. (7.16) is not strictly correct [2].

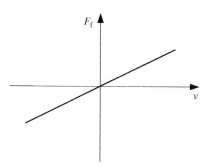

Figure 7.6   Viscous friction as a function of the cylinder speed

7.2.2.4   Combined Friction Model and Stribeck Friction

The combination of the Coulomb friction (Eq. (7.12)), the stiction (Eqs. (7.14)) and the viscous friction (Eq. (7.16)) results in the following equation for the friction force, F_f:

$$\begin{cases} F_f = F_C \ \text{sgn}(v) + F_v v, & \text{when } v \neq 0 \\ F_f = F_R, & \text{when } v = 0 \text{ and } |F_R| < F_S \\ F_f = F_S \ \text{sgn}(F_R), & \text{when } v = 0 \text{ and } |F_R| \geq F_S \end{cases} \qquad (7.17)$$

Equations (7.17) are graphically illustrated in Figure 7.7, where we observe a sudden change in the friction force in both directions at $v = 0$. This behaviour is related to the theoretical change from static to slip frictions, shown by the line AB, in Figure 7.5. In actual situations, however, the curve $F_f(v)$ is smooth in the region where slippage starts to happen. This phenomenon was described in a classic paper written by Stribeck [4] and is represented in the figure by dashed lines. The corresponding term that must be added in the first of Eqs. (7.17) has been designated as *Stribeck friction*.

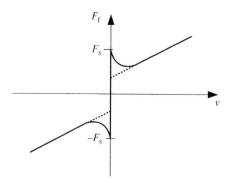

Figure 7.7   Combined dry, viscous and Stribeck friction

Considering the Stribeck friction, Eqs. (7.17) can be rewritten as [3]

$$
\begin{cases}
F_f = \text{sgn}(v)\left[F_C + (F_S - F_C)\,e^{-\left|\frac{v}{c_S}\right|}\right] + F_v v, & \text{when } v \neq 0 \\[3mm]
F_f = F_R, & \text{when } v = 0 \text{ and } |F_R| < F_S \\[3mm]
F_f = F_S\,\text{sgn}(F_R), & \text{when } v = 0 \text{ and } |F_R| \geq F_S
\end{cases}
\tag{7.18}
$$

where C_S is the *sliding speed coefficient* for the Stribeck friction term [1].

7.2.2.5 Velocity Threshold Approximation

The fact that F_f is discontinuous at $v = 0$ in Eqs. (7.18) makes these equations difficult to handle using numerical algorithms. A practical way to circumvent this problem is shown in Figure 7.8. In the figure, a sufficiently small threshold speed value, v_{th} ($v_{th} > 0$), is chosen so that the friction force, F_f, can be linearly approximated within the interval $-v_{th} \leq v \leq v_{th}$. As a result, instead of Eqs. (7.18), we can write the following approximate equations for the cylinder friction:[5]

$$
\begin{cases}
F_f = \dfrac{v}{v_{th}}\left[F_C + (F_S - F_C)\,e^{-\left|\frac{v_{th}}{c_S}\right|} + F_v v_{th}\right], & \text{when } |v| < v_{th} \\[4mm]
F_f = \text{sgn}(v)\left[F_C + (F_S - F_C)\,e^{-\left|\frac{v}{c_S}\right|}\right] + F_v v, & \text{when } |v| \geq v_{th}
\end{cases}
\tag{7.19}
$$

Since $v_{th} > 0$, there was no need to write the function $\text{sgn}(v_{th})$ multiplying the Coulomb and static friction terms in the first of the Eqs. (7.19). We also observe that the term within brackets in the first equation in (7.19) is constant, that is

$$
F_C + (F_S - F_C)e^{-\left|\frac{v_{th}}{c_S}\right|} + F_v v_{th} = F_{th} > 0
\tag{7.20}
$$

From Eqs. (7.19) and (7.20), we see that $F_f(v_{th}) = F_{th}$ and $F_f(-v_{th}) = -F_{th}$, as shown in Figure 7.8.

The value of the threshold speed, v_{th}, should be small enough to correctly represent the stiction phenomenon (i.e. F_{th} should be as close to F_S as possible) and large enough to prevent numerical instabilities. In our experience, the choice of $v_{th} = 1 \times 10^{-4}$ m/s has proven to be adequate for the numerical solution of Eqs. (7.10). The other terms F_C, F_S, F_v and C_S, in Eqs. (7.19), must be obtained experimentally. For instance, Table 7.1 has the data relative to a hydraulic test rig, described in Ref. [5], which have adopted for the examples given later in this chapter.

Figure 7.9 shows the curve $F_f(v)$ obtained for the cylinder described in Table 7.1, using Eqs. (7.19).

---

[5] See, for example the online documentation of SimHydraulics from MathWorks Inc., available at http://www .mathworks.com/help/physmod/hydro/ref/cylinderfriction.html (April 2014).

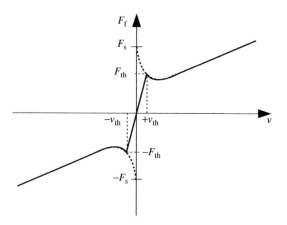

Figure 7.8 Friction model for the hydraulic cylinder

Table 7.1 Typical experimental data for the friction coefficients and cylinder dimensions [5]

Friction terms and cylinder data	Value
Coulomb friction, F_C	120 N
Static friction, F_S	880 N
Viscous friction coefficient, F_v	9570 N s/m
Sliding speed coefficient, C_S	0.02 m/s
Piston area, A_p	0.00312 m^2
Ring area, A_a	0.00153 m^2
Cylinder stroke, L	0.5 m

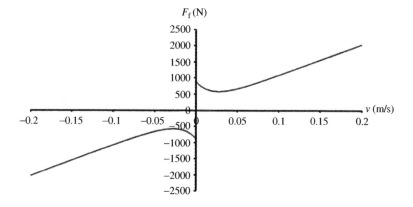

Figure 7.9 Friction force versus cylinder speed

7.2.3 Pilot-Operated Check Valves

The next elements that appear in Figure 7.1 are the pilot-operated check valves V_1 and V_2. The operational principle of these valves is illustrated in Figure 7.10, which shows a cross-sectional cut of a typical piloted check valve. Two situations are represented in the figure: one in which the valve opens by the action of the pressure p_1, when the flow direction is from port 1 to port 2 (normal operation mode, Figure 7.10(a)); and another where the flow direction can be either from port 1 to port 2 or from port 2 to port 1, and the valve opens by the action of the external pilot pressure, p_x (pilot operation mode, Figure 7.10(b)). In the first situation, p_1 is necessarily higher than p_2, whereas in the second situation, the pilot pressure, p_x, keeps the valve open so that no restriction is placed on the pressures p_1 or p_2.

The operation in normal mode is similar to the operation of a relief valve, where the valve aperture increases with the pressure differential. Observe, however, that the two valves V_1 and V_2 never operate in normal mode in the circuit shown in Figure 7.1, being always either closed or forcedly open by an external pilot pressure (see Figures 6.22–6.25). Therefore, we shall only consider the pilot operation mode in our mathematical model.

In pilot operation mode, the input pressure, p_1, has no influence on the position of the spool, which is kept at its rightmost position, x_{max}, by the action of an external pilot pressure, p_x. The valve can then be modelled by the orifice equation (2.79), given that the valve

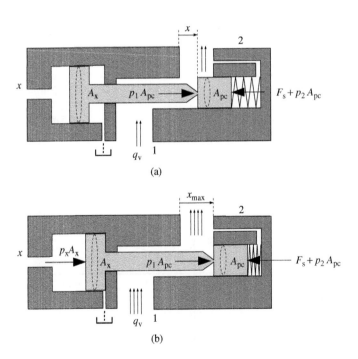

Figure 7.10 Piloted check valve: (a) partially-open during normal operation and (b) fully open during pilot operation

aperture is kept constant by an external force and not by the pressures p_1 and p_2 at the input and output ports.

7.2.3.1 Determination of the Minimum Pilot Pressure

In order to keep the valve spool open, the pilot pressure must overcome a minimum value that depends on the valve construction. For instance, considering that the rod-side of the pilot piston is connected to the tank, as shown in Figure 7.10, a force balance carried out over the valve spool gives[6]

$$x = \left(\frac{1}{k_s}\right)[A_{pc}(p_1 - p_2) + A_x p_x] - x_0 \tag{7.21}$$

where x_0 is the pre-tensioning spring deflection corresponding to the cracking pressure, p_c, and k_s is the valve spring constant (note that $F_s = k_s x + k_s x_0 = k_s x + p_c A_{pc}$).

From Eq. (7.21), we obtain the pilot pressure, p_x:

$$p_x = \left(\frac{A_{pc}}{A_x}\right)\left[\frac{k_s(x_0 + x)}{A_{pc}} - (p_1 - p_2)\right] \tag{7.22}$$

The pilot pressure, p_x, needs to be greater than a minimum value, $p_{x\min}$, for the check valve to start opening. Similarly, after p_x has reached a certain value $p_{x\max}$, the valve spool will be completely displaced to the right ($x = x_{\max}$), and the valve becomes similar to a fixed orifice. We can obtain the values of $p_{x\min}$ and $p_{x\max}$ by making $x = x_{\min} = 0$ and $x = x_{\max}$, respectively, in Eq. (7.22):

$$\begin{cases} p_{x\min} = \left(\dfrac{A_{pc}}{A_x}\right)\left[\dfrac{k_s x_0}{A_{pc}} - (p_1 - p_2)\right] \\[4mm] p_{x\max} = \left(\dfrac{A_{pc}}{A_x}\right)\left[\dfrac{k_s(x_0 + x_{\max})}{A_{pc}} - (p_1 - p_2)\right] \end{cases} \tag{7.23}$$

We usually do not have access to the value of $x_{\max}$. However, we can still have an idea of the magnitude of the pilot pressure from the first equation in Eq. (7.23), which can be written as follows:

$$p_{x\min} = \left(\frac{A_{pc}}{A_x}\right)(p_c - \Delta p_v) \tag{7.24}$$

where $\Delta p_v = p_1 - p_2$ and p_c is the cracking pressure, given by

$$p_c = \frac{k_s x_0}{A_{pc}} \tag{7.25}$$

---

[6] Here, we are disregarding the valve dynamics, which is an acceptable approach given that check valves usually respond very quickly when compared to the other hydraulic elements [6].

Remarks

1. Cracking pressures in piloted check valves are usually small. Typical values range from 1.5 to 10 bar [7].
2. If $p_1 > p_2$ in Eq. (7.24), we conclude that very little pressure is needed at the pilot port to open the valve. In fact, we observe from Eq. (7.24) that $p_{x\,min} < (p_c - \Delta p_v)$ in this case.[7] Given that $\Delta p_v > 0$, the minimum pilot pressure, $p_{x\,min}$, will be necessarily smaller than the cracking pressure, p_c (negative values of $p_{x\,min}$ indicate that the valve opens naturally by the action of the pressure differential, Δp_v).
3. If $p_1 < p_2$ in Eq. (7.24), the minimum pilot pressure, $p_{x\,min}$, will depend on the magnitude of the negative pressure differential, $-\Delta p_v$, and will be given by $p_{x\,min} = (A_{pc}/A_x)(p_c + |\Delta p_v|)$. In this case, the ratio A_{pc}/A_x plays a major role in reducing the magnitude or the pilot pressure in relation to the pressure differential between the input and output ports, Δp_v.

From Eq. (7.24), we see that the minimum pilot pressure depends on the internal dimensions of the valve, A_{pc} and A_x, which can be obtained from the manufacturer catalogues. As an example, consider the typical piloted check valve, shown in Figure 7.11 [7]. Assuming that the rod-end side of the pilot piston is vented through port Y, we observe that the valve configuration is similar to the valve schematically shown in Figure 7.10, so that Eq. (7.24) can be applied. Therefore, we can substitute the values of the areas, $A_{pc} = 0.42$ cm^2 and $A_x = 1.33$ cm^2, into Eq. (7.24) to obtain

$$p_{x\,min} = 0.32(p_c - \Delta p_v) \tag{7.26}$$

For the valve represented in Figure 7.11, four cracking pressures are available: 1.5, 3.0, 7.0 and 10.0 bar [7]. If we choose $p_c = 1.5$ bar, for example, the opening pilot pressure, $p_{x\,min}$,

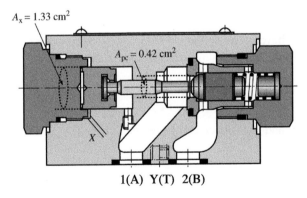

$A_x = 1.33$ cm^2

$A_{pc} = 0.42$ cm^2

X

1(A) Y(T) 2(B)

Figure 7.11 Cross-cut of a Bosch-Rexroth SL-type piloted check valve, series 6X (courtesy of Bosch-Rexroth Corp.)

[7] Note that $A_{pc}/A_x < 1$ in Figure 7.10.

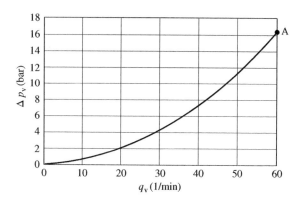

Figure 7.12 Typical pressure versus flow curve in piloted operational mode (reproduced courtesy of Bosch-Rexroth Corp.)

will be given by

$$p_{x\,min} = 0.525 - 0.35\Delta p_v \qquad (7.27)$$

The maximum pilot pressure, $p_{x\,max}$, relative to the valve operating at full aperture, should not be much higher than the minimum pilot pressure, $p_{x\,min}$, especially at small cracking pressures (e.g. 1.5 bar). Therefore, we can make the simplifying assumption that once the pilot pressure becomes higher than $p_{x\,min}$, given by Eq. (7.27), the check valve opens completely, reaching the point of maximum aperture right away. Such assumption allows us to perform one single check to verify whether the check valve is open or closed when a numerical solution is required.

In order to obtain a suitable mathematical model for the valve, we need to know the pressure versus flow curve, $\Delta p_v(q_v)$, for the valve operating in piloted mode. For instance, Figure 7.12 shows the curve $\Delta p_v(q_v)$ for the valve represented in Figure 7.11 [7] when the flow direction is from port 2 to port 1 (pilot operation mode). As will become clear shortly, only one point on the curve is needed to model the valve as a fixed orifice. For this purpose, we have randomly chosen point A (60.0, 16.3) on the extreme right of the figure.

7.2.3.2   Modelling Equation in Pilot Operation Mode

Once the piloted check valves are completely open, they can be viewed as simple orifices for which Eq. (2.79) can be applied. Therefore, the relation between the flow, q_v, and the pressure differential across the valve, Δp_v, in pilot operation mode can be generally written as

$$\begin{cases} q_v = 0, & \text{if } p_x \leq p_{x\,min} \\ q_v = C\,\text{sgn}\left(\Delta p_v\right)\sqrt{|\Delta p_v|}, & \text{if } p_x > p_{x\,min} \end{cases} \qquad (7.28)$$

where the coefficient, C, must be obtained from experimental data.

The second of Eqs. (7.28) is a more elaborated version of the orifice flow equation (2.79). Given that the valve is opened through an external pilot, there is no guarantee that the pressure p_1 will be greater than the pressure p_2, that is, the fluid can flow in either direction: from port 1 to port 2, or from port 2 to port 1 (see Figure 7.10). This is why we have used the module of Δp_v in the radicand. On the other hand, a good valve model should be able to correctly represent the flow direction; so, we have added the multiplier $\text{sgn}(\Delta p_v)$, to the equation, meaning that the flow is positive for a positive pressure differential $(p_1 > p_2)$ and the flow is negative for a negative pressure differential $(p_1 < p_2)$.

The coefficient, C, can be obtained from the characteristic curve in Figure 7.12, which shows the pilot operation mode of the check valve. If we consider a constant approximation for C, only one point in the characteristic curve is needed. Thus, if point A in Figure 7.12 is chosen, we have that

$$C = \frac{q_c}{\sqrt{p_A}} = \frac{60}{\sqrt{16.3}} = 14.86 \tag{7.29}$$

The valve represented by the characteristic curve shown in Figure 7.12 can then be modelled by the following equations when operating in pilot mode:

$$\begin{cases} q_v = 0, & \text{if } p_x \leq p_{x\min} \\ q_v = 14.86 \ \text{sgn}\left(\Delta p_v\right) \sqrt{|\Delta p_v|}, & \text{if } p_x > p_{x\min} \end{cases} \tag{7.30}$$

where p_1 and p_2 are given in bar and q_v in l/min.

7.3   Case Study

Now that we have all the necessary elements for modelling the circuit shown in Figure 7.1, we are in a position to perform some simulations. Because of the differential areas in the cylinder, the resulting mathematical model will be necessarily nonlinear, for which a numerical approach is appropriate, as will be seen shortly.

The following initial considerations are valid for this example:

1. The pump flow, q_{pt}, in Eqs. (7.10), changes in direction and magnitude accordingly to the following equation:

$$q_{pt} = q_{pmax} \sin\left(\frac{2\pi t}{T}\right) \tag{7.31}$$

where T is the oscillation period and q_{pmax} is the maximum pump flow. Note that $0 \leq t \leq T$.
2. The cylinder movement must happen within the stroke, L, in Figure 7.3. In other words, $0 < x < L$ in Eqs. (7.10). This poses a restriction on the value of the period, T, in Eq. (7.31).

It is convenient here to make a list of all the numerical data that will be used in this example. This is done in Table 7.2.

Table 7.2 Data used in the simulations

Circuit parameter	Value
Inner conduit volume, V	1.06×10^{-4} m^3
Piston area (single rod), A_p	0.00312 m^2
Ring area (single rod/double rod), A_a	0.00153 m^2
Cylinder stroke, L	0.5 m
Piston–rod–load mass, m	50 kg
Coulomb friction, F_C	120 N
Static friction, F_S	880 N
Viscous friction coefficient, F_v	9750 N s/m
Sliding speed coefficient, C_S	0.02 m/s
Effective bulk modulus, β_E	3.57×10^8 N/m^2 (Eq. (2.36))
Tank pressure, p_0	0 N/m^2 or 10×10^5 N/m^2
External load, F (quadrants I and IV)	+53,000 N
External load, F (quadrants II and III)	−26,500 N
Maximum pump flow, $q_{p\,\text{max}}$	3.397×10^{-4} m^3/s

7.3.1 Determination of the Pump Flow Period

In order to obtain the period, T, in Eq. (7.31), consider the curve $q_{pt}(t)$ shown in Figure 7.13. If we consider only the negative part of the $q_{pt}(t)$ curve where $T/2 < t < T$, we can write for the total cylinder displacement during retraction, x_a:

$$x_a = -\frac{q_{p\,\text{max}}}{A_a} \int_{T/2}^{T} \sin\left(\frac{2\pi t}{T}\right) dt \tag{7.32}$$

Similarly, the total displacement during extension, x_p, will be given by

$$x_p = \frac{q_{p\,\text{max}}}{A_p} \int_{0}^{T/2} \sin\left(\frac{2\pi t}{T}\right) dt \tag{7.33}$$

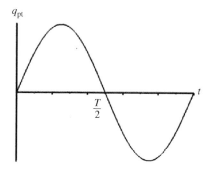

Figure 7.13 Pump output flow

From Eqs. (7.32) and (7.33), we obtain

$$x_p = \left(\frac{A_a}{A_p}\right) x_a \le \left(\frac{A_a}{A_p}\right) L \tag{7.34}$$

As an example, consider the operation in the second and third quadrants, where the average pump flow, q_p, is given by vA_a (Figures 6.23 and 6.24, respectively). The maximum value of the period, T, in this case can be obtained by solving the integral in Eq. (7.32):

$$T = \frac{x_a \pi A_a}{q_{p\,max}} = \frac{\varphi_T L \pi A_a}{q_{p\,max}} \tag{7.35}$$

where $0 < \varphi_T \le 1$. In this case study we make $\varphi_T = 0.9$ so that we can guarantee that the cylinder will move within the maximum allowed stroke.

The period T required for the cylinder to move a distance $\varphi_T L$ at the first and fourth quadrants, where $q_p = vA_p$ (Figures 6.22 and 6.25), is obviously longer than the period obtained for the second and third quadrants and is given by (see Eq. (7.33))

$$T = \frac{\varphi_T L \pi A_p}{q_{p\,max}} \tag{7.36}$$

Using the data from Table 7.2, and making $\varphi_T = 0.9$ in Eqs. (7.35) and (7.36), we obtain the values for the period T as shown in Table 7.3.

7.3.2 Numerical Simulation

In order to find a numerical solution to Eqs. (7.10), we apply the Runge–Kutta method introduced in Section 5.2.3. First, we observe that the system of equations (7.10) can be written as a single vector equation, as follows:

$$\frac{d\mathbf{y}}{dt} = f(\mathbf{y}, t) \tag{7.37}$$

Table 7.3 Extension and retraction periods used in our simulations

Operational quadrant (Figures 6.22–6.25)	T (s)
I and IV	12.99
II and III	6.37

where

$$
\mathbf{y} = \begin{bmatrix} p_{12} \\ p_{34} \\ v \\ x \end{bmatrix} \text{ and } f(\mathbf{y}, t) = \begin{bmatrix} \beta_E \left(\dfrac{q_p - A_p v - q_{V_1}}{V + A_p x} \right) \\ \beta_E \left[\dfrac{A_a v - q_p - q_{V_2}}{V + A_a (L - x)} \right] \\ \dfrac{(p_{12} A_p - p_{34} A_a) - (F + F_f)}{m} \\ v \end{bmatrix}
\tag{7.38}
$$

We remember here that the Runge–Kutta method is a marching method that calculates the approximate value of the solution vector, $\mathbf{y}$, at time t_{n+1}, $\mathbf{y}_{n+1}$ ($n = 1, 2, 3, \ldots$), from a known previous value of $\mathbf{y}$ at time t_n, $\mathbf{y}_n$. Therefore, for the process to be started, we need to know the initial value of the solution vector at time $t_0 = 0$. Given that the pump flow at time t_0 is zero and considering that the external force, F, at the cylinder changes direction depending on the operational quadrant (see Table 7.2), we can write for the initial solution vector, $\mathbf{y}_0$:

$$
\mathbf{y}_0 = \begin{bmatrix} |F/A_p| \, i \\ |F/A_a| \, j \\ 0 \\ \dfrac{L(1 - \varphi_T)}{2} \end{bmatrix}
\tag{7.39}
$$

where i and j depend on the initial direction of the force, F, as shown in Table 7.4.

The values of i and j given in Table 7.4 can also be obtained from the following equations:

$$
j = \frac{1}{2} \left(1 - \frac{F}{|F|} \right) \text{ and } i = 1 - \frac{1}{2} \left(1 - \frac{F}{|F|} \right)
\tag{7.40}
$$

With $\mathbf{y}_0$ defined by Eqs. (7.39) and (7.40) and T limited by Eqs. (7.35) and (7.36), we can proceed with the numerical simulation to obtain the solution vector, $\mathbf{y}$, for the time interval $0 < t < T$. The results are shown next ↓.

Table 7.4   Values of the parameters i and j in Eq. (7.39)

Sign of F	i	j
+	1	0
−	0	1

7.3.2.1 Results for the First and Fourth Quadrants

Figure 7.14 shows the curves obtained for the solution vector components considering a positive force, F, and a complete cycle of the pump flow (see Figure 7.14(a)). In these circumstances, the cylinder will operate in the first quadrant when expanding and in the fourth quadrant when retracting (see Figures 6.22 and 6.25).

From Figure 7.14(b), we see that the cylinder moves from $x = 0.025$ m to $x = 0.475$ m, leaving a space of 0.025 m between the piston and the heads. This guarantees that the fourth equation in (7.10) remains valid during the analysis. Observe that oscillations related to the fluid elasticity are present in the $v(t)$ curve (Figure 7.14(c)), as well as in the $p_{12}(t)$ and $p_{34}(t)$ curves (Figures 7.14(d) and (e)), near the points where the pump flow becomes zero.

Figure 7.14(e) shows the pressure at the low-pressure conduit 3–4 as a function of time when the tank pressure is zero bar gauge. Despite the usual oscillatory behaviour near the

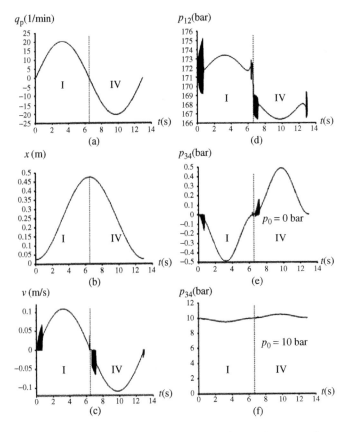

Figure 7.14 Numerical results for the first and fourth quadrants: (a) pump flow, q_p, (b) cylinder displacement, x, (c) cylinder speed, v, (d) pressure in line 1–2, p_{12}, (e) and (f) pressure in line 3–4, p_{34}

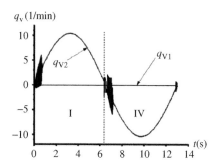

Figure 7.15   Piloted check-valve flows

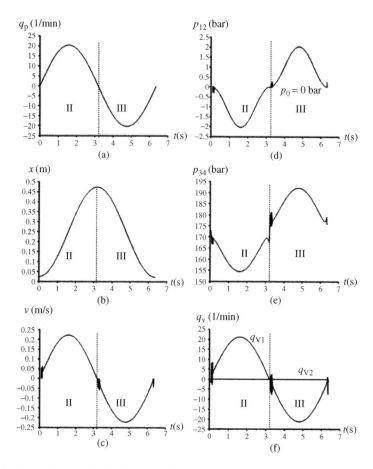

Figure 7.16   Numerical results for the second and third quadrants: (a) pump flow, q_p, (b) cylinder displacement, x, (c) cylinder speed, v, (d) pressure in line 1–2, p_{12}, (e) pressure in line 3–4, p_{34}, and (f) piloted check-valve flows, q_{v1} and q_{v2}

point where the pump flow changes sign, it is interesting to see that the pressure becomes lower than 1 atm (0 bar gauge) during the first quadrant. This suggests the possibility of 'vaporous cavitation' (see Section 5.1.1) in the circuit, which can be remedied by raising the tank pressure level, as indicated in Figure 7.14(f) where we have plotted the results for a tank pressure, p_0, of 10 bar. A practical means of raising the pressure p_0 is by using a charge circuit instead of a direct connection to the tank. We remember that this was the solution adopted in the original circuit shown in Figure 6.21.

The flows through the piloted check valves V_1 and V_2 are shown in Figure 7.15. Note that the transient model correctly represents the status of the check valves, as indicated in Figures 6.22 and 6.25, that is, only the check valve V_2 remains open during operation.

7.3.2.2 Results for the Second and Third Quadrants

Figure 7.16 shows the results obtained when the external force, F, is $-26,500$ N (Table 7.2). This corresponds to the situation in the second and third quadrants, as seen in Figures 6.23 and 6.24, respectively. Note that this time, the cylinder moves faster because of the differential areas.

As expected, the same observations concerning the possibility of cavitation remain valid in quadrant II. Interestingly, in this case, we observe that the pressure in line 1–2 reaches non-physical negative values[8] (Figure 7.16(d)). On the other hand, the pressure at line 3–4 is higher when compared to the corresponding situation shown in Figure 7.14(d), due to the smaller area influenced by the force F.[9] Note that this time, only the piloted check valve V_1 is open (Figure 7.16(f)), which confirms the situations shown in Figures 6.23 and 6.24.

Exercises

(1) Show that the friction force, F_f, exerted by the fluid onto the moving plate in Figure 7.17 is proportional to the speed, v, as expressed by Eq. (7.16).

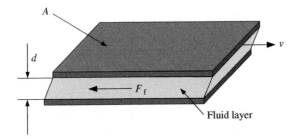

Figure 7.17 Viscous force between a stationary and a moving plate

---

[8] Assuming that the atmospheric pressure at sea level is approximately 1 bar, it is impossible to reach values below -1 bar in the circuit. This is, however, not seen by the numerical algorithm, which assumes that the pressures are defined as real numbers.

[9] In this case, the pressure reaches higher values in quadrant III, when the pump flow is directed against the cylinder load.

(2) Estimate the viscous friction, F_f(N), for the piston–cylinder shown in Figure 7.18 considering three different viscosity grades: ISO 22, 46 and 68 (Table 2.2). Assume the fluid temperature is kept at 40 °C in your calculations and that the oil density is 870 kg/m³.

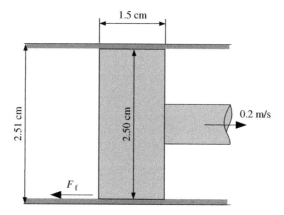

Figure 7.18   Viscous force in a cylinder

(3) The cylinder of the double-rod actuator shown in Figure 7.19 is held in the mid-stroke position by setting the pump displacement to zero. Starting from Eqs. (7.10), disregard the friction forces at the cylinder and the inner conduit volumes and show that the actuator in the figure can be modelled by the following equation:

$$\frac{d^2 v}{dt^2} + \left(\frac{2\beta_E A_a^2}{m V_0} \right) v + \left(\frac{1}{m} \right) \frac{dF}{dt} = 0$$

where m is the mass of the piston and the rod together.

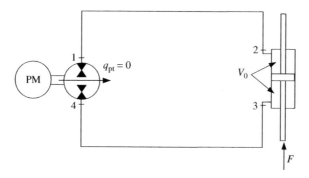

Figure 7.19   Simplified version of a double-rod actuator

(4) Start from the mathematical model given in Exercise 3 and obtain an expression for the natural frequency for the circuit shown in Figure 7.19.

(5) Obtain the solution for the mathematical model of the transmission shown in Figure 7.19 for F as given by

$$F = F_m \cos(\omega_f t)$$

where F_m and ω_f are constants.

References

[1] Andersson S, Söderberg A, Björklund S (2007). Friction models for sliding dry, boundary and mixed lubricated contacts. Tribology International, 40: 580–587.

[2] Guida D, Nilvetti F, Pappalardo CM (2010) Dry friction of bearings on dynamics and control of an inverted pendulum. Journal of Achievements in Materials and Manufacturing Engineering, 38(1): 80–94.

[3] Olsson H (1996) Control systems with friction, Dissertation. Lund Institute of Technology, Swiss.

[4] Stribeck R (1902) Die Wesentlichen Eigenschaften der Gleit- und Rollenlager – The key qualities of sliding and roller bearings. Zeitschrift des Vereines Seutscher Ingenieure, 46: 38–39, pp. 1342–1348, 1432–1437.

[5] Rahmfeld R (2002) Development and control of energy saving hydraulic servo drives for mobile systems, Dissertation. Hamburg University of Technology (TUHH), Germany.

[6] Williamson C, Ivantysynova M (2010) Stability and motion control of inertial loads with displacement controlled hydraulic actuators. Proceedings of the 6th FPNI – PhD Symposium, pp. 499–514, West-Lafayette, USA.

[7] Bosch-Rexroth (2002) Check valve: hydraulically pilot operated, types SV and SL series SX. Catalog – RE 21 460/06.02, Lohr am Main, Germany.

8

Practical Applications

We dedicate this chapter to explore some typical applications of hydrostatic transmissions and actuators. Our intention is not to cover all possible cases – since there are many – but to give an idea of the current and potential fields where hydrostatic transmissions and actuators can be applied. We begin by exploring the benefits of using hydrostatic transmissions as a natural choice for an infinitely variable transmission (IVT) in passenger cars and vehicles in general. We move on to show the applications of hydrostatic transmissions in heavy mobile machinery, such as cranes and excavators, which have already made use of this technology for a long time. Next, we explore the contemporary theme of hybrid vehicles. Then we show the utilization of hydrostatic transmissions and actuators in wind power generation. Finally, we close the chapter with an important aeronautical application of electrohydrostatic actuators.

8.1 Infinitely Variable Transmissions in Vehicles

To understand the advantages of IVTs and, in particular, hydrostatic transmissions in vehicular applications, we consider the way in which the engine power is transferred to the wheels in a passenger car. Figure 8.1 shows the schematics of a typical rear-driven car transmission, where the power coming from the engine, E, is transmitted to the wheels through the differential, D. In the figure, we only consider the case where the car is moving forward, accelerating from idle to maximum speed. As a result, the angular speed of the output shaft, ω_o, will vary from zero to a maximum value, ω_{omax}. Note that the engine in Figure 8.1 cannot operate under a minimum angular speed. Therefore, since the engine connects directly to the input shaft of the transmission, the input angular speed, ω_i, must fall within the interval $[\omega_{imin}, \omega_{imax}]$.

Now, suppose that we place a gearbox between the input and the output shafts. Because of the constant values of the transmission ratio, the output speed will vary from ω_{omin} to ω_{omax}

Hydrostatic Transmissions and Actuators: Operation, Modelling and Applications, First Edition.
Gustavo Koury Costa and Nariman Sepehri.
© 2015 John Wiley & Sons, Ltd. Published 2015 by John Wiley & Sons, Ltd.
Companion Website: www.wiley.com/go/costa/hydrostatic

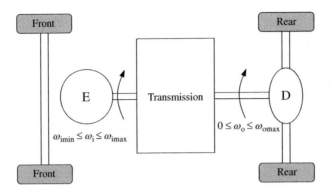

Figure 8.1   Schematics of a typical rear-driven vehicle

and not from 0 to ω_{omax} as required.[1] Given that it is not possible to have $\omega_o = 0$ without stalling the engine, alternative solutions must be adopted. Three possibilities are as follows:

1. Slowly connect the engine shaft to the transmission input by means of a clutch. This solution has been used for years and it is still popular among passenger cars (manual transmissions). The great advantage is the high transmission efficiency that is achieved when the gears are engaged.[2]
2. Couple the engine to the transmission through a torque converter. The automatic transmissions that equip passenger cars today make use of this technology. Comfort is a priority in these types of vehicles, and yet the actual energy efficiency is relatively low over the complete speed range of the vehicle when compared to using gearboxes for the transmission instead. There is, however, the advantage that part of the engine power converts into a higher torque at the wheels when the car stops.
3. Use an IVT between the engine and the differential. This solution is becoming popular in passenger cars as it allows for the angular speed of the output shaft to be continuously changed. An example of a mechanical IVT is seen in Figure 1.12, which shows a combination of a torque converter, a gearbox and a toroidal variator. Hydrostatic transmissions can be used as well. However, in this respect, they are overall less efficient than the mechanical IVT solutions [1]. Another alternative, which combines the flexibility of hydrostatic transmissions with the high efficiency of gearboxes, is to use a hydromechanical transmission (see Section 1.5).

In spite of the IVT technology being used, it is important to see how the vehicle will benefit from it. It is well known that both the power output and the fuel consumption of a spark-ignition engine depend on its angular speed [2]. For instance, Figure 8.2(a) shows the variation of the torque and power in a typical gasoline engine while Figure 8.2(b) gives the specific fuel consumption. Observe that as the engine speed changes from 1000

---

[1] Note that in a discretely variable transmission (e.g. a gearbox), $R_T \neq 0$, and as a consequence it is impossible to have $\omega_o = 0$ (see Eq. (1.1)).
[2] Note that this is not the case when there is still slippage within the clutch, that is, during the interval $0 \leq \omega_o \leq \omega_{omin}$, or when gears are being changed.

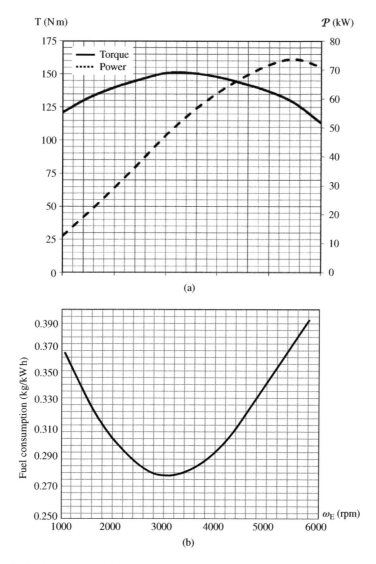

Figure 8.2 Typical (a) torque and power curves and (b) specific fuel consumption for an internal combustion engine

to 6000 rpm, torque and power reach their maximum values at different angular speeds (Figure 8.2(a)). Likewise, there is a certain engine speed for which fuel consumption is the lowest (Figure 8.2(b)).

One of the advantages of using an IVT connecting the engine to the differential is that it is possible to change the car speed and yet keep the engine running at a fixed rotation. For example, if the engine speed is maintained at approximately 3000 rpm, the lowest fuel consumption is obtained (Figure 8.2(b)). Another advantage is that the usual 'jolts' that occur when changing gears do not exist, which results in a smoother drive in the end.

8.2 Heavy Mobile Equipment

Hydrostatic and hydromechanical transmissions have been extensively used in tractors, cranes, bulldozers and other heavy-duty machines. The main reason is their ability to change the transmission ratio and to reverse the motion without the need of changing gears.[3] In the particular case of hydrostatic transmissions, there is also the spatial flexibility that allows the pump and motor to be arbitrarily placed in relation to one another. Other benefits include the possibility of energy regeneration through the use of accumulators and the capacity for keeping the engine speed within an optimal range, as seen in the previous section.

As a first example, consider the hydraulic bucket lift in Figure 8.3, where the cabin and the engine are placed on a turntable that is allowed to rotate freely around its vertical axis. There is absolutely no possible way of mechanically connecting the engine to the wheels in that case. This is one of the situations where hydrostatic transmissions are needed. Due to the possibility of connecting pump and motor through hoses the engine can be placed on the turning platform and still transmit power to hydraulic motors attached to the wheels, as shown on the right in the magnified excerpt from the portion of the picture labelled A. This spatial flexibility represents the main advantage in relation to mechanical transmissions in this case.

Hydromechanical transmissions (Section 1.5) have also been extensively used in heavy vehicles. One typical field of application is the transmission of wheel loaders (Figure 8.4). Wheel loaders must be precisely controlled both during pile operations and when they need

Figure 8.3 Hydraulic bucket lift: cabin and engine are mounted on a rotary platform (a), and hydraulic motor M is connected to the wheel (b)

---

[3] This comes in handy, for example in tractors, where hydrostatic transmissions allow for the speed of forward and reverse travels to be instantly and infinitely varied without any change in engine speed. As a result, the finish produced by the tractor (e.g. in agricultural applications) will be consistently homogeneous, given that there is no change in the rotational speed of the output shaft connected to the tractor implements. Moreover, it is easy to shift from forward to reverse by simply pushing a pedal without taking the hands off the wheel.

Figure 8.4   Wheel loader

to move quickly from place to place. Thus, when the loader is transporting (high speed and low torque), we can benefit from the higher mechanical efficiency of the gearbox. On the other hand, during pile operation, the hydrostatic transmission is preferred because of the ability to better control the loader speed.

Hydrostatic actuators can also be used in heavy machinery. If we recall from Figure 6.29, two or more cylinders can be controlled by their corresponding pumps and yet share one common charge circuit. For instance, Figure 8.5 shows a picture of a hydraulic excavator whose arm contains four sets of cylinders (labelled A through D). Because of the way that

Figure 8.5   Hydraulic excavator

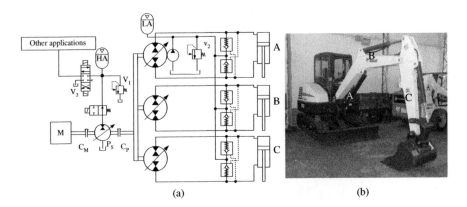

(a)   (b)

Figure 8.6   Hydraulic circuit for a mini-excavator

the excavator operates, the forces acting on each cylinder can vary in both direction and intensity, so that each cylinder can operate in any of the four quadrants.

It is possible, in principle, to replace the conventional valve-based circuits with hydrostatic actuators for every cylinder to increase cylinder control. Thus, all the features that were mentioned in Chapter 6 regarding hydrostatic actuators could be incorporated into the arm control of the excavator (see e.g. Refs. [3–5]. Although the use of hydrostatic actuators in heavy machines is still mostly experimental, some preliminary results about the substitution of hydrostatic actuators in place of a conventional hydraulic circuit using valves have shown that the efficiency of the latter can be almost 2.5 times greater and considerably less energy is consumed [4]. Moreover, depending on the circuit design, there exists the possibility of energy regeneration.

An example of a three cylinder regenerative hydrostatic circuit, which may be used in a mini-excavator, is given in Figure 8.6 (see Ref. [5] for a similar circuit).

The two clutches, C_M and C_P, in Figure 8.6, connect the internal combustion engine, M, to the energy storage circuit and to the three individual pumps that control each cylinder. The pump, P_S, located in the energy storage circuit, loads the high-pressure accumulator (HA) when storing energy at a pressure limited by the relief valve, V_1. The energy stored in the high-pressure accumulator may be used to power the cylinders or to supply energy to other hydraulic applications, depending on the position of the directional valve V_3, which can also be used to unload the accumulator HA, if necessary.

The use of the Common Pressure Rails (CPRs, see Section 6.3) in excavators and similar equipment has also been explored (see Refs. [6, 7]). For instance, Figure 8.7 shows a possible alternative circuit for the mini-excavator shown in Figure 8.6, following a similar solution presented in Ref. [6]. The circuit makes use of the Innas Hydraulic Transformer (IHT), introduced in Chapter 6 (Figure 6.52), to control each cylinder together with auxiliary directional valves.[4] The connection of the high- and low-pressure rails, R_1 and R_2, with the pump can be done as illustrated in Figure 6.47. Note the use of piloted check valves to avoid excessive load acceleration.

---

[4] The four-quadrant operation of the cylinder is detailed in Chapter 6 (see Figures 6.50 and 6.51).

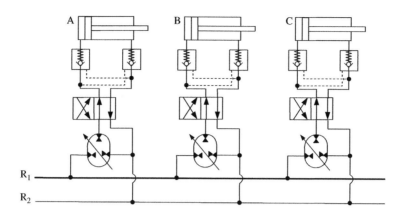

Figure 8.7   Circuit using hydraulic transformers

8.3   Hybrid Vehicles

It was back in 1963 when an article entitled 'Hydrostatic drive' was published in the magazine *Popular Mechanics*, in which it was said that 'tomorrow's cars may boost drive efficiency, yet discard the gearbox, clutch and even the brakes' [8]. The article made a reference to the experimental use of a hydrostatic transmission in a Triumph Herald Sedan in Scotland, UK.[5] Donald Firth, a designer of the Britain's National Engineering Laboratory where the experiments were conducted, predicted that hydrostatic cars would become popular 'perhaps in as few as three years'. However, more than five decades have passed, and we still do not see hydrostatic driven cars taking to the streets in large scale. In fact, despite the fact that it is perfectly conceivable to have a hydrostatic transmission (or *hydrostatic drive*, as it is referred to in the study of mobile hydraulics), issues such as noise, size and efficiency have hindered the use of hydrostatic transmissions in small vehicles [9].

The growing concern about environmental friendly engineering has cast some light on the idea of using hydrostatic transmissions in automobiles again [10]. In fact, the possibility of regenerating energy using hydraulic accumulators has stirred the interest on this matter, and we have seen a growth in dedicated research and prototyping of regenerative hydrostatic transmissions in vehicles. For instance, in 2006, the US Environmental Protection Agency (EPA) developed and launched an actual truck equipped with an energy-regenerative hydrostatic transmission circuit (Figure 8.8), with a reported fuel economy of up to 35% [11].

Other vehicles equipped with hydrostatic transmissions have also been developed and tested in recent years. In 2008, a hydrostatic transmission using digital displacement technology (Section 3.3.5) was successfully tested in a BMW 530i passenger car prototype by Artemis Intelligent Power [12]. The prototype is shown in Figure 8.9. With reference to Figure 8.9(a), the dual motor is about to be lifted into the packaging void previously occupied by the rear differential. The dual motor consists of two coaxial but completely independent digital displacement pump–motors. This arrangement easily allows for torque vectoring and traction control. The rest of the driveline comprises a digital displacement

---

[5] The video of this experiment can be seen on the Internet (see http://www.britishpathe.com/video/the-gearless-car-glasgow, accessed in May 2014).

Figure 8.8   Hydraulic truck equipped with an energy-regenerative hydrostatic transmission (courtesy of UPS – United Parcel Service)

(a)	(b)

Figure 8.9   Artemis BMW 350i prototype: (a) dual motor assembly and (b) high-pressure accumulator (courtesy of Artemis Intelligent Power Ltd)

pump (not shown in the figure) that is driven off the engine in place of the gearbox and the high-pressure accumulator, seen in Figure 8.9(b). The low-pressure side of the system is connected via a charge pump to a tank that is at atmospheric pressure. Tests showed an almost 50% reduction in fuel use on an urban driving cycle compared with earlier baseline tests with the standard 6-speed manual transmission.

Today, companies all over the world have been developing technology for automotive applications[6] and many publications exploring the theme of hydrostatic transmissions in vehicles exist (see Refs. [13–17]). It seems that the 1963 predictions are finally becoming a reality. In fact, large-scale production can already be foreseen. For instance, it has been

---

[6] For example, Parker-Hannifin (http://www.parker.com), Eaton (http://www.eaton.com) and Bosch-Rexroth (http://www.boschrexroth.com). Sites accessed in May, 2014.

recently reported that the auto manufacturer Peugeot Citroen has teamed up with Bosch to build a hydraulic hybrid car whose commercial launch is due in 2016.[7] We give a proper definition for *hybrid* vehicles in Section 8.3.1.

8.3.1 Definition

Generally speaking, a vehicle is called a 'hybrid' if it can be powered by two different energy sources. For example, a motorized bicycle that can still be powered by the rider when he or she desires is an example of a hybrid vehicle. On the other hand, diesel–electric loco-motives – in which the diesel engines are used to power electric generators, which in turn supply energy for the electric motors connected to the wheels – are not examples of hybrid vehicles since only one primary source of energy exists in the end. However, if the sources of energy for the locomotive were both diesel engines and electric motors, then the locomo-tive could be considered a hybrid [18]. Similarly, if the kinetic energy of a car, otherwise transformed into heat by the brakes, becomes a secondary source of energy along with the engine, then the car can be referred to as a hybrid. Therefore, we can generally state that

> A hybrid vehicle is a vehicle that can run on two or more energy sources, with one of them usually being reversible.

It is important to state that the definition above does not mean that the process of energy recovery must be thermodynamically reversible. Such process would be purely hypothetical, given that losses will always occur. Here, by 'reversible', we are referring to the parcel of energy that can be stored and later reused in the car (regenerated). For example, the recov-ered braking energy, which is ultimately released back to the wheels, can be considered 'reversible' in our definition.

8.3.2 Electric Hybrids

Electric hybrid vehicles make use of an electricity storage device (e.g. a battery pack) to subsequently store and release energy to the wheels as needed. For instance, while braking, an electricity generator can transform the kinetic energy of the car into electric energy and charge the battery. The stored energy can then be reused later when the car is accelerat-ing. Figure 8.10 shows the schematic configuration of a *series* electric hybrid [19]. In this figure, the engine is mechanically connected to an AC generator. The generated power is then rectified before being directed into the DC motor or the battery pack, depending on the desired operation mode (battery charge or traction). For instance, if the car is stopped with no power being required at the wheels, the engine can be used to charge the batteries. The electric motor, connected to the differential, D, can work as both motor and generator,

---

[7] See, for example http://www.extremetech.com/extreme/146450-peugeot-unveils-hydraulic-air-hybrid-80-mpg-car and http://www.psa-peugeot-citroen.com/en/featured-content/automotive-innovation/hybrid-air-engine-full-hybrid-gasoline. Both accessed in May, 2014.

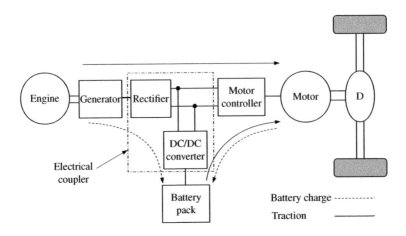

Figure 8.10   Series electric hybrid schematics

absorbing energy from the wheels (e.g. when the car brakes) and sending it back into the batteries. A motor controller and a power electronics DC/DC converter are therefore needed, as shown in the figure. Note that the energy accumulated in the batteries can be ultimately reused to drive the car wheels through the electric motor.

Another common hybrid configuration is shown in Figure 8.11 (*parallel* electric hybrid), where we observe that either the engine or the electric motor/generator can be mechanically connected to the differential. Note that a mechanical coupler[8] (MC) is needed in this case, and the function of the electric motor is restricted to energy regeneration. For instance, when the car is idle, the engine can connect to the electric motor through the MC. The motor then operates as a generator and charges the battery pack, as indicated by the dashed line. By disconnecting the engine from the wheels through the clutch, the energy previously stored in the batteries can be used to drive the differential, therefore regenerating energy.

8.3.3   Hydraulic Hybrids

Similar to electric hybrid vehicles, hydraulic hybrids are usually built in series or in parallel. Figure 8.12 illustrates a typical series configuration (pump and motor controllers are not shown for simplicity). The circuit uses two hydraulic accumulators as a means of storing and releasing hydraulic energy. Energy is stored by pumping oil from the low-pressure accumulator (LA) into the high-pressure accumulator (HA). This can be done either by the pump connected to the engine or by the pump/motor connected to the wheels through the differential, D (dashed lines). In regeneration mode, the pressure differential between the accumulators drives the hydrostatic pump/motor connected to the wheels (solid lines). The transmission operates as a normal IVT otherwise, connecting the engine shaft to the differential shaft.

---

[8] The role of the mechanical coupler is to carry the power coming from the engine and/or the motor into the differential. More information on some typical coupler designs can be found in Ref. [19].

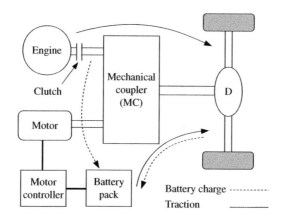

Figure 8.11   Parallel electric hybrid schematics

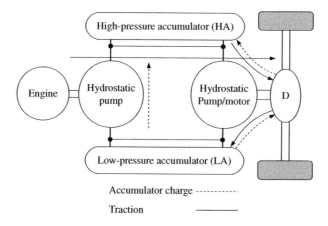

Figure 8.12   Series hydraulic hybrid schematics

Figure 8.13 illustrates the operation of a parallel hydraulic hybrid. In this case, the pump/motor and the high and low-pressure accumulators (HA and LA, respectively) are directly connected to the differential, D, through a MC. The same connection happens to the engine, which can be decoupled from the wheels through a clutch as indicated in the figure. When it is in storage mode, the hydrostatic pump/motor is driven either by the wheels or by the engine, delivering oil from the low-pressure accumulator into the high-pressure accumulator. The stored energy can be later reused to drive the wheels by disconnecting the engine from the differential. In direct-drive mode, the engine is mechanically connected to the differential and the pump/motor shaft is bypassed.

Note that the schemes shown in Figures 8.12 and 8.13 are conceptually identical to the schemes shown in Figures 8.10 and 8.11. For instance, in the series hydraulic configuration shown in Figure 8.12, the two accumulators play the role of the battery pack in the series electric scheme in Figure 8.10. Likewise, the pump plays the role of the electricity generator and the hydrostatic pump/motor corresponds to the electric motor/generator.

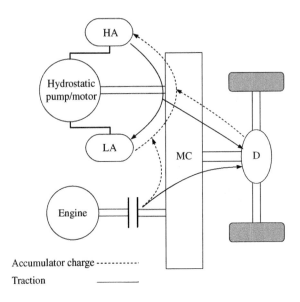

Figure 8.13   Parallel hydraulic hybrid schematics

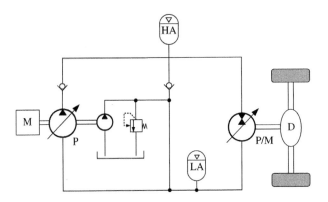

Figure 8.14   Simple series hybrid circuit

A simple example of a hydraulic circuit for a series hybrid vehicle can be seen in Figure 8.14, where the energy storage devices are the high- and low-pressure accumulators, HA and LA, respectively.

Figure 8.15 shows the operation with the vehicle moving forward (a), stopped (b) and reversing (c). One clear disadvantage of this circuit is that the accumulator HA, situated between the pump and the motor, drastically reduces the effective bulk modulus. In fact, depending on the size of the accumulator, there may exist a considerable delay of the motor response to a change in the pump displacement, which ultimately translates into a 'spongy' feeling to the driver [20].

In all three stages shown in Figure 8.15, the energy flow occurs from the engine to the wheels. Figure 8.16 shows two situations where energy is stored as a pressure differential

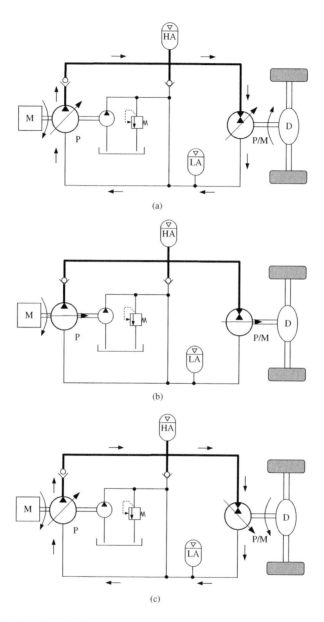

Figure 8.15   Operation in normal traction mode: (a) forward, (b) stop, (c) reverse

between the high and low-pressure accumulators, HA and LA. In the first case, energy from the engine flows into the high-pressure accumulator for a certain period of time while the car is idled.[9] In the second case, the pump/motor P/M, now operating as a pump, converts the kinetic energy from the wheels into hydraulic energy while charging the accumulator HA.

---

[9] Remember that the pressure inside the accumulator, HA, must remain below the safety limits.

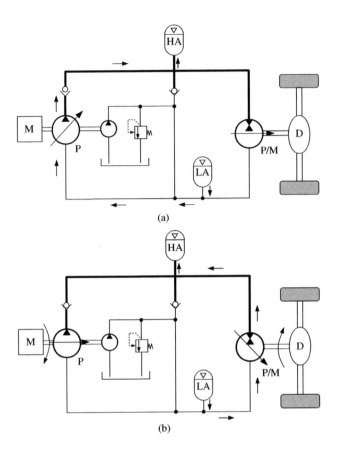

Figure 8.16 Energy storage modes: (a) energy storage (vehicle stopped), (b) regenerative braking

To reuse the energy stored in the high-pressure accumulator, we invert the displacement of the pump/motor, P/M, which then operates as a motor transferring the stored hydraulic energy into the wheels through the differential, as seen in Figure 8.17.

Some elements, such as a loop flush module and inter-conduit relief valves, that have not been shown in this circuit should be incorporated in a more complete version (see Figure 4.35). Other designs have also been proposed, including a blended series–parallel configuration [20], which among other things aim to solve the problem of the low-effective bulk modulus created by the high-pressure accumulator. In addition, advances have been made in the field of transmission control, as observed in Refs. [20, 21].

At this stage, it is interesting to show some actual fuel economy values obtained for a series and a parallel sport utility vehicle (SUV) prototype, according to Ref. [22]. Table 8.1 shows the results for mild and full hybrids. The terms 'mild' and 'full' are synonyms for 'parallel' and 'series' hybrids, respectively.

Two operating conditions have been analysed, namely 'engine off' and 'engine on' strategies. In the 'engine on' operation (EON), the engine is not turned off during energy recovery,

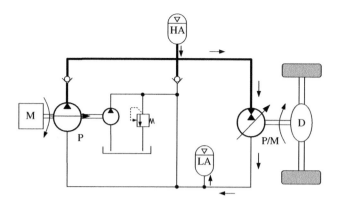

Figure 8.17   Regeneration mode

Table 8.1   Fuel economy improvement for hybrid SUV prototypes (gasoline engine)

Conventional (automatic)	Mild (EON)	Mild (EOFF)	Full (EON)	Full (EOFF)
Base	13%	17%	18%	34%

while in the 'engine off' strategy (EOFF), the engine is turned off whenever energy from the accumulators is being used to drive the wheels. We see that in spite of additional considerations such as maintenance cost, added weight and plant investment, the benefits of fuel economy make hydraulic hybrids a promising concept.

Hybrid technology can be applied to any type of vehicle. However, the benefits of energy regeneration are more evident in some particular applications. For instance, because heavy vehicles, such as buses, are prone to work on a stop and go basis for life, braking energy recovery makes more sense in their design when compared to the design of personal automobiles. In fact, this is one of the reasons why hydraulic hybrid technology is more frequently explored in this type of vehicle.

8.3.4   CPR-Based Hybrids

The 'CPR' concept, together with the use of efficient hydraulic transformers (see Section 6.3), can also be applied to hybrid vehicles [13, 23]. To understand the idea, we consider the use of dual-pump hydraulic transformers (see Figure 6.49) to control a hydrostatic pump/motor, as shown in Figure 8.18. Note that the displacements of the pump/motors P_1 and P_2, connected to the high-pressure and low-pressure rails, R_1 and R_2, are symmetrical in this case.[10]

The hydraulic motor M can be driven either by the pressure differential between the pressure rails R_1 and R_2 or by the pump/motors P_1 or P_2. When the motor is driven by one of

---

[10] This is true if we disregard the external losses at the motor and the fluid compressibility. In practice, a difference between the displacements of the pump/motors P_1 and P_2 is expected.

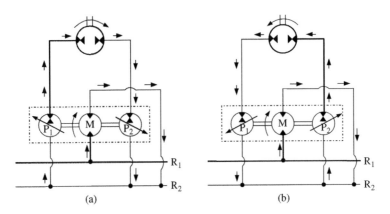

Figure 8.18 Dual-pump hydraulic transformers connected to a hydraulic motor with (a) clockwise rotation and (b) counter-clockwise rotation

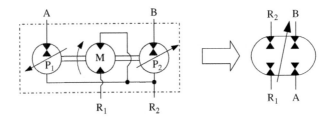

Figure 8.19 Equivalence between the IHT and the conventional design

the pumps, P_1 or P_2, energy is transferred from the wheels into the pressure rails. Clearly, the arrangement shown in Figure 8.18 is not practical because of the size of the individual components and the overall losses.

A compact dual-pump hydraulic transformer, based on the floating cup technology (see Figure 3.28), has been in development by the Dutch company Innas B.V. Figure 8.19 shows the port equivalence between the conventional dual-pump transformer and the IHT together with its hydraulic symbol.

Figure 8.20 shows how the IHT can be used in a hybrid car. The circuit shown in the figure is depleted of many of its elements, such as charge circuit, relief and check valves, for the sake of simplicity.[11] Basically, each wheel is driven by a fixed-displacement pump/motor whose ports are connected to an IHT. The circuit is configured in a way that each port, A and B, of each IHT connects to two motors at the same time through a T-joint so that the use of a differential is precluded.[12] The scheme shown in Figure 8.20 has been presented

[11] The design illustrated in Figure 8.20 has been named 'HyDrid' and is still under development. More information can be found in the Innas home page (http://www.innas.com).

[12] The T-joint is a natural flow divider. For example, if one of the wheels stops, all the flow coming from the IHT is automatically sent to the other wheel. We may, therefore, see the T-joint as a 'hydraulic differential' [23].

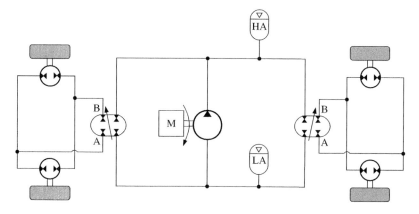

Figure 8.20   Innas hybrid scheme

Figure 8.21   Energy flow in an Innas hybrid vehicle

with two engines and two pumps, feeding the pressure rails [13]. However, given that the operational principle is not altered when a single engine is used, we have opted to use this simpler version as an illustration.

The concept behind the hybrid scheme shown in Figure 8.20 is different from the series or the parallel hybrids we saw in the previous section. Note that there is a disconnection between the energy source (engine and pump) and the consumption elements (hydraulic motors) so that energy never flows directly from the engine to the wheels as was the case in the hybrids represented by Figures 8.12 and 8.13. Figure 8.21 illustrates the energy flow for the circuit shown in Figure 8.20.

Tests carried on a 'HyDrid' prototype have resulted in an over 50% reduction of the fuel consumption with a considerable lowering of the CO_2 emission rate [23].

8.4   Wind Turbines

In order to understand the benefits of using hydrostatic transmissions in wind turbines, we must begin by briefly introducing the subject of wind power generation. Wind power input to the turbine blades must be transformed into electric power by an electricity generator. However, we know that wind blows at different speeds, and as a consequence, the turbine rotation does not remain constant during operation. The basic problem is therefore to extract alternate current electric energy at a constant frequency from a variable-speed input. Two types of electricity generators are usually employed to this end: asynchronous and synchronous generators.

8.4.1 Asynchronous Generators

An asynchronous generator is basically an 'inverted' electric motor. Therefore, a good way of understanding the main features of asynchronous generators is to first study the operational principle of asynchronous motors.

An asynchronous electric motor (Figure 8.22) consists of a stator and a rotor. The stator, also known as *armature*, is connected to the electric grid, that is, it receives electric energy from the grid. The rotor transforms the electric energy coming from the stator into mechanical energy (torque and angular speed) at its shaft.

Before we proceed to describe the operation of the asynchronous motor, it is important to remember the solenoid principle in which an electric current passing through a loop of wire (solenoid) creates a magnetic field whose intensity and orientation depend on the intensity and the direction of the electric current, I, as shown in Figure 8.23.

In an asynchronous motor, the stator consists of several electric coils, such as the solenoids shown in Figure 8.23. These coils produce a magnetic field when connected to the AC electric grid. Figure 8.24 shows the stator of a three-phase electric motor [24]. In this figure, the three electric phases are represented by the letters A, B and C and are connected to coils located around the stator in such a way that the electric current flowing in each pair of coils will, at different moments, produce differently oriented magnetic fields. For example, at time t_0, the voltage in phase A is at its peak and is responsible for the production of a strong magnetic field, indicated by the north pole, N. Phases B and C will produce weaker fields

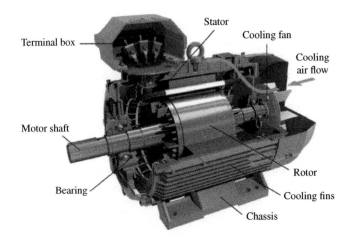

Figure 8.22 An asynchronous electric motor (courtesy of ABB motors)

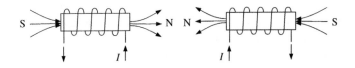

Figure 8.23 Magnetic field produced by an electric current in a solenoid

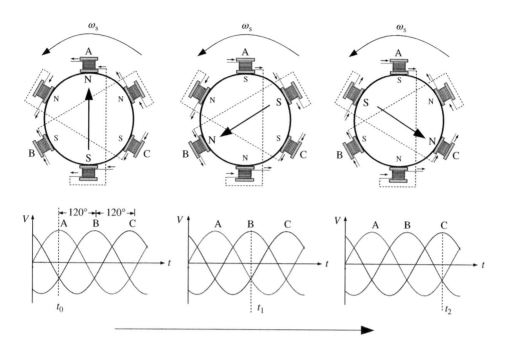

Figure 8.24   Rotating magnetic field in a typical stator

(represented in smaller letters), but they will combine to produce a 'resultant field' so that two magnetic poles will be formed around the stator. Because of the alternate nature of the voltage, the magnetic poles rotate counter-clockwise in time, as can be seen by the position of the resultant magnetic field at the subsequent instants t_1 and t_2. The angular speed of the rotating field, ω_s, is called *synchronous speed*.

For an arrangement in which N_p magnetic poles are created, it can be shown that the synchronous speed is given by [25]:

$$\omega_s \text{ (rpm)} = \frac{120f}{N_p} \tag{8.1}$$

where f is the grid frequency in hertz.

For example, the synchronous speed for the two-pole motor represented in Figure 8.24, assuming that the grid has a frequency of 60 Hz, is

$$\omega_s \text{ (rpm)} = \frac{120 \times 60}{2} = 3600 \text{ rpm}$$

From Eq. (8.1), we observe that the number of poles is determinant in the synchronous speed.

Suppose that we place a rectangular conductor inside the stator, as shown in Figure 8.25(a). Consider also that this rectangular rotor is connected to a shaft, as detailed in Figure 8.25(b), and is free to rotate around its bearings. The rotating magnetic

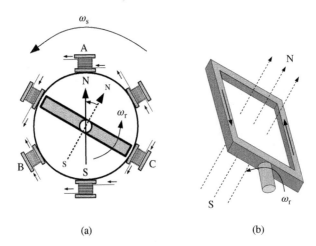

Figure 8.25   Asynchronous electric motor: (a) general schematics and (b) rotating coil detail

field in the armature will induce an electric current in the rectangular coil, creating another magnetic field (dashed arrows). As the induced magnetic field in the rotor tries to follow the rotating magnetic field of the stator, a torque is produced, causing the rotor to turn at an angular speed, ω_r. This is basically the way in which asynchronous electric motors operate. Note that in order to produce an electric current in the rotor, there must be always a relative speed between the rotor and the rotating magnetic field, that is, $\omega_s - \omega_r \neq 0$. In the particular case of asynchronous motors, the rotor must always rotate at a speed smaller than the synchronous speed ($\omega_s - \omega_r > 0$).

If a mechanical load is connected to the rotor shaft in Figure 8.25, electric power will be extracted from the electric grid and transformed into mechanical power. Now, imagine that an inverse situation happens and the rotor is connected to a source of mechanical power that forces it to rotate faster than the synchronous speed (e.g. the rotating shaft of a wind turbine). Again, there will be a difference between the rotor speed and the synchronous speed. However, this time the rotor speed will be higher than the synchronous speed ($\omega_s - \omega_r < 0$), and in order to decelerate the rotor, an 'inverted' electric current will be induced in the rotor. In the end, the difference between the mechanical energy input and the electrical energy supplied by the electric grid will be turned into electric energy at the rotor (minus the power losses), and the asynchronous motor will now operate as an asynchronous generator.

8.4.2   Synchronous Generators

We have seen that in asynchronous generators, the rotor needs to rotate at a higher speed in relation to the synchronous speed so that electricity can be generated into the grid. The inverse, that is, the rotor speed smaller than the synchronous speed, would result in an electric motor. As the name suggests, in synchronous motors and generators, there is no difference between the rotor speed and the synchronous speed.

In order to understand how a three-phase synchronous generator works, consider the simple scheme shown in Figure 8.26, where the rotor consists of three equally spaced rectangular coils as in Figure 8.25(b). The stator is a stationary magnet, which can be either a permanent magnet or an electromagnet. As the rotor revolves around its axis, an induced alternate voltage will appear in each coil. For a constant rotor speed, the induced electric tensions at phases A, B and C will be represented by sinusoid curves.

The electric frequency of each of the three phases shown in Figure 8.26 is dictated by rotor speed in such a way that no matter how slow or how fast the rotor turns, energy will always be generated. Another interesting characteristic of synchronous generators is that they do not depend on the existence of an electric grid to generate electricity. All that is required from the generator is the mechanical power input at its shaft. Figure 8.27 shows a 6.4 kW permanent magnetic generator for wind turbines.

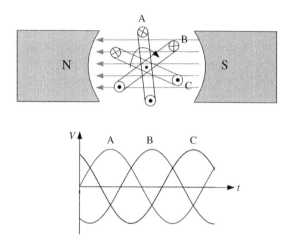

Figure 8.26   Simple three-phase synchronous generator

Figure 8.27   Permanent magnet synchronous generator (courtesy of The Switch)

Now that we have briefly discussed the way in which electrical energy is generated in wind turbines, we are ready to discuss the use of hydrostatic transmissions in wind generation. We begin with a brief introduction of the role of power transmissions in wind turbines.

8.4.3   General Aspects of Power Transmission in Wind Turbines

To understand the benefits of using hydrostatic transmissions in wind power plants, consider the situation illustrated in Figure 8.28. Wind blows at a variable speed, $v(t)$, and rotates the turbine. The turbine shaft then conveys the input power to the transmission as it rotates at a variable angular speed, $\omega_i(t)$. At the end of the chain, the input power reaches the electricity generator, whose shaft rotates at a speed $\omega_o(t)$, which may be variable or constant depending on the technology used in the transmission.

In typical wind turbines, the transmission in Figure 8.28 consists of a planetary gearbox [24], and the electricity generator is an asynchronous machine.[13] Given that turbine blades usually rotate at a low speed (especially in high-power units), the planetary gearboxes act as a speed amplifier so that $\omega_o(t)$ remains above the synchronous speed of the electricity generator.[14]

Now, considering that a gearbox is used for the transmission in Figure 8.28, the rotor speed, $\omega_o(t)$, will be given by a constant transmission ratio, R_T, multiplied by the input speed, $\omega_i(t)$, that is, $\omega_o(t) = R_T \omega_i(t)$. This may result in one of the following situations:

1. The rotor speed, $\omega_o(t)$, is lower than the synchronous speed, ω_s. In such a situation, the generator accelerates the rotor, acting as an electric motor and not a generator. In other words, the turbine operates as a big fan, extracting electric energy from the grid.

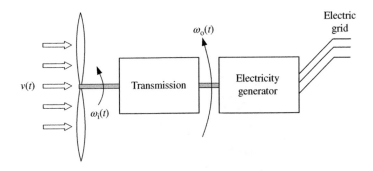

Figure 8.28   Power flow in a wind power generator

Clearly, this is an undesirable situation. The solution is to disconnect the turbine from the generator whenever the wind speed is below a minimal value, known as *cut-in speed* [24];

2. The rotor speed, $\omega_0(t)$, is higher than the synchronous speed, ω_s. In this situation, electricity is generated. However, the electric frequency output by the generator will not be constant because of the variable input speed. It is therefore necessary to adjust the generator output to the consuming grid. This is usually done by power electronic circuits placed between the generator and the grid. It is interesting to note that the electricity generator ends up acting as a brake, keeping the turbine speed within certain design limits.[15]

Another possible configuration is to use a synchronous generator. In such a case, no restriction is placed on the rotor speed, $\omega_0(t)$, in order for electricity to be generated. However, the generator output would still be variable, and a final adjustment to the consuming grid would need to be (electronically) done unless the transmission itself provides a means of making ω_0 constant.

8.4.4 Hydrostatic Transmission in Wind Turbines

The first benefit of using a hydrostatic transmission in the scheme shown in Figure 8.28 is the weight and size reduction under the turbine nacelle, especially in large turbines when the rotor speed is low and the input torque is high [29]. In addition, hydrostatic transmissions are 'spatially flexible', meaning that they can be more easily accommodated inside the turbine nacelle and even split between the turbine top and the ground level in a configuration where the pump would be connected to the turbine shaft and the motor would be connected to the electricity generator on the ground [30, 31]. In fact, it is even possible to transport the wind energy coming from several turbines into a single generator, as proposed by the 'Delft Offshore Turbine' (DOT)[16] project. The DOT concept is illustrated in Figure 8.29 for a single wind turbine. Note that the hydrostatic transmission units (pump and motor) are placed separately from one another. The pump is located inside the turbine nacelle, whereas the motor stays at ground level. Power coming from each turbine is then used to pump seawater into a Pelton turbine connected to a synchronous generator on the shore. In the end, the project can be seen as a combined wind farm and hydroelectric power plant.

Apart from the inherent versatility of hydrostatic transmissions, we may say that one of their greatest advantages in wind power applications comes from their infinitely variable transmission ratio [33]. By using a hydrostatic transmission between the turbine shaft and the electricity generator, it is possible to keep the generator rotor at a constant speed while the turbine speed changes. Therefore, it is possible to use a synchronous generator with a constant input speed, eliminating the need for a posterior electronic treatment of the generated electricity. As mentioned in Ref. [29]: ' ... the speed ratio of the hydrostatic

---

[15] Typically, the turbine speed must be such that the rotor speed remains at least 1% higher than the synchronous speed [24]. We also observe that very strong winds and gusts can dangerously accelerate the turbine. Safety-braking mechanisms must be activated in those cases (see Ref. [28]).

[16] The DOT project was launched in 2008 in the Delft University Wind Energy Research Institute (DUWIND) [32].

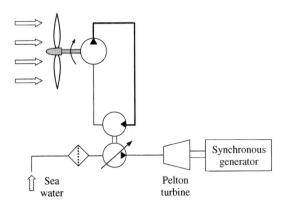

Figure 8.29   Schematics of the Delft Offshore Turbine

transmission can be adjusted by controlling the hydrostatic pressure and displacement ...
so we can use the versatile brushless synchronous generator at constant rotational speed
and enable good power controllability that meets grid requirement without the (electronic)
inverter/converter'.

The possibility of changing the turbine speed while keeping a constant speed at the generator shaft is also important when asynchronous generators are used. As seen earlier, in
gearbox-based wind turbines, the turbine speed is kept within a narrow range,[17] which may
reduce the aerodynamic efficiency because the turbine cannot adapt to changing wind conditions [27]. Using a hydrostatic transmission would allow the turbine to track the most
efficient operational regime while maintaining a constant difference between the rotor speed
and the synchronous speed at the generator.

One major disadvantage of hydrostatic transmissions in wind generation is their low overall efficiency [34], which can make their use uneconomical and prohibitive despite all the
positive characteristics mentioned so far. However, their reliability, simpler maintenance
and the already mentioned spatial flexibility and transmission ratio are advantages that may
outweigh the disadvantages. In recent years, new technologies have emerged that increase
the efficiencies of hydrostatic pumps and motors, paving the way for more regular use of
hydrostatic transmissions in wind turbines. Figure 8.30 shows an actual unit in development
by Artemis Intelligent Power. We observe in the nacelle detail that the hydrostatic transmission makes use of the digital displacement concept as a means of improving efficiency
(see Section 3.3.5).

---

[17] In other words, given that the transmission ratio between the turbine and the asynchronous generator is constant,
and given that the rotor speed at the generator must remain slightly above the synchronous speed, the turbine will
be naturally forced to operate at a nearly constant speed in the end, with little space for adjustments. One possible
solution is to decouple the generator from the grid and use power electronics to change the synchronous speed in the
armature. Power electronics would again be used to modify the output of the generator to match the frequency and
phase of the grid. In the end, this solution increases complexity, cost, weight and size, reduces efficiency and adds
sources of unreliability to the system [27].

Figure 8.30   A hydrostatic transmission turbine (courtesy of Artemis Intelligent Power Ltd)

8.5   Wave Energy Extraction

An interesting application of hydrostatic transmissions is wave energy extraction, where the mechanical energy of sea waves is converted into electrical energy. Figure 8.31 shows the schematics of a wave attenuator in operation. Attenuators lie parallel to the direction of the waves and absorb energy as they 'ride' the waves [35]. The converter consists of several tube sections linked by flexible joints. As waves pass down the length of the converter, the sections fold as seen in the figure. The mechanical energy associated with the movement of the tubes is then converted into hydraulic energy by the cylinders housed inside the joints as illustrated in the zoom-in diagram in Figure 8.31. Note that the rods and the cases of

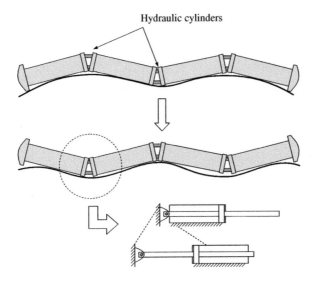

Figure 8.31   Wave converter in operation

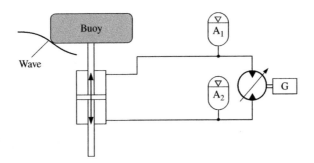

Figure 8.32   Wave energy extraction

the cylinders are mechanically coupled to the left and right tubes, such that they alternately expand and retract as the waves pass underneath the attenuator.

As an example of a hydraulic system to capture wave energy, consider the floating buoy converter shown in Figure 8.32. The buoy stands afloat on the water, where it is pushed upwards by the passing waves. The energy of the waves is then transferred to the buoy, whose column is connected to a hydraulic cylinder, which works as an energy conveyor by transferring energy from the ocean into the hydraulic circuit. The circuit represents a hydrostatic transmission, where the double-rod cylinder plays the role of the pump. Two

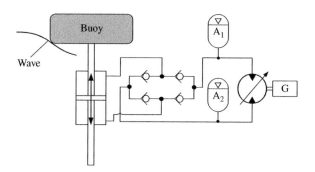

Figure 8.33 Circuit with flow rectifier

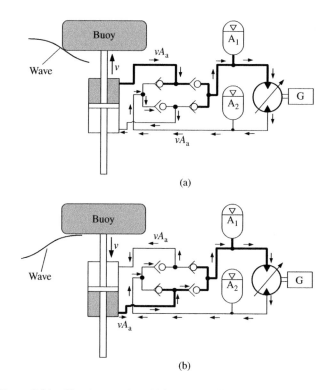

(a)

(b)

Figure 8.34 Circuit operation: (a) buoy ascending; (b) buoy descending

accumulators, A_1 and A_2, are placed in the circuit to provide for a smooth flow into the hydraulic motor, whose speed is controlled by its variable displacement. Note that the motor displacement must change from negative to positive whenever the buoy speed changes sign in order to keep the shaft of the electricity generator, G, rotating in the same direction.

Changing the motor displacement from a positive into a negative value is not simple, as it implies passing through the point when the displacement is zero. The problem is that as the motor displacement approaches zero, the transmission ratio approaches infinity, causing a huge increase in the motor speed. To solve this problem, a flow rectifier composed of four check valves, disposed as shown in Figure 8.33, can be used so that the flow through the variable displacement motor never changes its direction [35, 36]. The hydraulic circuit employed in the converter illustrated in Figure 8.31 essentially follows the design shown in Figure 8.33.

Figure 8.34 shows the operation of the hydraulic circuit illustrated in Figure 8.33. Note that the flow through the motor happens in one direction only and is unaffected by the piston speed.

8.6 Aeronautical Applications

The wings of an aeroplane contain movable aerodynamic surfaces that are usually controlled by hydraulic cylinders. As an example, Figure 8.35 shows the names of the aerodynamic surfaces and their distribution over the wings of a Boeing 767. For safety reasons, in a passenger aeroplane as the one shown in Figure 8.35, each aerodynamic surface is controlled by more than one cylinder [38]. In this particular aeroplane, three cylinders, connected to independent hydraulic circuits, are responsible for controlling the elevator and the rudder. The hydraulic

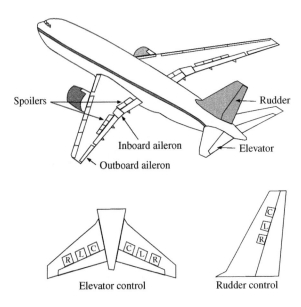

Figure 8.35   Aerodynamic surfaces on a Boeing 767 and details of the elevator and rudder control [37]

circuits to which these cylinders are connected are labelled as R (right hydraulic system), C (central hydraulic system) and L (left hydraulic system).

The hydraulic lines in a typical aeroplane constitute a considerable fraction of the plane weight. Here, the advantages of using electrohydrostatic actuators (EHAs) become more evident. Since they can be compacted into single units, they do not need to be hydraulically powered. Therefore, instead of having long hydraulic lines connecting the pumps to the actuators, power can be transferred through electric cables.[18] The use of electric wires instead of hydraulic conduits not only reduces the weight of the aeroplane but also makes the whole system less prone to leakages.

Another advantage of EHAs is that they are placed near the corresponding control surface. Therefore, any failure can be isolated and the unit can be rapidly turned off, whereas in a central hydraulic circuit, a major flaw in one conduit can potentially compromise the whole circuit, making it imperative to have more than one circuit in parallel for a fail-safe operation.

Figure 8.36(a) shows a typical example of a compact EHA that equips the Airbus A380 (Figure 8.36(b)). The aerodynamic control system of this aeroplane consists of four independent energy sources: two of them are hydraulic and the other two are electric. The electric systems power the EHAs connected to the aerodynamic surfaces and operate in standby, becoming operative in the event of a failure of the conventional hydraulic circuits [40].

An example of an EHA circuit for aileron control [41] is shown in Figure 8.37. In this figure, it is possible to identify the following components:

- The position sensor S, which is responsible for reading the cylinder displacement.
- A high-level power electronics module, which is responsible for the motor speed and torque control.
- A means of 'isolating' the cylinder in case of an electric failure (valve V_3).

The other elements – such as the 'peak protection' valves V_4 and V_5, and the valves to replenish the oil in the circuit, V_1 and V_2 – are common to any hydrostatic actuator.

We can understand the function of the bypass valve, V_3, if we consider that more than one EHA is connected to the aileron. If one of the actuators becomes defective due to an electric failure, the solenoid of the valve V_3 deactivates and the valve assumes its bypass position

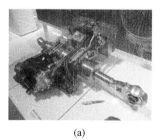

(a)   (b)

Figure 8.36   (a) Compact electrohydrostatic actuator used in the (b) Airbus A380 aeroplane

[18] Because of this particular characteristic, it is usual to refer to electrohydrostatic actuators in aeroplanes as *power-by-wire* (PBW) devices [39].

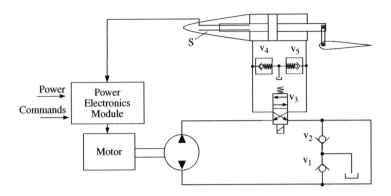

Figure 8.37   Typical circuit for an aileron EHA

automatically, letting the cylinder loose so that it does not interfere with the operation of the other actuators (fail-safe mode).

References

[1] Beachley NH, Frank AA (1979) Continuously variable transmissions: theory and practice. Lawrence Livermore Laboratory, University of California, Report UCRL-15037.

[2] Newton K, Steeds W, Garrett TK (1996) The motor vehicle, 12th Ed. SAE International, Warrendale, USA.

[3] Rahmfeld R, Ivantysynova M (2001) Displacement controlled linear actuator with differential cylinder – a way to save primary energy in mobile machines. 5th International Conference on Fluid Power Transmission and Control – ICFP 2001, pp. 296–301, Hangzhou, China.

[4] Williamson C, Zimmerman J, Ivantysynova M (2008) Efficiency study of an excavator hydraulic system based on displacement-controlled actuators. Bath ASME Symposium on Fluid Power and Motion Control – FPMC2008, pp. 291–307.

[5] Zimmerman J, Ivantysynova M (2011) Hybrid displacement controlled multi-actuator hydraulic systems. Proceedings of the Twelfth Scandinavian International Conference on Fluid Power, May 18–20, 2011, Tampere, Finland.

[6] Inderelst M, Sgro S, Murrenhoff H (2010) Energy recuperation in working hydraulics of excavators. Proceedings of the Fluid Power and Motion Control, FPMC 2010, Johnston DN and Plummer AR (Eds.), pp. 551–562.

[7] Vael G. Achten P, Fu Z (2000) The Innas hydraulic transformer the key to the hydrostatic common pressure rail. SAE Technical Paper 2000-01-2561.

[8] Arthur J (1963) Hydrostatic drive. Popular Mechanics, 120(2): 112–115.

[9] Halpert J (2011) Chrysler experiments with hydraulic hybrid minivans. MIT-Technology Review. http://m.technologyreview.com/energy/37230/. Accessed May 2014.

[10] Kargul J (2012) Affordable hybrids for a petroleum and carbon constrained world. 2012 – Clean Transportation Summit Clean, Environmental Protection Agency (EPA), USA. http://www.wwcleancities.org/documents/HydraulicHybrids_EPA_Kargul.pdf. Accessed 12 October 2012.

[11] UPS Press Release (2012) UPS to Add 40 hydraulic hybrid vehicles to its fleet. http://www.pressroom.ups.com/Press+Releases/Archive/2012/Q4/UPS+to+Add+40+Hydraulic+Hybrid+Vehicles+to+its+Fleet. Accessed May 2014.

[12] Artemis (2008) Hydraulic hybrid developed by Artemis. Intelligent Power Ltd. http://www.artemisip.com/news-media/videos/2008/05/01/artemis-hydraulic-hybrid-bmw. Accessed 28 August 2012.

[13] Achten PAJ (2007) The hybrid transmission, SAE Technical Paper 2007-01-4152. DOI: 10.4271/2007-01-4152.

[14] Deppen TO, Alleyne AG, Stelson KA, Meyer JJ (2010). Predictive energy management for parallel hydraulic hybrid passenger vehicle. Proceedings of the ASME 2010 Dynamic Systems and Control Conference, September 12–15, Cambridge, Massachusetts, USA.

[15] Rydberg K-E (2009). Energy efficient hydraulic hybrid drives. Proceedings of the 11th Scandinavian International Conference on Fluid Power, SICFP'09, June 2–4, 2009, Linköping, Sweden.

[16] Taylor J, Rampen W, Robertson A, Caldwell N (2011). Digital displacement hydraulic hybrids parallel hybrid drives for commercial vehicles. Proceedings of the JSAE Annual Congress, May 2011, Pacifico Yokohama, Japan.

[17] Zeljko S, Zoran J, Djordje D, Zlatomir Z, Dragan N (2013). Improving the fuel economy by using hydraulic hybrid powertrain in passenger cars. Machines, Technologies, Materials, (4): 24–26.

[18] Fuhs AE (2009) Hybrid vehicles and the future of personal transportation. CRC Press, Boca Raton, USA.

[19] Ehsani M, Gao Y, Emadi A (2010) Modern electric, hybrid electric and fuel cell vehicles – fundamentals, theory and design, 2nd Ed. CRC Press, Boca Raton, USA.

[20] Sprengel M, Ivantysynova M (2013) Investigation and energetic analysis of a novel hydraulic hybrid architecture for on-road vehicles. Proceedings of the 13th Scandinavian International Conference on Fluid Power, SICFP2013, June 3–5, 2013, Linköping, Sweden.

[21] Stelson KA, Meyer JJ, Alleyne AG, Hencey B (2008) Optimization of a passenger hydraulic hybrid vehicle to improve fuel economy. Proceedings of the 7th JFPS International Symposium on Fluid Power, TOYAMA September 15–18, 2008.

[22] EPA (2005) Progress report on clean and efficient automotive technologies under development at EPA – Interim technical report EPA420-R-04-002. United States Environmental Protection Agency.

[23] Achten P, Van den Brink T, Potma J, Schellekens M, Vael G (2009) A four-quadrant hydraulic transformer for hybrid vehicles. Proceedings of the 11th Scandinavian International Conference on Fluid Power, SICFP'09, June 2–4, Linköping, Sweden.

[24] Stiesdal H (1999) The wind turbine: components and operation. Bonus Energy A/S, Info Newsletter: Special Issue – Autumn 1999. http://users.wpi.edu/~cfurlong/me3320/DProject/BonusEnergy-1998.pdf. Accessed 05 March 2015.

[25] Klempner G, Kerszenbaum I (2004) Operation and maintenance of large turbo-generators. John-Wiley & Sons, USA.

[26] Green M (2010) Wind energy-background report, Yukon Energy Charrette. Natural Power, Vancouver, BC, CA.

[27] Thul B, Dutta R, Stelson KA (2011) Hydrostatic transmission for mid-size wind turbines. Proceedings of the 52nd National Conference on Fluid Power, Paper 40-2, pp. 1023–1038, LA, USA.

[28] El-Henaoui S (2008) When the wind blows pitch control systems turn. Moog Inc. http://www.moog.com/literature/ICD/Moog_WindArticle_ENGLISH_1July09.pdf. Accessed 29 May 2014.

[29] Sasaki M, Yuge A, Hayashi T, Nishino H, Uchida U, Noguchi T (2014). Large capacity hydrostatic transmission with variable displacement. Proceedings of the 9th International Fluid Power Conference, 9. IFK, March 24–26, 2014, Aachen, Germany.

[30] Ossyra JC, Patri S, Durbha D, Jankowski A, Stelson K, Wang F, Dutta R, Bohlman B, Marr J, Feist C, Lueker M (2012) Reliable lightweight transmission for off-shore, utility scale wind turbines. Technical report. Eaton Corporation Innovation Center, University of Minnesota & Clipper Windpower, LLC, USA.

[31] Dahlhaug O (2010) Potential top-mass reduction by hydraulic transmission. NTNU: http://www.sintef.no/project/Nowitech/Wind_presentations/Dahlhaug,%20O.G.,%20NTNU.pdf, Accessed 16 October 2011.

[32] Diepeveen NFB (2014) Seawater-based hydraulics for offshore wind turbines. Technical report, Delft University Wind Energy Research Institute (DUWIND).

[33] Diepeveen NFB, Laguna AJ (2011) Dynamic modeling of fluid power transmissions for wind turbines, Proceedings of the European Wind Energy Association – EWEA, November 21 –December 01, 2011, Amsterdam.

[34] Rampen W (2006) Gearless transmissions for large wind turbines – the history and future of hydraulic drives, Artemis Intelligent Power Ltd, Scotland.

[35] Drew B, Plummer AR, Sahinkaya MN (2009) A review of wave energy converter technology. Proceedings of the Institution of Mechanical Engineers, Part A: Journal of Power and Energy, 223: 223–887.

[36] Plummer AR, Cargo CJ, Hillis AJ, Schlotter M (2011) Hydraulic power transmission and control for wave energy converters. Proceedings of the 52nd National Conference on Fluid Power, Las Vegas, USA, Paper 17.2.

[37] Remudo T (2011) Boeing 767 systems review: schematics. Southeast Flight Support. http://www.remudoaviation.net/docs/B767%20Schematics.pdf. Accessed 26 July 2011.

[38] Geiger D (2011). Electrohydrostatic actuation modernizes fluid power. MSD-Motion System Design. http://motionsystemdesign.com/motors-drives/electrohydrostatic-actuation-modernizes-0111. Accessed 21 July 2011.

[39] Ramsey JW (2001) Power-by-wire. Avionics Today (Aviation Today Network). http://www.aviationtoday.com/av/military/Power-By-Wire_12671.html#.UPKdzaV8OlY. Accessed 13 January 2013.

[40] Van den Bossche D (2006) The A380 flight control electrohydrostatic actuators, achievements and lessons learnt. Proceedings of the 25th International Congress of the Aeronautical Sciences, September 3–8, 2006, Hamburg, Germany.

[41] Andersson J, Krus P, Nilsson K (1998) Optimization as a support for selection and design of aircraft actuation systems. Proceedings of the Seventh AIAA/USAF/NASA/ ISSMO Symposium on Multidisciplinary Analysis and Optimization, September 2–4, 1998, St. Louis, USA.

Appendix A

Hydraulic Symbols

Component name	Symbol
Pumps, prime movers, motors and actuators	
Fixed-displacement pump	
Fixed-displacement, bi-directional pump	
Variable-displacement pump	
Variable-displacement, bi-directional pump	
Fixed-displacement motor	

Hydrostatic Transmissions and Actuators: Operation, Modelling and Applications, First Edition.
Gustavo Koury Costa and Nariman Sepehri.
© 2015 John Wiley & Sons, Ltd. Published 2015 by John Wiley & Sons, Ltd.
Companion Website: www.wiley.com/go/costa/hydrostatic

Component name	Symbol
Fixed-displacement, bi-directional motor	
Variable-displacement motor	
Variable-displacement, bi-directional motor	
Fixed-displacement pump/motor	
Fixed-displacement, bidirectional pump/motor	
Variable-displacement pump/motor	
Variable-displacement, bidirectional pump/motor	
Single-acting (single-rod) cylinder	
Double-acting (double-rod) cylinder	

Component name	Symbol
Electric motor	
Non-electric prime mover	
Hydraulic power unit	
Clutch	
Hydraulic line elements	
Working line	
Drain line	
Line crossing	or
Junction	
Cooler	
Filter	
Tank	
Pressure and flow control	
Pressure gauge	

Component name	Symbol
Flow restrictor	
Flow control valve (adjustable flow restrictor)	
Shut-off valve	
Pressure relief valve	
Externally piloted pressure control valve	
Spring-loaded check valve	
Check valve	
Pilot operated check valve	
Directional valves	
Manual control	
Electrical control	

Component name	Symbol
Bidirectional electrical control	
Piloted	
Spring-return	
Two way, two position (2 × 2) valve	
Three way, two position (3 × 2) valve	
Four way, two position (4 × 2) valve	
Three way, three position (3 × 3) valve	
Four way, three position (4 × 3) valve	
Spring returned, 2 × 2 normally closed solenoid-activated valve	
Spring returned, 3 × 2 solenoid-activated valve	

Component name	Symbol
Spring centred, 4×3 solenoid-activated valve with closed centre	
Spring centred, 3×3 double piloted valve with closed centre	
Other symbols	
Hydraulic transformer	
Dual-pump hydraulic transformer	

Appendix B

Mathematics Review

B.1 The *Nabla* Operator ($\vec{\nabla}$)

In fluid mechanics, as well as in many branches of science, the following operation on a generic vector function **v**, whose components in $\mathfrak{R}^3$ are u, v and w, is usually found:

- *Divergent* of **v**:

$$\frac{\partial u}{\partial x} + \frac{\partial v}{\partial y} + \frac{\partial w}{\partial z}$$

Notice that the operation above is a transformation of the type $\mathfrak{R}^3 \to \mathfrak{R}$. Two other operations on scalar functions are also commonly found:

- *Gradient* of f:

$$\left(\frac{\partial f}{\partial x}, \frac{\partial f}{\partial y}, \frac{\partial f}{\partial z} \right) \text{ (an } \mathfrak{R} \to \mathfrak{R}^3 \text{ transformation)}$$

- *Laplacian* of f:

$$\frac{\partial^2 f}{\partial x^2} + \frac{\partial^2 f}{\partial y^2} + \frac{\partial^2 f}{\partial z^2} \text{ (an } \mathfrak{R} \to \mathfrak{R} \text{ transformation)}$$

There is another common transformation, from a vector in $\mathfrak{R}^3$ into another vector in $\mathfrak{R}^3$, called *rotational*. As we have not made use of this last transformation in this book, we will not present this operation here.[1]

The operations above can be made a lot easier if we use the vector operator *Nabla*, $\vec{\nabla}$, defined as a column matrix as follows:

$$\vec{\nabla}^T = \begin{bmatrix} \dfrac{\partial}{\partial x} & \dfrac{\partial}{\partial y} & \dfrac{\partial}{\partial z} \end{bmatrix} \tag{B.1}$$

---

[1] The interested reader can consult Ref. [1] for more a more detailed explanation.

Hydrostatic Transmissions and Actuators: Operation, Modelling and Applications, First Edition.
Gustavo Koury Costa and Nariman Sepehri.
© 2015 John Wiley & Sons, Ltd. Published 2015 by John Wiley & Sons, Ltd.
Companion Website: www.wiley.com/go/costa/hydrostatic

Note that we have used the transpose of $\vec{\nabla}$ (a line vector) in Eq. (B.1) for convenience. The choice of a column matrix to represent a vector is merely conventional, since either line or column matrices can be chosen to represent a vector.

Although $\vec{\nabla}$ is not a vector but a differential operator, we may equip it with some vector-like operations, such as multiplication by a scalar and inner product. For instance, let f be a function in $\mathfrak{R}$. We can 'multiply' the operator $\vec{\nabla}$ by the function f, obtaining the gradient of f, $\vec{\nabla}f$, as follows:

$$\vec{\nabla}f = \begin{bmatrix} \dfrac{\partial f}{\partial x} \\[2mm] \dfrac{\partial f}{\partial y} \\[2mm] \dfrac{\partial f}{\partial z} \end{bmatrix} \tag{B.2}$$

The divergent and Laplacian can also be defined in a similar way:

Divergent of **v**

$$\vec{\nabla} \cdot \mathbf{v} = \vec{\nabla}^T \mathbf{v} = \begin{bmatrix} \dfrac{\partial}{\partial x} & \dfrac{\partial}{\partial y} & \dfrac{\partial}{\partial z} \end{bmatrix} \begin{bmatrix} u \\ v \\ w \end{bmatrix} = \dfrac{\partial u}{\partial x} + \dfrac{\partial v}{\partial y} + \dfrac{\partial w}{\partial z} \tag{B.3}$$

Laplacian of f

$$\vec{\nabla} \cdot \left(\vec{\nabla}f\right) = \begin{bmatrix} \dfrac{\partial}{\partial x} & \dfrac{\partial}{\partial y} & \dfrac{\partial}{\partial z} \end{bmatrix} \begin{bmatrix} \dfrac{\partial f}{\partial x} \\[2mm] \dfrac{\partial f}{\partial y} \\[2mm] \dfrac{\partial f}{\partial z} \end{bmatrix} = \dfrac{\partial^2 f}{\partial x^2} + \dfrac{\partial^2 f}{\partial y^2} + \dfrac{\partial^2 f}{\partial z^2} \tag{B.4}$$

Because of the appearance of the second derivatives in the Laplacian of the function f, it is also usual to write $\vec{\nabla}^2 f = \vec{\nabla} \cdot \left(\vec{\nabla}f\right)$.

B.2   Ordinary Differential Equations (ODEs)

Ordinary differential equations are basic to the dynamic analysis of hydrostatic transmissions and actuators. In this section, we present some introductory notes on this vast and complex theme, which we believe are sufficient for what is needed to understand the concepts explored in this book. The interested reader can consult Refs. [2, 3] for a more complete exposition on the subject.

B.2.1 General Aspects and Definitions for ODEs

B.2.1.1 Linear and Non-Linear Equations

Consider the following second-order[2] ODE:

$$a_0 \frac{d^2y}{dt^2} + a_1 \frac{dy}{dt} + a_2 y = f \tag{B.5}$$

where, in the most general scenario, each coefficient a_i, $i = 0, \ldots, 2$, may be a function of y and its derivatives as well as a function of time. Similarly, the right-hand side term, f, may be a function of y and its derivatives as well as of time.

Equation (B.5) is said to be linear if all the coefficients a_i are constants or at most functions of the independent variable, t. On the other hand, if any of the coefficients a_i is a function of the dependent variable, y, and/or its derivatives, the equation is non-linear. Linear differential equations usually admit an analytical solution while non-linear equations are more difficult to be solved, given that there is no general theory that can be applied to them [2]. Therefore, we usually recur to numerical methods when the solution of a non-linear differential equation is needed.

The function, f, on the right-hand side of Eq. (B.5), is sometimes known as a 'forcing or excitation term' [4], a term inherited from mechanical vibrations or 'input function', as it is usually referred to in control engineering [5]. f is always related to external energy inputs in actual engineering problems. When $f = 0$, the differential equation (B.5) is called *homogeneous*.

Two different definitions are employed for the solution of the differential equation (B.5). First, a *general solution* is a function $y = h(t)$ that satisfies the equality (B.5) for whatever initial conditions are applied at $t = 0$. If the solution also requires that a specific set of initial conditions be satisfied, we use the term *particular solution* instead [6]. There may be more than one general solution for an ODE, and in the particular case of linear homogeneous ODEs, different solutions can be combined to form another solution, as will be seen next.

B.2.1.2 The Superposition Principle for Linear Homogeneous ODEs

Let h_1 and h_2 be two general solutions of the linear homogeneous ODE:

$$a_0 \frac{d^2y}{dt^2} + a_1 \frac{dy}{dt} + a_2 y = 0 \tag{B.6}$$

The superposition principle (or linearity principle) states that the linear combination of h_1 and h_2, $h = k_1 h_1 + k_2 h_2$ (k_1 and k_2 are constants), is also a solution of (B.6). To prove it, we

---

[2] The term 'second-order' relates to the order of the higher derivative that appears in the equation. Higher order differential equations are not used in this book and will not be studied here.

substitute $h = k_1 h_1 + k_2 h_2$ into Eq. (B.6) and obtain

$$k_1 \left(a_0 \frac{d^2 h_1}{dt^2} + a_1 \frac{dh_1}{dt} + a_2 h_1 \right) + k_2 \left(a_0 \frac{d^2 h_2}{dt^2} + a_1 \frac{dh_2}{dt} + a_2 h_2 \right) = 0 \qquad \text{(B.7)}$$

Since h_1 and h_2 are solutions of (B.6), both terms within parentheses in Eq. (B.7) are zero, which proves that (B.7) also constitutes an equality for every constant k_1 and k_2. This completes the proof of the superposition principle. We will make use of this property when we find a general solution for a homogeneous ODE with constant coefficients.

B.2.1.3   General Solution of Homogeneous ODEs with Constant Coefficients

Consider the following particular homogeneous ODE with constant coefficients a and b:

$$\frac{d^2 y}{dt^2} + a \frac{dy}{dt} + by = 0 \qquad \text{(B.8)}$$

Equation (B.8) is representative of natural mechanical vibrations where no exciting force is present, such as a mass–spring system that remains oscillating when the spring is compressed and subsequently released or a pendulum that keeps on swinging after the initial push. Wherever there is elasticity or any other form of energy storage, free oscillations can be present and this is why Eq. (B.8) is so significant.

In this section, we obtain a general solution for Eq. (B.8). The main idea is to formally define two properties associated with natural oscillations: the natural frequency,[3] ω, and the damping ratio, ζ. These properties constitute important parameters for the dynamic analysis of any hydraulic circuit.

Through simple observation, we note that the function $y = ke^{\lambda t}$, where k ($k \neq 0$) and λ are constants, constitutes a general solution to Eq. (B.8). In fact, if we substitute $y = ke^{\lambda t}$ into Eq. (B.8), we obtain

$$\frac{d^2 \left(ke^{\lambda t} \right)}{dt^2} + a \frac{d \left(ke^{\lambda t} \right)}{dt} + bke^{\lambda t} = \left(\lambda^2 + a\lambda + b \right) ke^{\lambda t} = 0 \qquad \text{(B.9)}$$

Given that $k \neq 0$ in Eq. (B.9), $y = ke^{\lambda t}$ will be a general solution of (B.8) as long as the characteristic equation $\lambda^2 + a\lambda + b = 0$ is satisfied. This gives the following possible values for the constant λ:

$$\lambda_1, \lambda_2 = \frac{-a \pm \sqrt{a^2 - 4b}}{2} \qquad \text{(B.10)}$$

Let us make $a = 2\zeta\omega$ and $b = \omega^2$ in Eqs. (B.8) and (B.9) where ζ and ω are two newly defined coefficients (the reason for that will become clear as we progress). We can then rewrite Eq. (B.10) in a different way:

$$\lambda_1, \lambda_2 = -\zeta\omega \pm \frac{1}{2} \sqrt{(2\zeta\omega)^2 - 4\omega^2} = \omega \left(-\zeta \pm \sqrt{\zeta^2 - 1} \right) \qquad \text{(B.11)}$$

---

[3] Strictly speaking, the natural 'angular' frequency would be a preferred term to avoid confusion with the natural frequency f, measured in s^{-1} (ω is given in rad/s). However, in order to avoid unnecessarily long phrases, we prefer to make clear from the context and from the symbol used (f or ω) to which frequency we are referring.

Note that the coefficient ζ in Eq. (B.11) defines the nature of the roots λ_1 and λ_2. Four cases can be identified:

1. If $\zeta = 0$, the two roots λ_1 and λ_2 are imaginary, that is, $\lambda_1, \lambda_2 = \pm\omega j$, where $j = \sqrt{-1}$. The general solution of Eq. (B.8) is then given by

$$y = k_1 e^{\omega j t} + k_2 e^{-\omega j t} \tag{B.12}$$

Equation (B.12) can be written in a different manner if we apply the well-known Euler's formula $e^{jx} = \cos(x) + j\sin(x)$:

$$y = (k_1 + k_2)\cos(\omega t) + (k_1 - k_2)j\sin(\omega t) = A\cos(\omega t) + B\sin(\omega t) \tag{B.13}$$

where A and B are constants ($A = k_1 + k_2$ and $B = k_1 j - k_2 j$). Equation (B.13) describes an undamped harmonic oscillation whose natural frequency is ω. We therefore say that when $\zeta = 0$, the physical system represented by Eq. (B.8) is *undamped*.

2. If $\zeta = 1$, the characteristic equation $\lambda^2 + a\lambda + b = 0$ has one real root, that is, $\lambda_1 = \lambda_2 = -\omega$. The system described by Eq. (B.8) is now defined as *critically damped* and its general solution is given by

$$y = k_1 e^{-\omega t} + k_2 e^{-\omega t} = A e^{-\omega t} \tag{B.14}$$

Note that now the solution, y, is non-oscillatory and decays exponentially as time progresses.

3. If $0 < \zeta < 1$, the roots of the characteristic equation $\lambda^2 + a\lambda + b = 0$ are complex conjugates. In this case, the general solution becomes

$$y = k_1 e^{\omega(-\zeta + \gamma j)t} + k_2 e^{\omega(-\zeta - \gamma j)t} \tag{B.15}$$

where $\gamma = \sqrt{1 - \zeta^2}$. Using Euler's formula, Eq. (B.15) can be rewritten as

$$y = e^{(-\zeta\omega t)}[A\cos(\omega_\gamma t) + B\sin(\omega_\gamma t)] \tag{B.16}$$

where $\omega_\gamma = \gamma\omega$. From Eq. (B.16) we see that the solution is oscillatory with a modified (damped) natural frequency, ω_γ, and a progressively reducing amplitude due to the exponential factor $e^{(-\zeta\omega t)}$. The corresponding physical system is referred to as *underdamped*.

4. If $\zeta > 1$, the characteristic equation $\lambda^2 + a\lambda + b = 0$ has two real roots and we say that the system is *overdamped*. Similar to the critically damped situation, the solution of Eq. (B.8) is non-oscillatory; however, it presents a faster damping ratio, as can be seen by the solution equation, which in this case becomes

$$y = k_1 e^{\omega\lambda_1 t} + k_2 e^{-\omega\lambda_2 t} \tag{B.17}$$

where

$$\begin{cases} \lambda_1 = \omega\left(\sqrt{\zeta^2 - 1} - \zeta\right) = \omega\left(\sqrt{\zeta^2 - 1} - \sqrt{\zeta^2}\right) < 0 \\ \lambda_2 = \omega\left(-\zeta - \sqrt{\zeta^2 - 1}\right) < 0 \end{cases} \tag{B.18}$$

Three definitions have been implicitly given in the previous discussion. We repeat them here because of their importance:

1. The undamped natural frequency, ω:

$$\omega = \sqrt{b} \tag{B.19}$$

2. The damping ratio, ζ:

$$\zeta = \frac{a}{2\omega} = \frac{a}{2\sqrt{b}} \tag{B.20}$$

3. The damped natural frequency, ω_γ:

$$\omega_\gamma = \sqrt{b}\sqrt{1 - \zeta^2} = \sqrt{b - \frac{a^2}{4}} \tag{B.21}$$

Figure B.1 illustrates the four different types of solution of Eq. (B.8) according to the value of the damping ratio, ζ.

The time lapse between two consecutive values of the function y, corresponding to a complete revolution of the angle $\omega_\gamma t$ in Eq. (B.16) where $\omega_\gamma t$ changes from $2\pi N$ to $2\pi(N + 1)$ radians ($N = 0, 1, 2, \ldots$), is called the period, T, of the correlated sinusoidal function. The frequency, f, is defined as the number of cycles per unit time. Since one complete cycle is performed during the period, T, we have that

$$f = \frac{1}{T} \tag{B.22}$$

The relation between the frequency, f, and the angular frequency, ω_γ, is straightforward and comes from the fact that one cycle corresponds to a 2π revolution in the angle $\omega_\gamma t$, that is

$$\omega_\gamma = \frac{2\pi}{T} = 2\pi f \tag{B.23}$$

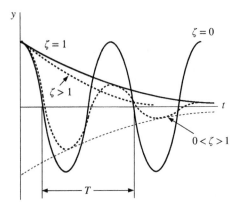

Figure B.1   Solutions to the second-order homogeneous ODE with constant coefficients

Equation (B.23) is valid for the undamped natural frequency, ω, as well. Usual units for ω_γ and f are rad/s and s^{-1} (Hz), respectively.

B.2.1.4 Non-Homogeneous Linear Equations

Non-homogeneous equations are not as simple as homogeneous equations, even in the constant coefficient case. We are not going to deal with this subject in detail here, but we shall present one particular method for solving initial value problems in the following section. Before we proceed, let us state an important theorem concerning linear, non-homogeneous ODEs [2]:

Let

$$\ddot{y} + a(t)\dot{y} + b(t)y = f(t) \tag{B.24}$$

be a given linear ODE whose homogeneous counterpart is[4]

$$\ddot{y} + a(t)\dot{y} + b(t)y = 0 \tag{B.25}$$

Now let $h_{\mathrm{h}} = k_1 y_1 + k_2 y_2$ be a general solution of Eq. (B.25) and h_{p} a particular solution satisfying Eq. (B.24). It is possible to show that the general solution of Eq. (B.24), h, can be written as

$$h = h_{\mathrm{h}} + h_{\mathrm{p}} \tag{B.26}$$

In other words, the general solution of a linear ODE is always composed of the corresponding homogeneous solution plus one particular solution that satisfies the equation as a whole. This theorem guarantees that concepts such as natural frequency and damping ratio remain valid even in the presence of a forcing term, $f(t)$ on the right-hand side of the ODE.

B.2.2 The Laplace[5] Transform Method

In this section, we introduce a classic method of solving linear ODEs that has been widely used due to its relative simplicity. We begin by defining the Laplace transform integral and showing how to obtain it for some usual functions.

B.2.2.1 The Laplace Transform Integral

The Laplace transform of a function $f(t)$ is defined by the following improper integral[6] [3, 6]:

$$\mathcal{L}(f) = F(s) = \int_0^\infty e^{-st} f(t)\, dt \tag{B.27}$$

---

[4] Here we are using the upper dot notation for the time derivatives, that is $\ddot{y} = \dfrac{d^2 y}{dt^2}$ and $\dot{y} = \dfrac{dy}{dt}$.

[5] In honour of the French mathematician Pierre Simon Marquis de Laplace (1749–1827).

[6] An integral where one of the limits of integration (or both) is set to infinity is called an improper integral.

where s is a complex variable. The following properties are immediately deductible (a prime, following the function f, denotes time differentiation):

1. *Linearity*: Let k_1 and k_2 be constants and f_1 and f_2 be two arbitrary functions:

$$\mathcal{L}(k_1 f_1 + k_2 f_2) = k_1 \int_0^\infty e^{-st} f_1 \, dt + k_2 \int_0^\infty e^{-st} f_2 \, dt = k_1 \mathcal{L}(f_1) + k_2 \mathcal{L}(f_2)$$

2. *Laplace transform of the first and second derivatives*: If the function f is twice differentiable and has a Laplace transform, along with its first and second derivatives, f' and f'', after integrating by parts, we obtain

$$\mathcal{L}(f') = \int_0^\infty e^{-st} \left(\frac{df}{dt} \right) dt = e^{-st} f(t)|_0^\infty + s \int_0^\infty e^{-st} f(t) \, dt = s\mathcal{L}(f) - f(0)$$

$$\mathcal{L}(f'') = \int_0^\infty e^{-st} \left(\frac{d^2 f}{dt^2} \right) dt = e^{-st} f'(t)|_0^\infty + s \int_0^\infty e^{-st} f'(t) \, dt = s^2 \mathcal{L}(f) - sf(0) - f'(0)$$

3. *Attenuation property*: Let α be a real constant and let the Laplace transform of the function $f(t)$ be $F(s)$, then

$$\mathcal{L}[e^{-\alpha t} f(t)] = \int_0^\infty e^{-st} e^{-\alpha t} f(t) \, dt = \int_0^\infty e^{-(s+\alpha)t} f(t) \, dt = \mathcal{L}(s + \alpha) = F(s + \alpha)$$

As an example of the attenuation property, Table B.1 gives the Laplace transform of $\cos(\omega t)$.[7] The Laplace transform of $e^{-\alpha t} \cos(\omega t)$ will then be given by

$$\mathcal{L}[e^{-\alpha t} f(t)] = F(s + a) = \frac{s + a}{(s + a)^2 + \omega^2} \tag{B.28}$$

Table B.1 Some selected Laplace transforms

f	$\mathcal{L}(f) = F(s)$
1	$\dfrac{1}{s}$
$A \times 1(t)$, or A	$\dfrac{A}{s}$
t^n (n is an integer)	$\dfrac{n!}{s^{(n+1)}}$
$\sin(\omega t)$	$\dfrac{\omega}{s^2 + \omega^2}$
$\cos(\omega t)$	$\dfrac{s}{s^2 + \omega^2}$

[7] All the transformations in Table B.1 have been obtained through the Laplace integral (B.27).

Using the aforementioned properties, we can obtain the Laplace transform of practically any function for which the Laplace integral converges.[8] For instance, the Laplace transform of $f(t) = t$ will be given by

$$\mathcal{L}t| = \int_0^\infty e^{-st} t\, dt = -\frac{te^{-st}}{s}\Big|_0^\infty + \frac{1}{s}\int_0^\infty e^{-st}\, dt = -\frac{e^{-st}}{s^2}\Big|_0^\infty = \frac{1}{s^2} \qquad (B.29)$$

As another example, let us determine $\mathcal{L}(0)$. If we pick an arbitrary function of time, f, and use the linearity property of the Laplace transform, we can write that

$$\mathcal{L}(0) = \mathcal{L}(0 \times f) = 0 \times \mathcal{L}(f) = 0 \qquad (B.30)$$

One particular case of interest is the function f, which is discontinuous at $t = 0$, represented in Figure B.2. For the particular case shown in the figure, we have that $f(-0) = 0$ and[9] $f(0+) = f_0$. In situations similar to this, it is possible to write the Laplace transform of the function f as[10]:

$$\mathcal{L}(t) = \int_{0-}^\infty e^{-st} f(t)\, dt \qquad (B.31)$$

As an example of the application of Eq. (B.31), consider the *step function* defined as

$$f(t) = \begin{cases} 0, & t < 0 \\ A, & t > 0 \end{cases} \qquad (B.32)$$

A usual representation of the function (B.32) is [7]:

$$f(t) = A \times 1(t) \qquad (B.33)$$

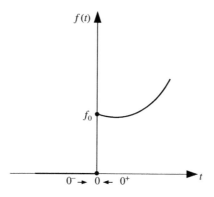

Figure B.2   Example of a discontinuous function at $t = 0$

---

[8] The interested reader can find more on the convergence of Laplace transforms in Ref. [7].
[9] If we skip the mathematical formalities, we can say that -0 and $+0$ indicate a point on the t-axis immediately to the left and to the right of point $t = 0$, respectively.
[10] Actually, Eq. (B.31) has been referred to as the Laplace transform definition by some authors (see Ref. [5] and the references therein).

where $1(t)$ is the unit step function:

$$1(t) = \begin{cases} 0, & t < 0 \\ 1, & t > 0 \end{cases} \tag{B.34}$$

Using the definition given by Eq. (B.31), the Laplace transform of the function $A \times 1(t)$ becomes

$$\mathcal{L}(t) = \int_{0^-}^{\infty} e^{-st}[A \times 1(t)]\,dt = \int_{0^-}^{0} e^{-st}(0)\,dt + \int_{0}^{\infty} Ae^{-st}\,dt = \frac{A}{s} \tag{B.35}$$

B.2.2.2 Solution of Linear ODEs

In order to exemplify the use of the Laplace transform for solving linear ODEs, consider the following equation with constant coefficients:

$$\frac{d^2y}{dt^2} + a\frac{dy}{dt} + by = f(t) \quad \text{for } t \in \,]0, \infty[\tag{B.36}$$

subject to the following initial conditions:

$$y(0^-) = 0 \text{ and } \frac{dy}{dt}(0^-) = 0 \tag{B.37}$$

An equation of this type was found in Case Study 1, Chapter 5 (see Eq. 5.27). To apply the Laplace transform method to Eqs. (B.36) and (B.37), we start by taking the transforms of both sides. Using the linearity property, we have that

$$\mathcal{L}\left(\frac{d^2y}{dt^2}\right) + a\mathcal{L}\left(\frac{dy}{dt}\right) + b\mathcal{L}(y) - \mathcal{L}(f) = \mathcal{L}(0) \tag{B.38}$$

We then expand the transforms of the first and second derivatives and make $\mathcal{L}(0) = 0$:

$$s^2\mathcal{L}(y) - sy(0) - y'(0) + a[s\mathcal{L}(y) - y(0)] + b\mathcal{L}(y) - \mathcal{L}(f) = 0 \tag{B.39}$$

If we assume that the solution function, y, is such that $y = 0$ for $t < 0$, not being defined at $t = 0$, as illustrated in Figure B.2, we can write Eq. (B.39) as

$$s^2\mathcal{L}(y) - sy(0^-) - y'(0^-) + a[s\mathcal{L}(y) - y(0^-)] + b\mathcal{L}(y) - \mathcal{L}(f) = 0 \tag{B.40}$$

where the Laplace transform, $\mathcal{L}(f)$, is now taken between the limits 0^- and $+\infty$.

Substituting the initial conditions given by Eqs. (B.37) into Eq. (B.40), we can solve for $\mathcal{L}(y)$:

$$\mathcal{L}(y) = \frac{\mathcal{L}(f)}{s^2 + as + b} \tag{B.41}$$

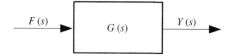

Figure B.3   Block diagram representation of a linear differential equation

Equation (B.41) does not give the solution of Eqs. (B.36) and (B.37) in a direct manner. Rather, it shows a relationship between the Laplace transform of the functions y and f instead. This leads to the block diagram representation shown in Figure B.3, where we can identify the Laplace transform of the input function, $F(s) = \mathcal{L}(f)$, the Laplace transform of the output function, $Y(s) = \mathcal{L}(y)$, and the transfer function $G(s) = (s^2 + as + b)^{-1}$. Equation (B.41) can be conveniently rewritten as

$$Y(y) = G(s)F(s) \tag{B.42}$$

Once the original differential equation has been written in the form of Eq. (B.42), we can obtain the output, y, by using the inverse of the Laplace transform.

B.2.2.3   Case Study 1: Step Input

Suppose that f, in Eq. (B.41), has been defined as

$$f(t) = c \times 1(t) \tag{B.43}$$

where c is constant.

Equation (B.41) then becomes

$$\mathcal{L}(y) = \frac{\mathcal{L}[c \times 1(t)]}{s^2 + as + b} = \frac{c}{s(s^2 + as + b)} \tag{B.44}$$

Generally speaking, if $Y(y) = \mathcal{L}(y)$, we have that $y = \mathcal{L}^{-1}[Y(y)]$. In other words, if we are able to find the inverse of the Laplace transform of the right-hand side of Eq. (B.44), the output function, y, will be given by

$$y = \mathcal{L}^{-1}\left[\frac{c}{s\left(s^2 + as + b\right)}\right] \tag{B.45}$$

A practical means of finding the Laplace transform inverse is by a direct comparison with the transform of known functions. For instance, from Table B.1, we know that

$$\mathcal{L}[\cos(\omega t)] = \frac{s}{s^2 + \omega^2} \tag{B.46}$$

Therefore, we have

$$\mathcal{L}^{-1}\left[\frac{s}{s^2 + \omega^2}\right] = \cos(\omega t) \tag{B.47}$$

To find the inverse of the Laplace transform of more complex functions, such as the one given by Eq. (B.44), we must break it into functions for which the Laplace transforms are known. The linearity property of the inverse transform is very helpful at this stage and is given here without a formal proof. In this aspect, let F_1 and F_2 be two functions of a complex variable, s, and let k_1 and k_2 be two constants. It can be shown that

$$\mathcal{L}^{-1}[k_1 F_1 + k_2 F_2] = k_1 \mathcal{L}^{-1}(F_1) + k_2 \mathcal{L}^{-1}(F_2) \tag{B.48}$$

In order to use the linearity property of the inverse Laplace transform in the solution of Eq. (B.45), we first expand the expression within brackets into partial fractions,[11] as follows:

$$\frac{c}{s(s^2 + as + b)} = \frac{A}{s} + \frac{Bs + C}{s^2 + as + b} \tag{B.49}$$

where

$$A = \frac{c}{b}, \quad B = -\frac{c}{b} \quad \text{and} \quad C = -\frac{ac}{b} \tag{B.50}$$

Now, using the linearity property, we obtain

$$y = \mathcal{L}^{-1}\left[\frac{A}{s} + \frac{Bs + C}{s^2 + as + b}\right] = \mathcal{L}^{-1}\left(\frac{A}{s}\right) + \mathcal{L}^{-1}\left(\frac{Bs + C}{s^2 + as + b}\right) \tag{B.51}$$

The first inversion follows immediately from Table B.1, and we are left with

$$y = A + \mathcal{L}^{-1}\left(\frac{Bs + C}{s^2 + as + b}\right) \tag{B.52}$$

Finding the inverse Laplace transform of the function within parentheses in Eq. (B.52) is not initially straightforward. However, we can still make some modifications. We begin by completing the squares in the denominator:

$$\frac{Bs + C}{s^2 + as + b} = \frac{Bs + C}{\left(s + \dfrac{a}{2}\right)^2 + \omega_\gamma^2} \tag{B.53}$$

where

$$\omega_\gamma^2 = b - \frac{a^2}{4} \tag{B.54}$$

We may now rewrite the numerator of Eq. (B.53) in a slightly different way:

$$\frac{Bs + C}{s^2 + as + b} = \frac{B\left(s + \dfrac{a}{2}\right) + C - \left(\dfrac{a}{2}\right)B}{\left(s + \dfrac{a}{2}\right)^2 + \omega_\gamma^2} \tag{B.55}$$

---

[11] For more information about partial fraction decomposition, consult Refs. [5, 6].

The right-hand side of Eq. (B.55) can then be expanded as follows:

$$\frac{Bs + C}{s^2 + as + b} = B\left[\frac{\left(s + \frac{a}{2}\right)}{\left(s + \frac{a}{2}\right)^2 + \omega_\gamma^2}\right] + \left[\frac{C - \left(\frac{a}{2}\right)B}{\left(s + \frac{a}{2}\right)^2 + \omega_\gamma^2}\right] \qquad (B.56)$$

By applying the inverse of the Laplace transform to both sides of Eq. (B.56) and using the linearity property, we obtain

$$\mathcal{L}^{-1}\left(\frac{Bs + C}{s^2 + as + b}\right) = B\mathcal{L}^{-1}\left[\frac{\left(s + \frac{a}{2}\right)}{\left(s + \frac{a}{2}\right)^2 + \omega_\gamma^2}\right] + \mathcal{L}^{-1}\left[\frac{C - \left(\frac{a}{2}\right)B}{\left(s + \frac{a}{2}\right)^2 + \omega_\gamma^2}\right] \qquad (B.57)$$

The inversion of the first term on the right-hand side of Eq. (B.57) is readily obtained as (see Eq. (B.28)):

$$B\mathcal{L}^{-1}\left[\frac{\left(s + \frac{a}{2}\right)}{\left(s + \frac{a}{2}\right)^2 + \omega_\gamma^2}\right] = Be^{-\left(\frac{a}{2}\right)t}\cos(\omega_\gamma t) \qquad (B.58)$$

In order to obtain the inverse of the Laplace transform of the second term on the right-hand side of Eq. (B.57), we first note that

$$\mathcal{L}^{-1}\left[\frac{C - \left(\frac{a}{2}\right)B}{\left(s + \frac{a}{2}\right)^2 + \omega_\gamma^2}\right] = \left[\frac{C - \left(\frac{a}{2}\right)B}{\omega_\gamma}\right]\mathcal{L}^{-1}\left[\frac{\omega_\gamma}{\left(s + \frac{a}{2}\right)^2 + \omega_\gamma^2}\right] \qquad (B.59)$$

The inverse of the Laplace transform in Eq. (B.59) can be obtained from Table B.1 and the attenuation property:

$$\left[\frac{C - \left(\frac{a}{2}\right)B}{\omega_\gamma}\right]\mathcal{L}^{-1}\left[\frac{\omega_\gamma}{\left(s + \frac{a}{2}\right)^2 + \omega_\gamma^2}\right] = \left[\frac{C - \left(\frac{a}{2}\right)B}{\omega_\gamma}\right]e^{-\left(\frac{a}{2}\right)t}\sin(\omega_\gamma t) \qquad (B.60)$$

The solution of the differential equations (B.36) and (B.37) with the input function, f, given by Eq. (B.43) can then be written as

$$y = A + Be^{-\left(\frac{a}{2}\right)t}\cos(\omega_\gamma t) + \left[\frac{C - \left(\frac{a}{2}\right)B}{\omega_\gamma}\right]e^{-\left(\frac{a}{2}\right)t}\sin(\omega_\gamma t) \qquad (B.61)$$

Substituting the values for A, B and C, given by Eqs. (B.50), into Eq. (B.61), we finally obtain

$$y = \left(\frac{c}{b}\right)\left[1 - e^{-\left(\frac{a}{2}\right)t}\cos(\omega_\gamma t)\right] - \left(\frac{ac}{2b\omega_\gamma}\right)e^{\left(-\frac{a}{2}\right)t}\sin(\omega_\gamma t) \qquad (B.62)$$

Equation (B.62) can be cast into a slightly different form:

$$y = \left(\frac{c}{b}\right)\left\{1 - e^{-\left(\frac{a}{2}\right)t}\left[\cos(\omega_\gamma t) + \left(\frac{a}{2\omega_\gamma}\right)\sin(\omega_\gamma t)\right]\right\} \qquad (B.63)$$

Remarks

1. We can quickly conclude by comparing Eqs. (B.21) and (B.54) that the term, ω_γ, in Eq. (B.63) corresponds to the damped natural angular frequency of the physical system represented by Eqs. (B.36), (B.37) and (B.43).
2. Observe that the damping ratio, ζ, given by Eq. (B.20), is not readily identifiable within Eq. (B.63).

B.2.2.4   Case Study 2: Non-dissipative Equation and Harmonic Input

In many situations, it is important to predict the behaviour of a hydraulic system when dissipative forces are not present. In this case, the mathematical model becomes

$$\frac{d^2y}{dt^2} + by = f(t), \quad \text{for } t \in \,]0, \infty[\qquad (B.64)$$

For homogeneous initial conditions, applying the Laplace transform method to Eq. (B.64) yields

$$s^2 \mathcal{L}(y) + b\mathcal{L}(y) - \mathcal{L}(f) = 0 \qquad (B.65)$$

Solving Eq. (B.65) for $\mathcal{L}(y)$, we obtain

$$\mathcal{L}(y) = \frac{\mathcal{L}(f)}{s^2 + b} \qquad (B.66)$$

Now, consider that the input function, f, in Eq. (B.66), is given by

$$f = A_m\sin(\omega_f t) \qquad (B.67)$$

where A_m is the function amplitude and ω_f is the input frequency.

Equation (B.66) then becomes

$$\mathcal{L}(y) = \frac{\mathcal{L}[A_m\sin(\omega_f t)]}{s^2 + b} = \frac{A_m\mathcal{L}[\sin(\omega_f t)]}{s^2 + b} = \frac{A_m\omega_f}{(s^2 + \omega_f^2)(s^2 + b)} \qquad (B.68)$$

The right-hand side of Eq. (B.68) can be expanded into partial fractions as follows:

$$A_m \left[\frac{\omega_f}{\left(s^2 + \omega_f^2\right)\left(s^2 + b\right)} \right] = A_m \left[\frac{As + C}{s^2 + \omega_f^2} + \frac{Bs + D}{s^2 + b} \right] \tag{B.69}$$

Solving Eq. (B.69) for the constants A through D, we obtain (we leave the demonstration as an exercise):

$$A = B = 0, \quad C = \frac{\omega_f}{b - \omega_f^2} \quad \text{and} \quad D = \frac{\omega_f}{\omega_f^2 - b} \tag{B.70}$$

We can therefore rewrite Eq. (B.68) as

$$\mathcal{L}(y) = A_m \left[\frac{C}{s^2 + \omega_f^2} + \frac{D}{s^2 + b} \right] \tag{B.71}$$

The solution function, y, can then be found by taking the inverse of the Laplace transform at both sides of Eq. (B.71):

$$y = A_m C \mathcal{L}^{-1} \left(\frac{1}{s^2 + \omega_f^2} \right) + A_m D \mathcal{L}^{-1} \left(\frac{1}{s^2 + b} \right) \tag{B.72}$$

In order to solve Eq. (B.72), we first rewrite it in a slightly different form:

$$y = \left(\frac{A_m C}{\omega_f} \right) \mathcal{L}^{-1} \left(\frac{\omega_f}{s^2 + \omega_f^2} \right) + \left(\frac{A_m D}{\sqrt{b}} \right) \mathcal{L}^{-1} \left(\frac{\sqrt{b}}{s^2 + b} \right) \tag{B.73}$$

From Table B.1, we immediately recognize the Laplace inverses as sine functions. Therefore, we have

$$y = \left(\frac{A_m C}{\omega_f} \right) \sin(\omega_f t) + \left(\frac{A_m D}{\sqrt{b}} \right) \sin(t\sqrt{b}) \tag{B.74}$$

Substituting the coefficients C and D, given by Eqs. (B.70) into Eq. (B.74), we finally obtain

$$y = \frac{A_m \omega_f}{(b - \omega_f^2)} \left[\left(\frac{1}{\omega_f} \right) \sin(\omega_f t) - \left(\frac{1}{\sqrt{b}} \right) \sin(t\sqrt{b}) \right] \tag{B.75}$$

From Eq. (B.19), we have that $\omega = \sqrt{b}$. Therefore, we can rewrite Eq. (B.75) in terms of the natural frequency, ω:

$$y = \frac{A_m \omega_f}{(\omega^2 - \omega_f^2)} \left[\frac{\sin(\omega_f t)}{\omega_f} - \frac{\sin(\omega t)}{\omega} \right] \tag{B.76}$$

References

[1] Fox RW, McDonald BT, Pritchard PJ (2004) Introduction to fluid mechanics, 6th Ed., John Wiley & Sons, USA.

[2] Boyce WE, DiPrima RC (2001) Elementary differential equations and boundary value problems. John Wiley & Sons, USA.

[3] Butkov E (1973) Mathematical physics. Addison-Wesley, USA.

[4] Kelly SG (2012) Mechanical vibrations – theory and applications. Cengage Learning, USA.

[5] Nise NS (2011) Control systems engineering, 6th Ed., John Wiley & Sons, USA.

[6] Kreyszig E (2006) Advanced engineering mathematics, 9th Ed., John Wiley & Sons, USA.

[7] Ogata K (1997) Modern control engineering, 3rd Ed., Prentice-Hall, USA.

Appendix C

Fluid Dynamics Equations

C.1   Introduction

Consider an infinitesimal particle with a fixed mass dm, moving along with the flow with a speed $\mathbf{v}$, as shown in Figure C.1.

As the fluid element dm moves along with the flow, it becomes subject to forces acting on every face of the infinitesimal cube due to the interaction with its neighbour particles. Figure C.2 shows the forces acting on the lateral faces of the element, perpendicular to the x-axis. The analysis that follows must be also applied to the other four faces of the infinitesimal cube.

Observing Figure C.2, we see that the force acting on the left face is $\mathbf{F}_x$. As we move from the left face to the right face of the cube, in the direction of the x-axis, the force $\mathbf{F}_x$ is changed from $\mathbf{F}_x$ into $\mathbf{F}_{x+\Delta x}$. We may then perform a first-order Taylor expansion of $\mathbf{F}_x$ to obtain

$$\mathbf{F}_{x+\Delta x} = \mathbf{F}_x + \frac{\partial \mathbf{F}_x}{\partial x} \Delta x \qquad (C.1)$$

The resultant force, $\mathbf{R}_x$, can be easily obtained from Eq. (C.1) as

$$\mathbf{R}_x = \mathbf{F}_{x+\Delta x} - \mathbf{F}_x = \frac{\partial \mathbf{F}_x}{\partial x} \Delta x \qquad (C.2)$$

We observe that the components of the forces $\mathbf{F}_x$ and $\mathbf{F}_{x+\Delta x}$ in Eq. (C.2) can be obtained by multiplying the elementary area $dydz$ by the correspondent tension: τ_{xx}, τ_{xy} or τ_{xz}. The tension notation adopted here follows the conventional rule in which the first sub-index indicates the plane and the second sub-index indicates the direction. Therefore, τ_{xz}, for example indicates that the tension is taken in a plane perpendicular to x and oriented along z.

Hydrostatic Transmissions and Actuators: Operation, Modelling and Applications, First Edition.
Gustavo Koury Costa and Nariman Sepehri.
© 2015 John Wiley & Sons, Ltd. Published 2015 by John Wiley & Sons, Ltd.
Companion Website: www.wiley.com/go/costa/hydrostatic

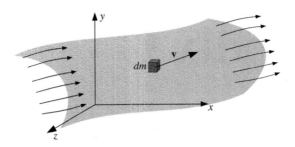

Figure C.1   Particle tracking on a flow

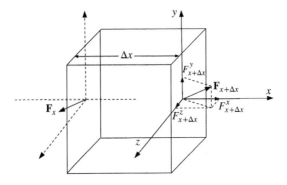

Figure C.2   Equilibrium of forces in a fluid element in the x-direction

We therefore have

$$
\mathbf{F}_x = \begin{bmatrix} \tau_{xx}\,dy\,dz \\[4pt] \tau_{xy}\,dy\,dz \\[4pt] \tau_{xz}\,dy\,dz \end{bmatrix}, \mathbf{F}_{x+\Delta x} = \begin{bmatrix} \left[\left(\tau_{xx} + \dfrac{\partial \tau_{xx}}{\partial x}\Delta x\right) dy\,dz\right] \\[10pt] \left(\tau_{xy} + \dfrac{\partial \tau_{xy}}{\partial x}\Delta x\right) dy\,dz \\[10pt] \left[\left(\tau_{xz} + \dfrac{\partial \tau_{xz}}{\partial x}\Delta x\right) dy\,dz\right] \end{bmatrix} \tag{C.3}
$$

Substituting $\mathbf{F}_x$ and $\mathbf{F}_{x+\Delta x}$, given by Eq. (C.3), into Eq. (C.2), we obtain the following expression for $\mathbf{R}_x$, in the limit where $\Delta x \rightarrow dx$:

$$
\mathbf{R}_x = \begin{bmatrix} \left[\left(\dfrac{\partial \tau_{xx}}{\partial x}dy\,dz\right) dx\right] \\[10pt] \left(\dfrac{\partial \tau_{xy}}{\partial x}dy\,dz\right) dx \\[10pt] \left[\left(\dfrac{\partial \tau_{xz}}{\partial x}dy\,dz\right) dx\right] \end{bmatrix} \tag{C.4}
$$

If we proceed with the determination of the resultant force for each face of the element, we arrive at the following matrix equation:

$$
\begin{bmatrix} \mathbf{R}_x^T \\ \mathbf{R}_y^T \\ \mathbf{R}_z^T \end{bmatrix} = \begin{bmatrix} \dfrac{\partial}{\partial x} & \dfrac{\partial}{\partial y} & \dfrac{\partial}{\partial z} \end{bmatrix} \begin{pmatrix} \tau_{xx} & \tau_{xy} & \tau_{xz} \\ \tau_{yx} & \tau_{yy} & \tau_{yz} \\ \tau_{zx} & \tau_{zy} & \tau_{zz} \end{pmatrix} \begin{bmatrix} \mathbf{i}^T \\ \mathbf{j}^T \\ \mathbf{k}^T \end{bmatrix} dx\,dy\,dz
\tag{C.5}
$$

where $\mathbf{i}$, $\mathbf{j}$ and $\mathbf{k}$ are unit vectors along the x, y and z directions. In Eq. (C.5), we also identify the transpose of the Nabla operator (more on the Nabla operator can be found in Appendix B):

$$
\vec{\nabla}^T = \begin{bmatrix} \dfrac{\partial}{\partial x} & \dfrac{\partial}{\partial y} & \dfrac{\partial}{\partial z} \end{bmatrix}
\tag{C.6}
$$

C.2 Fluid Stresses and Distortion Rates

Equation (C.5) contains a very important entity in fluid mechanics: the *viscous tensor* $[\tau]$, which is the 3×3 matrix made up of the nine viscous tensions τ_{xx}, τ_{xy}, $\dots$, τ_{zz}. In this aspect, we observe that the viscous tension τ_{yx} that appears in the Newton's law of viscosity in Eq. (2.1) is just a part of the bigger picture regarding the more general force balance, illustrated in Figure C.2. In fact, the cubic element undergoes a generalized distortion in the edges and the faces, which depends on the stresses acting on it. For instance, consider the bidimensional view of the fluid element dm (Figure C.1), now represented in the xy plane for simplicity as shown in Figure C.3. In the figure, we can identify three different types of distortion:

- Elongation (or contraction) along the x-axis (Δx),
- Elongation (or contraction) along the y-axis (Δy),
- Angular distortion (shearing) in the plane ($\alpha_x + \alpha_y$).

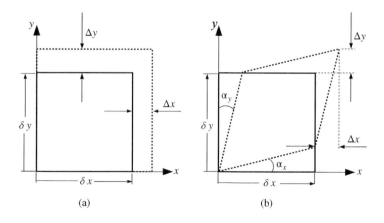

(a) (b)

Figure C.3 Strain state of a fluid element in the plane xy. (a) Elongation, (b) shearing

The linear (E) and angular (Γ) deformations of the element in the plane xy (Figure C.3) are defined as follows:

$$\begin{cases} E_{xx} = \dfrac{\Delta x}{\delta x} \\[2mm] E_{yy} = \dfrac{\Delta y}{\delta y} \\[2mm] \Gamma_{xy} = \Gamma_{yx} = \alpha_x + \alpha_y = \tan^{-1}\left(\dfrac{\Delta y}{\delta x} + \dfrac{\Delta x}{\delta y}\right) \cong \dfrac{\Delta y}{\delta x} + \dfrac{\Delta x}{\delta y} \end{cases} \tag{C.7}$$

Similar to Hooke's law[1] in solid elasticity, it is possible to relate the element deformations[2] to the components of the stress tensor $[\tau]$ (Eq. C.5). However, we observe that fluid elements are constantly changing their shape as they move. Because of that, it is appropriate to relate the element stresses to the deformation rates instead. The linear and angular deformation rates (ε and γ, respectively) can be obtained by dividing the linear and angular deformations by the corresponding time interval, Δt, in which they take place. In the limit when $\Delta t \to 0$, we have

$$\begin{cases} \varepsilon_{xx} = \lim_{\Delta t \to 0}\left(\dfrac{E_{xx}}{\Delta t}\right) = \lim_{\Delta t \to 0}\left(\dfrac{\Delta x/\Delta t}{\delta x}\right) = \dfrac{\partial u}{\partial x} \\[3mm] \varepsilon_{yy} = \lim_{\Delta t \to 0}\left(\dfrac{E_{yy}}{\Delta t}\right) = \lim_{\Delta t \to 0}\left(\dfrac{\Delta y/\Delta t}{\delta y}\right) = \dfrac{\partial v}{\partial y} \\[3mm] \gamma_{xy} = \gamma_{yx} = \lim_{\Delta t \to 0}\left(\dfrac{\alpha_x + \alpha_y}{\Delta t}\right) = \lim_{\Delta t \to 0}\left(\dfrac{\Delta y/\Delta t}{\delta x} + \dfrac{\Delta x/\Delta t}{\delta y}\right) = \dfrac{\partial v}{\partial x} + \dfrac{\partial u}{\partial y} \end{cases} \tag{C.8}$$

where u and v are the xy components of the velocity vector $\mathbf{v}^T = \begin{bmatrix} u & v & w \end{bmatrix}$.

The relations between the linear and angular deformation rates and the components of the stress tensor are not easy to be developed and will be given here without a formal proof (the interested reader can consult Ref. [1] for a comprehensive discussion):

$$\tau_{xx} = -p + \lambda \vec{\nabla} \cdot \mathbf{v} + 2\mu \frac{\partial u}{\partial x}, \quad \tau_{yy} = -p + \lambda \vec{\nabla} \cdot \mathbf{v} + 2\mu \frac{\partial v}{\partial y},$$

$$\tau_{zz} = -p + \lambda \vec{\nabla} \cdot \mathbf{v} + 2\mu \frac{\partial w}{\partial z}, \quad \tau_{xy} = \tau_{yx} = \mu\left(\frac{\partial u}{\partial y} + \frac{\partial v}{\partial x}\right),$$

$$\tau_{xz} = \tau_{zx} = \mu\left(\frac{\partial w}{\partial x} + \frac{\partial u}{\partial z}\right), \quad \tau_{yz} = \tau_{zy} = \mu\left(\frac{\partial v}{\partial z} + \frac{\partial w}{\partial y}\right) \tag{C.9}$$

In Eqs. (C.9), we identify two coefficients: the *absolute* (or *dynamic*) *viscosity*, μ, which is related to the shear stresses, and the *bulk viscosity*, λ, which has to do with the fluid

---

[1] In honour of the English scientist Robert Hooke (1635–1703). Hooke's law relates the inner stresses in a solid body to their relative strains in the elastic regime.

[2] Note that there will also be translation and rotation of the fluid element as it travels along the flow. However, none of these movements influences the stresses on the fluid element.

compressibility along each of the Cartesian axes. We observe that while it is relatively easy to make sense of the off-diagonal terms (τ_{ij}, $i \neq j$), given that the fluid viscosity, μ, is related to the shearing of the element (angular distortion), the same cannot be said about the normal viscous tensions (diagonal components of the tensor $[\tau]$), τ_{ii}, which are composed of three distinct terms:

1. The first term is the hydrostatic pressure, p, which appears as negative, indicating that it is always directed towards the centre of the element. The name 'hydrostatic' comes from the fact that while all the other tension components vanish when the fluid is at rest, the pressure p is the only one that will remain.
2. The second term only exists if the fluid is being compressed or elongated as it flows. This comes from the fact that if the velocity divergent, $\vec{\nabla} \cdot \mathbf{v}$, is not zero, the immediate implication is that there must be a relative motion between at least two parallel faces of the cubic element.
3. The third term also has to do with fluid elasticity, but it only accounts for the elastic distortion along the axis where the tension is being considered.

Considering the observations 2 and 3 above, we assume that an incompressible fluid would reduce the diagonal terms of the stress tensor to $-p$. In fact, if the fluid is incompressible, there cannot be any difference in speed between two parallel faces of the cube, which leads to $\vec{\nabla} \cdot \mathbf{v} = 0$. The same conclusion can be drawn from the mass conservation equation, as will be seen in the following section.

C.3　Differential Fluid Dynamics Equations

C.3.1　Conservation of Mass

In what follows, we are considering the fluid as a continuous media, that is its physical properties are well defined in each and every point in space and vary continuously all through the volume occupied by the fluid.[3]

Under the continuum hypothesis, let us consider the plane representation of the infinitesimal control volume (CV) through which the fluid flows, as shown in Figure C.4 (only the balance along the x-axis is represented). Assuming a homogeneous fluid, where the density does not vary in space, a mass balance over the CV gives

$$\frac{\partial \rho}{\partial t} dx\,dy\,dz = \rho \left(u - \frac{\partial u}{\partial x} \frac{dx}{2} \right) dy\,dz + \rho \left(v - \frac{\partial v}{\partial y} \frac{dy}{2} \right) dx\,dz$$

$$+ \rho \left(w - \frac{\partial w}{\partial z} \frac{dz}{2} \right) dx\,dz - \rho \left(u + \frac{\partial u}{\partial x} \frac{dx}{2} \right) dy\,dz$$

$$- \rho \left(v + \frac{\partial v}{\partial y} \frac{dy}{2} \right) dx\,dz - \rho \left(w + \frac{\partial w}{\partial z} \frac{dz}{2} \right) dx\,dz \qquad (C.10)$$

---

[3] Derivatives and first-order (truncated) Taylor series expansions, which will be needed in the development of the fluid dynamic equations, are only defined for functions that are continuous at least up to the first derivative.

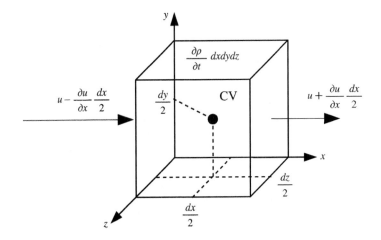

Figure C.4   Infinitesimal control volume and mass balance along the *x*-axis

where ρ is the fluid density, measured at the centre of the control volume, and u, v and w are the x, y and z components of the velocity vector $\mathbf{v}^T = \begin{bmatrix} u & v & w \end{bmatrix}$.

Equation (C.10) can be further simplified to the following expression:

$$\frac{\partial \rho}{\partial t} = -\rho \left(\frac{\partial u}{\partial x} + \frac{\partial v}{\partial y} + \frac{\partial w}{\partial z} \right) \tag{C.11}$$

We can write Eq. (C.11) in a more usual form:

$$\frac{\partial \rho}{\partial t} + \rho \vec{\nabla} \cdot \mathbf{v} = 0 \tag{C.12}$$

Before we move on to the next section, let us slightly modify Eq. (C.12) and write it as a function of pressure. By definition, the density ρ is the inverse of the specific volume, v. Using this fact together with the definition of the fluid bulk modulus given by Eq. (2.4), and considering that the temperature of the fluid does not change as it flows, we obtain for the mass balance:

$$\left(\frac{\rho}{\beta} \right) \frac{\partial p}{\partial t} + \rho \nabla \cdot \mathbf{v} = 0 \tag{C.13}$$

Given that the compressibility of liquids is very small (β is very large), the term containing the time derivative in Eq. (C.13) is approximately zero.[4] In the limit situation where the fluid is incompressible, $\beta \to \infty$, and Eq. (C.13) becomes

$$\nabla \cdot \mathbf{v} = 0 \tag{C.14}$$

Equation (C.14) confirms the statement made at the end of last section that in incompressible fluids the divergent of the velocity vector is zero.

---

[4] Fluid flows, which exhibit this characteristic, are sometimes referred to as *quasi-incompressible* flows.

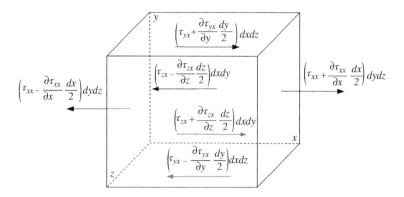

Figure C.5   Force balance in a fluid element in the x-direction

C.3.2   Conservation of Momentum

In order to obtain the momentum equations, we follow an infinitesimal particle with mass dm as it moves with the flow instead of analysing the flow through a fixed control volume. Figure C.5 shows the forces acting on the fluid element where only the tension components along the x-axis are shown for clarity. The resultant force, F_x, will then be given by[5]

$$F_x = \left(\tau_{xx} + \frac{\partial \tau_{xx}}{\partial x} \frac{dx}{2} \right) dy\,dz - \left(\tau_{xx} - \frac{\partial \tau_{xx}}{\partial x} \frac{dx}{2} \right) dy\,dz$$

$$+ \left(\tau_{yx} + \frac{\partial \tau_{yx}}{\partial y} \frac{dy}{2} \right) dx\,dz - \left(\tau_{yx} - \frac{\partial \tau_{yx}}{\partial y} \frac{dy}{2} \right) dx\,dz$$

$$+ \left(\tau_{zx} + \frac{\partial \tau_{zx}}{\partial z} \frac{dz}{2} \right) dx\,dy - \left(\tau_{zx} - \frac{\partial \tau_{yx}}{\partial z} \frac{dz}{2} \right) dx\,dy \qquad (C.15)$$

Equation (C.15) simplifies to

$$F_x = \left(\frac{\partial \tau_{xx}}{\partial x} + \frac{\partial \tau_{yx}}{\partial y} + \frac{\partial \tau_{zx}}{\partial z} \right) dx\,dy\,dz \qquad (C.16)$$

Similarly, we obtain the following expressions for the resultant forces, F_y and F_z, along the y- and z-axis:

$$F_y = \left(\frac{\partial \tau_{xy}}{\partial x} + \frac{\partial \tau_{yy}}{\partial y} + \frac{\partial \tau_{zy}}{\partial z} \right) dx\,dy\,dz \qquad (C.17)$$

$$F_z = \left(\frac{\partial \tau_{xz}}{\partial x} + \frac{\partial \tau_{yz}}{\partial y} + \frac{\partial \tau_{zz}}{\partial z} \right) dx\,dy\,dz \qquad (C.18)$$

---

[5] Here, we disregard the body forces due to gravity and any possible electromagnetic effects.

Now, the second law of Newton states that the mass of the fluid element times its acceleration equals the resultant force $\mathbf{F}^T = \begin{bmatrix} F_x & F_y & F_z \end{bmatrix}$. On the other hand, the acceleration of the fluid element is given by

$$\mathbf{a} = \frac{\partial \mathbf{v}}{\partial t} = \frac{\partial}{\partial t} \begin{bmatrix} u(x,y,z,t) \\ v(x,y,z,t) \\ w(x,y,z,t) \end{bmatrix} = \begin{bmatrix} \dfrac{\partial u}{\partial x}\dfrac{dx}{dt} + \dfrac{\partial u}{\partial y}\dfrac{dy}{dt} + \dfrac{\partial u}{\partial z}\dfrac{dz}{dt} + \dfrac{\partial u}{\partial t} \\ \dfrac{\partial v}{\partial x}\dfrac{dx}{dt} + \dfrac{\partial v}{\partial y}\dfrac{dy}{dt} + \dfrac{\partial v}{\partial z}\dfrac{dz}{dt} + \dfrac{\partial v}{\partial t} \\ \dfrac{\partial w}{\partial x}\dfrac{dx}{dt} + \dfrac{\partial w}{\partial y}\dfrac{dy}{dt} + \dfrac{\partial w}{\partial z}\dfrac{dz}{dt} + \dfrac{\partial w}{\partial t} \end{bmatrix} \tag{C.19}$$

where the chain rule was used for the derivatives of the velocity components.

In Eqs. (C.19), we observe that the spatial derivatives in relation to time dx/dt, dy/dt and dz/dt give the velocity components u, v and w, respectively. Therefore, we can rewrite Eq. (C.19) as

$$\mathbf{a} = \begin{bmatrix} \vec{\nabla} u \cdot \mathbf{v} + \dfrac{\partial u}{\partial t} \\ \vec{\nabla} v \cdot \mathbf{v} + \dfrac{\partial v}{\partial t} \\ \vec{\nabla} w \cdot \mathbf{v} + \dfrac{\partial w}{\partial t} \end{bmatrix} \tag{C.20}$$

The Second Law of Newton states that

$$\mathbf{F} = (\rho \, dx \, dy \, dz)\mathbf{a} \tag{C.21}$$

Substituting $\mathbf{F} = \begin{bmatrix} F_x & F_y & F_z \end{bmatrix}^T$ with F_x, F_y and F_z, given by Eqs. (C.16)–(C.18), and the acceleration, $\mathbf{a}$, given by Eq. (C.20) into Eq. (C.21), we finally obtain

$$\rho \begin{bmatrix} \vec{\nabla} u \cdot \mathbf{v} + \dfrac{\partial u}{\partial t} \\ \vec{\nabla} v \cdot \mathbf{v} + \dfrac{\partial v}{\partial t} \\ \vec{\nabla} w \cdot \mathbf{v} + \dfrac{\partial w}{\partial t} \end{bmatrix} = \begin{bmatrix} \dfrac{\partial \tau_{xx}}{\partial x} + \dfrac{\partial \tau_{yx}}{\partial y} + \dfrac{\partial \tau_{zx}}{\partial z} \\ \dfrac{\partial \tau_{xy}}{\partial x} + \dfrac{\partial \tau_{yy}}{\partial y} + \dfrac{\partial \tau_{zy}}{\partial z} \\ \dfrac{\partial \tau_{xz}}{\partial x} + \dfrac{\partial \tau_{yz}}{\partial y} + \dfrac{\partial \tau_{zz}}{\partial z} \end{bmatrix} \tag{C.22}$$

Equation (C.22) can be split into three equations, one for each force component. If the relationships between the fluid viscosity and the viscous tensions given in Eq. (C.9) are also inserted into Eq. (C.22), we obtain the *Navier–Stokes equations*.[6] The Navier–Stokes

---

[6] In honour of the French and British scientists Claude-Louis Navier (1785–1836) and George Gabriel Stokes (1819–1903).

equations constitute a second-order non-linear set of partial differential equations and are not easily solved. In fact, analytical solutions are only possible for some simplified cases.[7]

The Navier–Stokes equations for a fluid, where the density is not a function of the spatial coordinates x, y and z, in the absence of body (gravitational and electromagnetic) forces, are the following[8] (we leave the demonstration as an exercise to the student):

$$\begin{cases} \rho\vec{\nabla}u \cdot \mathbf{v} + \rho\dfrac{\partial u}{\partial t} = -\dfrac{\partial p}{\partial x} + \mu\left(\dfrac{\partial^2 u}{\partial x^2} + \dfrac{\partial^2 u}{\partial y^2} + \dfrac{\partial^2 u}{\partial z^2}\right) \\[2mm] \rho\vec{\nabla}v \cdot \mathbf{v} + \rho\dfrac{\partial v}{\partial t} = -\dfrac{\partial p}{\partial y} + \mu\left(\dfrac{\partial^2 v}{\partial x^2} + \dfrac{\partial^2 v}{\partial y^2} + \dfrac{\partial^2 v}{\partial z^2}\right) \\[2mm] \rho\vec{\nabla}w \cdot \mathbf{v} + \rho\dfrac{\partial w}{\partial t} = -\dfrac{\partial p}{\partial z} + \mu\left(\dfrac{\partial^2 w}{\partial x^2} + \dfrac{\partial^2 w}{\partial y^2} + \dfrac{\partial^2 w}{\partial z^2}\right) \end{cases} \qquad \text{(C.23)}$$

The steady-state version of the incompressible Navier–Stokes equations can be obtained from Eqs. (C.23) by making the time derivatives equal to zero:

$$\begin{cases} \rho\vec{\nabla}u \cdot \mathbf{v} = -\dfrac{\partial p}{\partial x} + \mu\left(\dfrac{\partial^2 u}{\partial x^2} + \dfrac{\partial^2 u}{\partial y^2} + \dfrac{\partial^2 u}{\partial z^2}\right) \\[2mm] \rho\vec{\nabla}v \cdot \mathbf{v} = -\dfrac{\partial p}{\partial y} + \mu\left(\dfrac{\partial^2 v}{\partial x^2} + \dfrac{\partial^2 v}{\partial y^2} + \dfrac{\partial^2 v}{\partial z^2}\right) \\[2mm] \rho\vec{\nabla}w \cdot \mathbf{v} = -\dfrac{\partial p}{\partial z} + \mu\left(\dfrac{\partial^2 w}{\partial x^2} + \dfrac{\partial^2 w}{\partial y^2} + \dfrac{\partial^2 w}{\partial z^2}\right) \end{cases} \qquad \text{(C.24)}$$

Equations (C.24) become even more simplified for some specific types of flows. For instance, for 1D flows along the x-axis, v and w become zero. The steady-state Navier–Stokes equations are then reduced to

$$\begin{cases} \rho\dfrac{\partial u}{\partial x}u = -\dfrac{\partial p}{\partial x} + \mu\left(\dfrac{\partial^2 u}{\partial x^2} + \dfrac{\partial^2 u}{\partial y^2} + \dfrac{\partial^2 u}{\partial z^2}\right) \\[2mm] \dfrac{\partial p}{\partial y} = 0, \ \dfrac{\partial p}{\partial z} = 0 \end{cases} \qquad \text{(C.25)}$$

Equations (C.25) imply that for 1D incompressible laminar flows at steady-state regime, the pressure can only vary along the flow direction, that is $p = p(x)$. In addition, given that

---

[7] The numerical solution of the Navier–Stokes equations, however, has been the focus of intensive research lately. Finite Element and Finite Volume commercial software, for example are already available for a large spectrum of fluid flows.

[8] Embedded in Eqs. (C.23) is the simplifying assumption that the fluid viscosity, μ, is constant. One situation in which such an assumption applies is the laminar flow of a Newtonian fluid.

$v = w = 0$, Eq. (C.14) simplifies to $\nabla \cdot \mathbf{v} = \partial u / \partial x = 0$. Using this fact in Eqs. (C.25), we obtain

$$\frac{\partial p}{\partial x} = \mu \left(\frac{\partial^2 u}{\partial y^2} + \frac{\partial^2 u}{\partial z^2} \right) \tag{C.26}$$

A further simplification can be made if we consider that the speed does not vary in the z-direction (see Figure 2.1 for an example). In this case, Eq. (C.26) becomes

$$\frac{\partial p}{\partial x} = \mu \left(\frac{\partial^2 u}{\partial y^2} \right) \tag{C.27}$$

C.3.3   Navier–Stokes Equations in Cylindrical Coordinates

There are moments when it is more convenient to write the Navier–Stokes equations in cylindrical coordinates. This is the case with flows inside conduits, where there is a cylindrical symmetry.

Figure C.6 shows the graphical relation between cylindrical and Cartesian coordinates. From Figure C.6, the following relations between the Cartesian system (x, y, z) and the cylindrical system (r, θ, z) can be written as

$$\begin{cases} x = r\cos(\theta) \\ y = r\sin(\theta) \\ z = z \end{cases} \tag{C.28}$$

We will not give the complete mass conservation and Navier–Stokes equations in cylindrical coordinates here since they are very lengthy and have not been used in this book (the student can find the complete expressions in Ref. [2]). Instead, we limit ourselves to

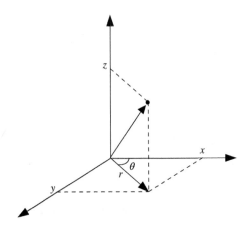

Figure C.6   Cartesian and cylindrical coordinates

show the cylindrical equivalent of Eq. (C.27), considering a unidimensional flow oriented along the z-axis (see, e.g. Figure 2.13):

$$\mu \frac{\partial}{\partial r} \left(r \frac{\partial u}{\partial r} \right) = r \frac{\partial p}{\partial z} \tag{C.29}$$

C.4 Control Volume Analysis

In the previous section, we studied the differential form of the governing fluid dynamics equations. The conservation of mass and momentum resulted in a system or partial differential equations that could only be solved for some particular cases. In several situations, however, we can obtain a great deal of information from a fluid dynamics problem by performing a global analysis where we analyse the flow from a macroscopic point of view. In this section, we show how this can be done for a general fluid flow.

C.4.1 The Reynolds Transport Theorem

Although we could proceed to develop the conservation equations over a control volume independently, the best way to proceed is to formulate a general equation for the balance of any physical property of the fluid as it crosses a determined region in space (control volume). For instance, let N be a physical property measured in a determined bulk of fluid, taken arbitrarily in the flow. By a 'bulk of fluid', we mean a system composed of the same particles in the flow so that the mass of the system does not change. The boundaries of the system, on the other hand, deform as the flow progresses. Figure C.7 exemplifies the situation where we see a system of mass m (grey area) moving in space at two instants of time, t and $t + \Delta t$. As the system moves from left to right, it crosses through a fixed control volume in space, CV (dashed circle). We assume that at time t, both the system and the control volume are overlapped, as shown in Figure ure C.7(a).

The variation of a given system property N, between times t and $t + \Delta t$, can be written as

$$\left. \frac{\Delta N}{\Delta t} \right|_{\text{sys}} = \frac{N_{\text{sys}}(t + \Delta t) - N_{\text{sys}}(t)}{\Delta t} \tag{C.30}$$

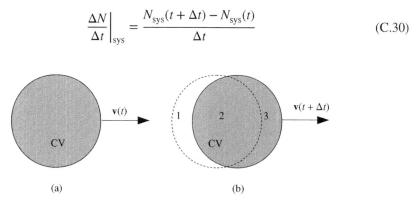

Figure C.7 Flow through a control volume

Observing the geometry of Figure C.7(b), it is easy to conclude that

$$N_{\text{sys}}(t + \Delta t) = (N_2 + N_3)|_{t+\Delta t} = (N_{\text{CV}} - N_1 + N_3)|_{t+\Delta t} \qquad \text{(C.31)}$$

Substituting $N_{\text{sys}}(t + \Delta t)$, given by Eq. (C.31), into Eq. (C.30), we obtain

$$\left. \frac{\Delta N}{\Delta t} \right|_{\text{sys}} = \frac{N_{\text{CV}}(t + \Delta t) - N_{\text{sys}}(t)}{\Delta t} + \left. \frac{N_3 - N_1}{\Delta t} \right|_{t+\Delta t} \qquad \text{(C.32)}$$

Since the control volume and the system coincide at time t, for the limit when $\Delta t \to 0$, we have

$$\lim_{\Delta t \to 0} \left[\frac{N_{\text{CV}}(t + \Delta t) - N_{\text{sys}}(t)}{\Delta t} \right] = \left. \frac{\partial N}{\partial t} \right|_{\text{CV}} = \frac{\partial}{\partial t} \int_{\text{CV}} \rho \eta \, dV \qquad \text{(C.33)}$$

where ρ is the density of the fluid inside the control volume and η is the physical property, N, per unit mass, that is

$$\eta = \frac{N}{m} \qquad \text{(C.34)}$$

Now, observe that the second term on the right-hand side of Eq. (C.32) represents the net rate of the property N into the control volume. This can be easily concluded if we note that $N_3/\Delta t$ is the flow rate of N that leaves the control volume and $N_1/\Delta t$ is the flow rate of N entering the control volume. Therefore, if we consider the whole surface over the control volume as S, for the limit when $\Delta t \to 0$, we can write

$$\lim_{\Delta t \to 0} \left. \frac{N_3 - N_1}{\Delta t} \right|_{t+\Delta t} = \int_S \eta \rho \mathbf{v} \cdot \mathbf{n} \, dS \qquad \text{(C.35)}$$

where $\mathbf{v}$ is the fluid velocity crossing the boundary of the control volume and $\mathbf{n}$ is the unit vector normal to the elementary area, dS.

Equation (C.35) can be better understood with the help of Figure C.8. In the figure, we see a small portion of fluid coming out of the CV. The elementary volume is $dS\cos(\alpha)v\,dt$ and the amount of the physical property, N, displaced during the time lapse dt, is dN.

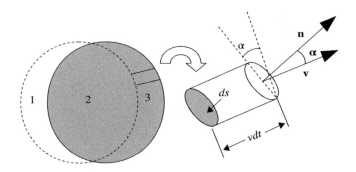

Figure C.8 Elementary flow through the area dS

From Figure C.8, we have that

$$dN = \rho\eta[v\cos(\alpha)dS]\,dt = (\rho\eta\mathbf{v}\cdot\mathbf{n})\,dS\,dt \tag{C.36}$$

where we have made $v\cos(\alpha) = \mathbf{v}\cdot\mathbf{n}$ (remember that $\mathbf{n}$ is a unit vector).

The 'flow rate' of the property, N, across the elementary boundary, dS, is then given by

$$\frac{dN}{dt} = \rho\eta\mathbf{v}\cdot\mathbf{n}\,dS \tag{C.37}$$

Integration of Eq. (C.37) over the boundary, S, of the CV results in the integral on the right-hand side of Eq. (C.35). The Reynolds transport theorem is obtained by combining Eqs. (C.32), (C.33) and (C.35), for the limit when $\Delta t \to 0$:

$$\left.\frac{dN}{dt}\right|_{\text{sys}} = \frac{\partial}{\partial t}\int_{\text{CV}}\rho\eta\,dV + \int_S \eta\rho\mathbf{v}\cdot\mathbf{n}\,dS \tag{C.38}$$

If the flow develops in a steady-state regime, there is no variation of the fluid property, N inside the control volume, and the Reynolds transport theorem becomes

$$\left.\frac{dN}{dt}\right|_{\text{sys}} = \int_S \eta\rho\mathbf{v}\cdot\mathbf{n}\,dS \tag{C.39}$$

The Reynolds transport theorem is one of the most important theorems in fluid dynamics and makes it simpler to develop balanced equations through an arbitrary volume in space. We apply this theorem to study the conservation of mass, momentum and energy in the following sections.

C.4.2 Mass and Momentum Conservation

As an example of application of the Reynolds transport theorem to an actual fluid dynamics problem, first consider the mass and linear momentum conservation applied to a moving system in relation to a fixed control volume in space. Since the physical properties being studied are the mass, m, and the linear momentum, $m\mathbf{v}$, in each case Eq. (C.38) becomes

$$\begin{cases} \left.\dfrac{dm}{dt}\right|_{\text{sys}} = \dfrac{\partial}{\partial t}\displaystyle\int_{\text{CV}}\rho\,dV + \int_S \rho\mathbf{v}\cdot\mathbf{n}\,dS \\[4mm] \left.\dfrac{d\,(m\mathbf{v})}{dt}\right|_{\text{sys}} = \dfrac{\partial}{\partial t}\displaystyle\int_{\text{CV}}\rho\mathbf{v}\,dV + \int_S \rho\mathbf{v}(\mathbf{v}\cdot\mathbf{n})\,dS \end{cases} \tag{C.40}$$

In Eq. (C.40), we observe that η (Eq. (C.34)) becomes 1 when $N = m$ and $\mathbf{v}$ when $N = m\mathbf{v}$. We also note that the conservation of momentum is in fact a vector equation and must be equally satisfied for each of the three components of the velocity vector, u, v and w. Thus, we can write the conservation of linear momentum as

$$
\begin{cases}
\left.\dfrac{d\,(mu)}{dt}\right|_{\text{sys}} = \dfrac{\partial}{\partial t} \displaystyle\int_{\text{CV}} \rho u\,dV + \int_S \rho u (\mathbf{v} \cdot \mathbf{n})\,dS \\[2em]
\left.\dfrac{d\,(mv)}{dt}\right|_{\text{sys}} = \dfrac{\partial}{\partial t} \displaystyle\int_{\text{CV}} \rho v\,dV + \int_S \rho v (\mathbf{v} \cdot \mathbf{n})\,dS \\[2em]
\left.\dfrac{d\,(mw)}{dt}\right|_{\text{sys}} = \dfrac{\partial}{\partial t} \displaystyle\int_{\text{CV}} \rho w\,dV + \int_S \rho w (\mathbf{v} \cdot \mathbf{n})\,dS
\end{cases}
\tag{C.41}
$$

Let us now apply some physical rules to the left-hand sides of Eqs. (C.40) and (C.41) and extract some valuable information from the Reynolds transport theorem. For instance, we know that the mass of the system cannot change (by definition). Therefore, we have

$$
\left.\frac{dm}{dt}\right|_{\text{sys}} = 0
\tag{C.42}
$$

From the Second Law of Newton, we know that the variation of linear momentum in a system equals the resultant of the forces acting on it, $\sum \mathbf{F}$. Thus, we can write that

$$
\left.\frac{d\,(m\mathbf{v})}{dt}\right|_{\text{sys}} = \sum \mathbf{F} =
\begin{pmatrix}
\sum F_x \\
\sum F_y \\
\sum F_z
\end{pmatrix}
\tag{C.43}
$$

where $\sum F_x$, $\sum F_y$ and $\sum F_z$ are the resultant components in the x-, y- and z-direction, respectively.

Substitution of Eqs. (C.42) and (C.43) into Eqs. (C.40) and (C.41) gives

$$
\begin{cases}
0 = \dfrac{\partial}{\partial t} \displaystyle\int_{\text{CV}} \rho\,dV + \int_S \rho \mathbf{v} \cdot \mathbf{n}\,dS \\[2em]
\displaystyle\sum F_x = \dfrac{\partial}{\partial t} \int_{\text{CV}} \rho u\,dV + \int_S \rho u\,(\mathbf{v} \cdot \mathbf{n})\,dS \\[2em]
\displaystyle\sum F_y = \dfrac{\partial}{\partial t} \int_{\text{CV}} \rho v\,dV + \int_S \rho v (\mathbf{v} \cdot \mathbf{n})\,dS \\[2em]
\displaystyle\sum F_z = \dfrac{\partial}{\partial t} \int_{\text{CV}} \rho w\,dV + \int_S \rho w (\mathbf{v} \cdot \mathbf{n})\,dS
\end{cases}
\tag{C.44}
$$

The steady-state version of the mass and momentum conservation equations can be obtained from Eqs. (C.44) by assigning the value zero to the time derivatives:

$$\begin{cases} 0 = \int_S \rho \mathbf{v} \cdot \mathbf{n} \, dS \\[2mm] \sum F_x = \int_S \rho u \, (\mathbf{v} \cdot \mathbf{n}) \, dS \\[2mm] \sum F_y = \int_S \rho v (\mathbf{v} \cdot \mathbf{n}) \, dS \\[2mm] \sum F_z = \int_S \rho w (\mathbf{v} \cdot \mathbf{n}) \, dS \end{cases} \qquad (\text{C.45})$$

C.4.3 Conservation of Energy

In order to write the energy conservation equations for a control volume, we first need to give a correct definition of the property N. Observe that every moving system carries an amount of energy with it. Such energy can be usually divided into potential, kinetic and internal (temperature related), as follows:

$$N = U + \frac{mv^2}{2} + mgh \qquad (\text{C.46})$$

where m is the mass of the system, U is the internal energy of the fluid, g is the acceleration of gravity and h is the height of the centre of mass of the system, in relation to a convenient referential.

Substituting Eq. (C.46) into the Reynolds transport equation (C.38) and making use of Eq. (C.34), we have

$$\frac{dN}{dt}\bigg|_{\text{sys}} = \frac{\partial}{\partial t} \int_{\text{CV}} \rho \left(u + \frac{v^2}{2} + gh \right) dV + \int_S \rho \left(u + \frac{v^2}{2} + gh \right) \mathbf{v} \cdot \mathbf{n} \, dS \qquad (\text{C.47})$$

where u is the internal energy per unit mass ($u = U/m$).

The first law of thermodynamics states that the variation of energy of a system equals the heat power supplied to it, dQ/dt, minus the rate of work done by the system on its surroundings, dW/dt. Therefore, we can rewrite Eq. (C.47) as

$$\frac{dQ}{dt} - \frac{dW}{dt} = \frac{\partial}{\partial t} \int_{\text{CV}} \rho \left(u + \frac{v^2}{2} + gh \right) dV + \int_S \rho \left(u + \frac{v^2}{2} + gh \right) \mathbf{v} \cdot \mathbf{n} \, dS \qquad (\text{C.48})$$

In Eq. (C.48), the rate of work, dW/dt, done by the system (or to the system) can be better understood with the help of Figure C.9, where we see the system being deformed by the action of the surroundings (Figure C.9(a)). Figure C.9(b) shows the details of the action of an elementary force, $d\mathbf{F}$, pushing the system boundary inwardly at time t, when both the system and the control volume are overlapped (Figure C.7).

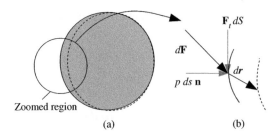

Figure C.9 Work done by an elementary force, $d\mathbf{F}$, on a moving system

In the most general scenario, the force $d\mathbf{F}$ (Figure C.9(b)) is inclined in relation to the normal vector, $\mathbf{n}$, so that both tangential and normal forces are present at the system boundary. We see that in the figure where the force $d\mathbf{F}$ decomposes into a tangential[9] and a normal component, $\mathbf{F}_t dS$ and $p\mathbf{n} dS$ (p is the pressure).[10] Therefore, in order to obtain the rate of work, dW/dt, done on the system by the boundary forces, we integrate the work of the infinitesimal force, $d\mathbf{F}$, over the boundary of the control volume,[11] S, as follows:

$$\frac{\partial W}{\partial t} = \frac{\partial}{\partial t}\int_S d\mathbf{F} \cdot d\mathbf{r} = \int_S \mathbf{v} \cdot d\mathbf{F} = \int_S \mathbf{v} \cdot (p\mathbf{n} + \mathbf{F}_t)dS = \int_S p\mathbf{v} \cdot \mathbf{n}\, dS + \int_S \mathbf{v} \cdot \mathbf{F}_t dS \quad (C.49)$$

Substituting dW/dt, given by (C.49), into Eq. (C.48), we arrive at the usual form of the energy conservation equation:

$$\frac{dQ}{dt} - \int_S p\mathbf{v} \cdot \mathbf{n}\, dS - \int_S \mathbf{v} \cdot d\mathbf{F}_t\, dS = \frac{\partial}{\partial t}\int_{CV} \rho\left(u + \frac{v^2}{2} + gh\right) dV +$$
$$\int_S \rho\left(u + \frac{v^2}{2} + gh\right)\mathbf{v} \cdot \mathbf{n} \cdot dS \quad (C.50)$$

For steady-state flows, Eq. (C.50) becomes

$$\frac{dQ}{dt} - \int_S p\mathbf{v} \cdot \mathbf{n}\, dS - \int_S \mathbf{v} \cdot d\mathbf{F}_t\, dS = \int_S \rho\left(u + \frac{v^2}{2} + gh\right)\mathbf{v} \cdot \mathbf{n} \cdot dS \quad (C.51)$$

A very interesting application of Eq. (C.51) is to use it for a hypothetical non-viscous flow, where no heat is generated and no tangential forces are present. Consider, for instance, the flow through an imaginary duct with no friction happening at the walls, as shown in Figure C.10. Assume also that the flow has reached the steady-state regime and that the fluid density, ρ, is constant. Under these circumstances, Eq. (C.51) becomes

$$-\int_S p\mathbf{v} \cdot \mathbf{n}\, dS = \int_S \rho\left(u + \frac{v^2}{2} + gh\right)\mathbf{v} \cdot \mathbf{n}\, dS \quad (C.52)$$

---

[9] Observe that $\mathbf{F}_t$ is, actually, a force per unit area.
[10] Note that in a hypothetical non-viscous flow, the force $d\mathbf{F}$ would necessarily act along the vector $\mathbf{n}$, given that no tangential stresses would exist (see Eqs. (C.9)).
[11] Remember that the boundary of the system coincides with the boundary of the control volume at time t.

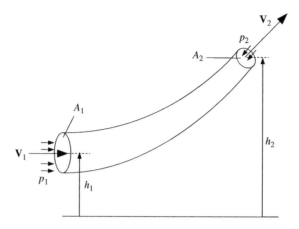

Figure C.10 Ideal flow through a hypothetical tube

Developing the integrals in Eq. (C.52), we obtain

$$p_1 v_1 A_1 - p_2 v_2 A_2 = v_2 A_2 \left(\rho u_2 + \frac{\rho v_2^2}{2} + \rho g h_2 \right) - v_1 A_1 \left(\rho u_1 + \frac{\rho v_1^2}{2} + \rho g h_1 \right) \qquad \text{(C.53)}$$

The mass conservation equation in Eqs. (C.45) gives

$$0 = \int_S \rho \mathbf{v} \cdot \mathbf{n} \, dS = \rho v_2 A_2 - \rho v_1 A_1 \qquad \text{(C.54)}$$

From Eq. (C.54), we know that $v_2 A_2 = v_1 A_1$. Using this fact in Eq. (C.53), we obtain

$$p_1 v_1 A_1 - p_2 v_1 A_1 = v_1 A_1 \left(\rho u_2 + \frac{\rho v_2^2}{2} + \rho g h_2 \right) - v_1 A_1 \left(\rho u_1 + \frac{\rho v_1^2}{2} + \rho g h_1 \right) \qquad \text{(C.55)}$$

Dividing Eq. (C.55) by $v_1 A_1$ results in

$$\rho u_1 + p_1 + \frac{\rho v_1^2}{2} + \rho g h_1 = \rho u_2 + p_2 + \frac{\rho v_2^2}{2} + \rho g h_2 \qquad \text{(C.56)}$$

If the flow is isothermal, $u_2 = u_1$, and Eq. (C.56) becomes

$$p_1 + \frac{\rho v_1^2}{2} + \rho g h_1 = p_2 + \frac{\rho v_2^2}{2} + \rho g h_2 \qquad \text{(C.57)}$$

Equation (C.57) is known as the *Bernoulli*[12] *equation* and is very important for fluid dynamics in general. It is understood that whenever losses are not present in a flow, energy is

---

[12] In honour of the Dutch–Swiss mathematician Daniel Bernoulli (1700–1782).

conserved. Equation (C.57) states that the energy per unit volume at the flow entrance (section 1), which is composed of pressure, kinetic energy per unit volume and potential gravitational energy per unit volume, must be equal to the energy per unit volume at the flow exit (section 2). It follows naturally from Eq. (C.57) that when energy is not conserved, the following expression holds

$$p_1 + \frac{\rho v_1^2}{2} + \rho g h_1 = p_2 + \frac{\rho v_2^2}{2} + \rho g h_2 + \Delta e \tag{C.58}$$

where Δe represents the energy losses per unit volume between sections 1 and 2 (Figure C.10).

References

[1] Schlichting H (1979) Boundary-layer theory. McGraw-Hill, USA.
[2] Fox RW, McDonald AT, Pritchard PJ (2004) Introduction to fluid mechanics, 6th Ed., John Wiley & Sons, USA.

Index

Hydrostatic Transmissions and Actuators: Operation, Modelling and Applications, First Edition.
Gustavo Koury Costa and Nariman Sepehri.
© 2015 John Wiley & Sons, Ltd. Published 2015 by John Wiley & Sons, Ltd.
Companion Website: www.wiley.com/go/costa/hydrostatic

Printed and bound by CPI Group (UK) Ltd, Croydon, CR0 4YY

16/04/2025

14658560-0002